数据驱动鲁棒选址优化

王曙明　著

科学出版社
北　京

内 容 简 介

本书从数据驱动决策与最优化角度出发，深入细致地阐述不确定环境下选址决策与优化问题中面临的一系列挑战，并提出相应的数据驱动鲁棒选址优化建模方法．主要内容包括基于有限概率分布信息的数据驱动（分布）鲁棒选址优化方法、基于决策依赖型结构化数据驱动（分布）鲁棒选址优化方法以及基于时间序列预测模型的多阶段（分布）鲁棒选址优化方法等等．本书以相关企业实际供应链决策问题为背景，全面阐述不确定环境下数据驱动鲁棒选址优化建模框架及求解算法，不仅丰富了选址优化建模理论，而且为企业提升选址与供应链运营决策质量提供了切实可行的数据驱动方法．

本书既可以供运筹与优化、物流与供应链管理等相关企业参考，同时也可以作为高校及科研院所相关专业的教师、研究生及研究人员的研读书籍或教学用书．

图书在版编目(CIP)数据

数据驱动鲁棒选址优化/王曙明著. -- 北京：科学出版社，2025.3

ISBN 978-7-03-077977-9

Ⅰ. ①数… Ⅱ. ①王… Ⅲ. ①鲁棒控制 Ⅳ. ①TP273

中国国家版本馆 CIP 数据核字(2024)第 002701 号

责任编辑：郝　悦　孙翠勤／责任校对：杨聪敏

责任印制：张　伟／封面设计：有道设计

科学出版社出版

北京东黄城根北街 16 号

邮政编码：100717

http://www.sciencep.com

涿州市般润文化传播有限公司印刷

科学出版社发行　各地新华书店经销

*

2025 年 3 月第　一　版　开本：720×1000　1/16

2025 年 3 月第一次印刷　印张：15 3/4

字数：316 000

定价：178.00 元

（如有印装质量问题，我社负责调换）

前　言

供应链韧性一直是国家现代流通体系、物流体系以及综合交通运输系统建设的重要一环. 2022 年 10 月 16 日, 习近平总书记在中国共产党第二十次全国代表大会上做报告时, 强调 “加快发展物联网, 建设高效顺畅的流通体系, 降低物流成本”“优化基础设施布局、结构、功能和系统集成, 构建现代化基础设施体系”, 对新时代国家流通体系建设提出了新要求. 2024 年 3 月 5 日, 在第十四届全国人民代表大会第二次会议上, 《政府工作报告》对供应链体系和交通基础设施建设工作提出了具体要求, 如 “加强充电桩、冷链物流、寄递配送设施建设”“推动产业链供应链优化升级”“增强产业链供应链韧性和竞争力” 等. 在供应链体系建设中, 物流中心、仓库以及工厂等基础设施的选址决策, 一直是保证供应链体系韧性与安全的关键. 而在企业 “数智化” 转型的大背景下, 设施选址的决策环境发生了巨大变化, 进而带来了以下新的挑战与机遇.

需求高度不确定性与选址决策的鲁棒性. 新零售时代, 商品多元化销售渠道的建立以及商品品类的快速更新, 导致基于历史销售数据对未来长期需求进行准确预测已经越来越困难. 这使得企业在确定设施选址决策时面临需求高度不确定性的挑战. 因此, 如何针对未来需求的高度不确定性, 形成更为鲁棒的设施选址决策, 以提升供应链网络的韧性, 成为选址优化领域的重要课题.

需求协变量信息与选址决策的有效性. 在企业 “数智化” 转型的过程中, 越来越多的企业开始使用信息技术和数字化工具来采集、分析与需求相关的协变量信息, 如消费价格指数、商品特征以及气候环境等. 在 “数智化” 背景下, 企业的数据环境发生了变革, 而如何全面有效地利用这些协变量信息来更加精准地刻画未来需求状态, 以提升选址决策的有效性, 也成为重要课题.

大规模 (高维) 需求与选址决策计算的高效性. 在新零售时代, 商品品类数量呈指数增长. 同时, 现代物流网络的发展以及直播带货的兴起引领了新的线上消费模式, 促使商品需求规模剧增. 这些因素导致现代供应链企业往往需要针对大规模需求网络进行设施选址决策优化.

针对上述挑战, 本书重点介绍不确定环境下设施选址问题的数据驱动优化方法. 按模型复杂度, 从基于有限概率分布信息的优化模型, 到基于时间序列预测信息的优化模型进行介绍. 本书主要分为三个部分: 基于有限概率分布信息的数据驱动 (分布) 鲁棒选址优化方法、基于决策依赖型结构化数据驱动 (分布) 鲁棒选址优化方法以及基于时间序列预测模型的多阶段 (分布) 鲁棒选址优化方法. 第一部分主要介绍了在需求和成本不确定环境下, 如何基于有限概率分布信息构建分布鲁棒枢纽选址优化模型. 进一步地, 结合需求协变量信息, 建立了状态依赖数据驱动型分布鲁棒设施选址优化模型, 并设计了模型的精确求解算法. 具体内容呈现于本书第 2、3、4 章. 第二部分详细介绍了设施选址问题的结构化鲁棒优化模型, 涵盖了针对资源回收系统运营环境不确定性的多目标鲁棒性分析框架构建, 以及在公私合营框架下, 考虑地方政府与资源回收系统运营商决策依赖型的资源回收系统选址优化模型. 该部分内容见于本书第 5、6 章. 第三部分主要从预测驱动决策的角度, 阐述了基于时间序列预测集的多阶段 (分布) 鲁棒选址优化方法. 内容包括基于预测不确定集的多阶段鲁棒资源回收设施选址优化模型, 以及基于沃瑟斯坦 (Wasserstein) 距离嵌套型分布不确定集的预算驱动型多阶段枢纽选址优化模型. 该部分内容见于本书第 7、8 章.

本书作者长期从事随机鲁棒优化、统计学习、模型不确定性及其在选址、物流与供应链管理、交通等领域的应用研究, 本书内容反映了作者及其研究团队近年来在该领域的主要研究成果. 在本书编写过程中, 作者详细查阅了大量参考文献, 总结分析了该领域的研究现状, 力图用言简意赅的语言介绍作者在数据驱动鲁棒选址优化领域的研究成果. 然而, 由于本书所介绍的内容为选址优化领域的研究热点, 书中难免存在疏漏之处, 在此恳请广大读者批评指正.

本书的研究工作得到了国家自然科学基金优秀青年科学基金项目 (编号: 71922020)、面上项目 (编号: 72171221、72471224) 以及中央高校基本科研业务经费专项资金 (编号: UCAS-E2ET0808X2) 的资助. 同时, 本书也得到了中国科学院大学经济与管理学院、中国科学院大学数字经济监测预测预警与政策仿真教育部哲学社会科学实验室 (培育) 以及教育部哲学社会科学创新团队 "数智时代经济管理复杂系统建模创新团队" 的大力支持. 此外, 在本书内容的组织过程中, 首都经济贸易大学的刘天奇助理教授、中国科学院大学的科研助理毛宇晨、北京交通大学的胡杰副教授、中国科学院大学的博士生刘丰、赵嘉、赵熠以及硕士生张嘉辰等均参与了本书的撰写工作, 在此一并表示感谢!

王曙明

2024 年 11 月 30 日

目　录

第 1 章 引　　言

1.1 设施选址问题简介

设施选址问题的起源可以追溯到 17 世纪法国数学家费马 (Fermat) 提出的一个几何问题: 在欧几里得平面上, 哪个点到三个给定点的距离之和最小? 该问题对选址科学的发展产生了深远影响, 并且至今仍有许多学者致力于相关问题研究 (Benko and Coroian, 2018; Görner and Kanzow, 2016). 在 20 世纪初, 朗哈特 (Launhardt) 和韦伯 (Weber) 在 3-节点 Weber 问题的研究中首次提出了规范的 (normative) 设施选址模型 (Fearon, 2006; Launhardt, 1900). 该模型的目标是为一个炼钢设施选择最优的地址, 使得钢铁供应链的总运输成本最小. 随后几十年中, 选址科学不断发展, 并成功地应用于多种实际问题. 直到 20 世纪 60 年代, 米勒 (Miehle) 和库珀 (Cooper) 将 Weber 问题拓展到多个设施选址的情况, 这标志着现代选址科学的产生 (Cooper, 1963; Miehle, 1958). 特别是, Cooper 提出了 p-中位工厂 (设施) 选址问题 (p-median facility location problem), 即在多个设施备选地址中选择 p 个地址, 以使设施到需求点之间的运输成本最小. 这一问题是现代选址科学的核心问题, 也是当前研究的热点 (Brimberg and Drezner, 2013; Drezner and Salhi, 2017; Croci et al., 2023).

设施选址问题根据选址决策的性质可以分为连续、网络或离散三类. 此外, 也可根据设施选址问题的目标、约束条件或设施的类型进行分类, 如 p-中位设施选址问题、p-中心 (p-center) 设施选址问题、工厂设施选址问题、枢纽设施选址问题、多层级设施选址问题以及多阶段设施选址问题等 (Laporte et al., 2019). 本书主要讨论两种具有离散选址决策的设施选址问题, 即工厂 (设施) 选址问题 (facility location problem) 和枢纽设施选址问题 (hub location problem).

工厂设施选址问题是选址科学的一类重要问题. 在这一问题中, 决策者需要在有限的备选地址中找到最优的建厂地址, 以充分满足客户的需求. 工厂设施选址包含两个关键决策: 设施选址决策和商品运输决策. 设施选址决策确定了工厂设施应该建在何处, 而运输决策决定如何从工厂将商品运送至客户, 以满足其需求. 该问题的目标是确定最优的工厂建设和商品运输决策, 以最小化建厂和运输成本 (Fernández and Landete, 2019).

工厂设施选址模型的应用范围极为广泛, 它在供应链管理、分布式系统、人道主义救援、应急系统、路径规划、货物运输以及资源回收管理等实际问题中发挥着至关重要的作用. Melo 等 (2009) 详细总结了工厂设施选址模型在供应链管理问题中的应用. Klose 和 Drexl (2005) 探讨了工厂设施选址模型在分布式系统设计中的应用. Balcik 和 Beamon (2008) 的研究是将人道主义救援与设施选址模型相结合的一个重要成果. Erkut 等 (2008) 运用设施选址模型解决了希腊城市废物管理问题. 其他相关应用研究可参考 (Daskin et al., 2002; Jia et al., 2007) 以及 Nagy 和 Salhi (2007). 此外, 工厂设施选址模型还被应用于机器调度、集群分析及组合拍卖等领域 (Drezner and Hamacher, 2004; Escudero et al., 2009; Klose and Drexl, 2005; Singh, 2008).

枢纽设施选址问题是交通运输、电信通信和计算机系统网络设计的核心问题. 该问题通过设计枢纽-辐射型网络 (hub-and-spoke network) 来实现在始发地与目的地之间高效运输资源. 在枢纽选址问题中, 始发地相同但目的地不同的商品或货物在抵达枢纽后进行分拣, 然后与其他目的地相同的商品一同运输. 这有助于降低运输成本, 并通过货物集中运输实现规模经济效益. 枢纽设施选址问题的决策主要包括枢纽设施选址和商品运输 (或网络设计) 决策, 其目标为最小化建设枢纽和运输成本 (Campbell and O'Kelly, 2012; Contreras and O'Kelly, 2019).

枢纽设施选址模型在交通运输领域应用十分广泛, 包括快递包裹运输、航空货运和客运、卡车运输和快速运输系统 (Campbell and O'Kelly, 2012; Farahani et al., 2013). 对于这些问题, 枢纽设施选址模型中的商品或货物是指由各种交通运输工具在公路、铁路、水路和空中航线运输的商品, 如快递包裹、乘客、邮件和货物等. 枢纽设施通常为商品分拣中心或运输终端 (O'Kelly and Bryan, 1998; Contreras and O'Kelly, 2019). Bryan 和 O'Kelly (1999) 总结了枢纽设施选址模型在航空运输中的应用. Kuby 和 Gray (1993) 研究了 Federal Express 公司如何基于枢纽设施选址模型设计其快递运输网络. Limbourg 和 Jourquin (2009) 研究了枢纽设施选址模型在欧洲公路和铁路货物运输中的应用. 更多相关研究可以参考 (Campbell et al., 2005; Gelareh and Pisinger, 2011; Çetiner et al., 2010; Gelareh and Nickel, 2008).

此外, 枢纽设施选址模型在电信领域的应用主要体现在各种分布式数据网络设计问题中, 模型涉及的对象是通过各种物理链路 (光纤或同轴电缆) 或空中链路 (卫星或微波链路) 所传输的电子数据. 枢纽设施是各种交换机、多路复用器和路由器等硬件. 同样地, 在电信领域, 枢纽设施选址模型通过降低数据传输成本来实现规模经济效应 (Klincewicz, 1998). 关于枢纽设施选址模型在电信领域的其他应用, 读者可以参考 (Alumur and Kara, 2008; Carello et al., 2004; Kim and O'Kelly, 2009; Yaman and Carello, 2005).

1.2 设施选址问题与不确定性

自 20 世纪 60 年代以来, 选址科学受到了越来越多学者和业界专家的关注, 已成为一个热点研究领域. 作为该领域的一类核心问题, 设施选址问题的目标为确定一个或多个设施的 "最佳" 位置以便为一组需求点提供服务. 其中, "最佳" 的含义取决于所研究问题的性质, 即所考虑的约束条件和优化目标. 随着选址科学与其他应用科学的交互发展, 设施选址已被广泛地应用于供应链管理、交通运输、通信网络、应急管理以及城市固体垃圾管理等众多实际应用领域中 (Dönmez et al., 2021; Farahani et al., 2019; Gollowitzer and Ljubić, 2011; Melo et al., 2009; Saldanha-da-Gama, 2022).

设施选址问题中的 "设施" 具有广泛的含义, 在生产与运营环境下, 一般指物流中心、仓库、分销中心、工厂和废物处理厂等设施, 这些设施具有固定成本高和运行寿命长的特点. 所以, 设施选址决策成本高昂且一般很难被改变, 而且对运营有长期影响. 在设施运营期间, 其运营环境可能会发生很大的变化. 比如, 成本、需求、运输时间以及设施选址模型的其他参数可能会具有高度的不确定性. 因此, 决策环境的不确定性是设施选址问题中的一个重要挑战 (Snyder, 2006). 这使得不确定环境下的设施选址模型成为选址科学和不确定优化领域的研究热点. 在设施选址问题中, 环境的不确定性通常可以根据其来源被分为三类: 供给方或服务方的不确定性、需求方的不确定性, 以及供给方与需求方之间的不确定性 (Shen et al., 2011). 供给方的不确定性包括不确定的供给能力或产能、交货时间和设施状态 (Cui et al., 2010; Yu et al., 2017; Afify et al., 2019). 需求方的不确定性主要是指需求的不确定性 (Atamtürk and Zhang, 2007; Baron et al., 2011; Gülpınar et al., 2013). 两者之间的不确定性包括不确定的运输成本、运输时间和运输路径状态 (Gao and Qin, 2016; Mišković et al., 2017; Nikoofal and Sadjadi, 2010).

进一步, 根据参数的不确定性特征, 我们又可以将不确定性分为三类情况: 不确定参数概率分布已知、不确定参数概率分布未知, 以及不确定参数概率分布部分可知. 不确定参数概率分布已知的情况是指不确定参数可由已知的概率分布刻画. 这种情况下的设施选址问题属于随机优化问题, 这类问题的决策目标一般是优化目标函数的期望值 (Louveaux, 1993). 不确定参数概率分布未知的情况是指不确定环境中参数的概率分布信息不可获得, 这时可用不确定集对其进行刻画. 这种情况下的设施选址问题是一个鲁棒优化问题, 这类问题一般优化不确定集中的极值 (worst-case value) 目标 (Bertsimas and Sim, 2004). 介于上述两种不确定性环境之间的情况为不确定参数概率分布部分可知也就是分布不确定性 (ambiguity), 即不确定参数的概率分布属于一个分布不确定集中, 决策者优化极值概率分布 (worst-case probability distribution 或 extreme probability distribution) 下的期望目标值 (Delage and Ye, 2010). 这种情况下的设施选址问题是一个分布鲁棒优化问题 (Saif and Delage, 2021).

1.2.1 随机设施选址问题

在随机设施选址模型中, 决策者通常假设不确定参数的概率分布是已知的, 并将模型的目标定为最小化设施选址系统的期望成本, 或最大化期望利润 (Snyder, 2006). Cooper (1974) 研究了需求点位置不确定下的随机 Weber 模型. Mirchandani 等 (1985) 首次提出了运输路径距离不确定下的随机设施选址模型, 并以最小化其期望成本为目标. Louveaux (1986) 研究了具有不确定需求、生产成本和价格的随机 p-中位容量限制型设施选址问题. Listeş 和 Dekker (2005) 构建了需求不确定下反向物流网络设施选址问题的随机规划模型. Sim 等 (2009) 研究了运输时间不确定且服从正态分布下的枢纽设施选址问题, 其目标是最小化期望运输时间. Yang (2009) 在不确定需求服从一个已知离散概率分布的假设下, 提出了一个两阶段随机规划航空运输枢纽设施选址模型. Contreras 等 (2011) 在商品需求和运输成本都存在不确定性的情况下, 研究了容量充足型设施选址问题, 并构建了两阶段随机规划模型. Alumur 等 (2012) 对容量充足型设施选址问题进行了研究, 并提出了两个随机规划模型, 来分别处理设施建设成本和需求不确定性的情况. Sun 等 (2012) 以及 Srivastava 和 Nema (2012) 在假设垃圾产生率、单位成本和收益等参数均服从均匀分布的前提下, 分别研究了单目标和多目标随机能源回收设施选址和容量规划问题.

上述研究大多以选址系统的期望成本或收益为目标, 还有一些研究在期望目标值的基础上, 同时考虑了系统的风险规避水平, 构建了均值-方差设施选址模型. 例如, Jucker 和 Carlson (1976) 在销售价格具有不确定性的情况下, 研究了随机均值-方差容量充足型设施选址模型. Hodder 和 Jucker (1985) 将上述模型推广到考虑不确定参数相关的情况. Verter 和 Dincer (1992) 总结了与随机设施选址和产能扩张问题相关的研究, 重点关注均值-方差模型在全球制造问题中的应用. Wagner 等 (2009) 将金融中的 “在险价值” 的概念引入需求不确定下设施选址问题的研究中, 并构建了基于 “在险价值” 的随机设施选址模型.

除了均值-方差随机设施选址模型, 还有大量的研究以设施选址系统满足某个约束的概率为目标 (Baron et al., 2011). 代表性的研究包括 (Frank, 1966; Carbone, 1974; Shiode and Drezner, 2003; Berman and Wang, 2006; Xiong et al., 2016). 此外, 多阶段随机设施选址问题也是一个重要研究方向. Correia 等 (2018) 开发了基于给定需求分布的多阶段容量限制型枢纽选址模型. Yu 等 (2021) 研究了一个多阶段随机仓库选址问题, 其需求分布的支撑集为有限需求场景 (scenarios), 并推导出一个多阶段模型和两阶段模型之间的差距下界. 关于随机设施选址问题的相关研究可以参考 (Snyder, 2006).

1.2.2 鲁棒设施选址问题

鲁棒优化模型通常构建一个不确定集来刻画参数的不确定性, 然后优化不确定集极值情况下的目标值, 以获得鲁棒解 (Baron et al., 2011). 鲁棒优化建模方法主要分为两类: 静态鲁棒优化和 (两阶段) 自适应鲁棒优化. 这两种鲁棒优化方法已广泛应用于设施选址问题 (Ben-Tal et al., 2009, 2004; Bertsimas and Sim, 2004; Yanikoglu et al., 2019). 静态鲁棒设施选址模型将选址决策和运输决策视为第一阶段决策或现时决策 (here-and-now decision). 在不确定参数的实现值被准确观测之前, 这些决策就已经确定. Baron 等 (2011) 利用区间 (interval) 和椭球 (ellipsoid) 不确定集来刻画客户需求的不确定性, 并基于此开发了鲁棒设施选址模型. Meraklı 和 Yaman (2017) 考虑了一个不确定需求服从管式 (hose) 约束的容量限制型设施选址问题, 并构建了基于多面体不确定集的鲁棒设施选址模型. Zetina 等 (2017) 同时考虑了成本和需求的不确定性, 并采用了 Bertsimas 和 Sim (2004) 提出的预算不确定集 (budget uncertainty set), 构建了鲁棒设施选址模型. 关于静态鲁棒设施选址问题的其他研究可参考 (Boukani et al., 2016; Meraklı and Yaman, 2016; Shahabi and Unnikrishnan, 2014; Zhu et al., 2018).

自适应鲁棒设施选址模型假定选址决策是在不确定性观测之前做出的, 而运输或运营决策则是在观测到不确定参数实现值后做出的. 尽管自适应鲁棒优化模型被证明可以有效解决静态鲁棒优化模型决策过于保守的问题, 但是一般来说它难以求解 (Ben-Tal et al., 2004). Zeng 和 Zhao (2013) 在本德斯 (Benders) 分解算法框架下, 提出了一种列和约束生成 (column-and-constrain generation, CCG) 方法, 用于求解基于多面体不确定集的自适应鲁棒优化模型, 并将该 CCG 方法应用于自适应鲁棒容量限制型设施选址模型的求解. Gabrel 等 (2014) 采用了类似于 Zeng 和 Zhao (2013) 的方法研究了容量限制型设施选址问题, 并通过在每次迭代中求解两次第二阶段子问题, 推导出有效的不等式以加速算法. Wang 等 (2016) 研究了原料产量不确定的多阶段资源回收设施选址问题, 并分别设计了仿射决策规则方法和 Benders 分解方法来近似和精确求解所构建的多阶段鲁棒设施选址模型. Ardestani-Jaafari 和 Delage (2018) 研究了多阶段鲁棒设施选址-运输问题, 并构建了多阶段自适应鲁棒设施选址模型, 且提出了一个高效计算的逼近模型. Wang 和 Ng (2019) 研究了原料产量和组成成分同时具有不确定性的两阶段自适应鲁棒资源回收设施选址问题, 并构建了预算不确定集来刻画两种不确定性. 为了求解该问题, 作者提出了一种融合二分搜索、线性化与 Benders 分解的精确算法. Wang 等 (2021a) 基于需求的均值和支撑集信息提出了目标驱动型自适应鲁棒设施选址优化问题.

1.2.3 分布鲁棒设施选址问题

分布鲁棒优化结合了随机规划与鲁棒优化的建模思想, 可以有效处理参数概率分布不确定的设施选址问题. 分布鲁棒优化模型通过构建包含不确定参数概率分布的分布不确定集来优化极值概率分布下的期望损失 (Bertsimas et al., 2019; Delage and Ye, 2010; Wiesemann et al., 2014; Gao and Kleywegt, 2023). 在假定不确定参数的均值和方差属于给定的分布不确定集的情况下, Gülpınar 等 (2013) 研究了单阶段分布鲁棒设施选址问题. Lu 等 (2015) 研究了一个具有不确定中断概率的可靠性设施选址问题, 并考虑了不同设施中断概率分布之间的相关性. Liu 等 (2019) 研究了一个具有分布鲁棒机会约束的应急医疗服务站设施选址和路径规划问题, 所提出的模型可以等价地转化为一个混合整数二阶锥规划问题. Wang 等 (2020) 在运输成本和商品需求同时具有不确定性的情况下, 研究了枢纽设施选址问题, 并构建了自适应分布鲁棒枢纽设施选址模型, 其中分布不确定集包含均值、边际离差和绝对离差等概率分布信息. Saif 和 Delage (2021) 利用 Givens 和 Shortt (1984) 提出的 Wasserstein 球构建需求分布不确定集, 研究了分布鲁棒容量限制型设施选址问题, 其中 Wasserstein 距离由 ℓ_1 范数定义. 在单阶段情况下, 作者将模型等价地转化为混合整数线性规划问题. 对于两阶段问题, 由于 Wasserstein 距离中使用了 ℓ_1 范数, 该问题可以转化为基于多面体不确定集的自适应鲁棒优化问题. 最近, Shehadeh 和 Sanci (2021) 构建了基于双峰 (bimodal) 分布不确定集的分布鲁棒设施选址模型. 此外, Basciftci 等 (2021) 研究了基于决策依赖分布不确定集的分布鲁棒设施选址问题, 并将所提出的分布鲁棒设施选址模型转化为一个混合整数线性规划问题. 基于 ∞ 型 Wasserstein 分布不确定集, Wang 等 (2021b) 研究了多阶段分布鲁棒仓库选址模型的构建. Liu 等 (2022) 研究了不确定商品需求与社会经济因素、时间、季节等状态信息相关下的设施选址问题; 作者基于一般锥结构的状态依赖分布不确定集, 构建了两阶段分布鲁棒设施选址模型, 并重点分析了供应链网络的韧性以及模型的精确求解算法. 他们提出了一种嵌套次梯度分解算法, 该算法可用于求解带有一般锥结构不确定性的两阶段随机鲁棒选址优化问题, 并从算法结构上统一了两阶段动态随机鲁棒优化算法与求解两阶段动态随机优化问题的 L 型 (L-shaped) 算法.

上述多种不确定环境往往同时存在, 并且不确定性参数与决策变量可能交互影响 (Basciftci et al., 2021; Cheng et al., 2021; Wang et al., 2020). 同时, 随着大数据技术的发展, 海量数据的采集、清洗和处理变得简便, 这使得我们可以获得更多以往无法准确获得的参数信息, 进而形成各种数据驱动型优化建模方法 (Bertsimas et al., 2018, 2023; Delage and Ye, 2010; Mohajerin Esfahani and

Kuhn, 2018). 基于上述两点, 本书主要介绍如何构建复杂不确定环境下的数据驱动型设施选址优化模型. 具体而言, 第 3 章和第 4 章分别介绍需求和运输成本同时具有不确定性以及不确定性参数具有状态依赖特性下的数据驱动型分布鲁棒优化设施选址模型的构建及求解. 第 5 章以能源回收系统设施选址问题为背景, 考虑了原料量和成分同时具有不确定性下的数据驱动型自适应鲁棒优化设施选址模型的构建, 并提出了评估系统经济可行性的鲁棒性指标. 进一步, 第 6 章主要探讨原料不确定性具有分类决策依赖的公私合作下能源回收设施选址问题. 针对不确定环境下多阶段设施选址问题, 第 7 章和第 8 章分别探讨基于预测模型的数据驱动型多阶段鲁棒优化及分布鲁棒优化设施选址模型的构建.

本书以实际问题为背景探讨了不确定环境下多种数据驱动型设施选址优化模型的构建及其求解方法, 不但丰富了设施选址建模理论, 而且为决策者提供了更具实用价值的数据驱动建模方法, 提升了选址决策的质量和企业对抗不确定性影响的能力.

1.3 分布鲁棒优化与数据驱动方法

分布鲁棒优化 (distributionally robust optimization) 作为不确定决策与优化问题最重要的建模范式之一, 在过去十年已经取得了重要的发展与应用 (Delage and Ye, 2010; Mohajerin Esfahani and Kuhn, 2018; Rahimian and Mehrotra, 2019; Gao and Kleywegt, 2023). 粗略地讲, 分布鲁棒优化是将基于给定概率分布的随机规划模型推广至不确定概率分布情形下的一种抵抗分布不确定性的最优化方法. 因此, 从建模的角度, 分布鲁棒优化方法最大的优势在于其不但利用了已有的概率分布信息, 而且同时具备对抗分布不确定性的鲁棒性机制 (Delage and Ye, 2010; Wiesemann et al., 2014; Bertsimas et al., 2019).

在数据科学不断发展的大趋势下, 数据驱动鲁棒优化模型的搭建正在成为分布鲁棒优化理论的重要研究热点. Hu 和 Hong (2013) 首先基于 ϕ-散度 (ϕ-divergence) 函数构建了具有一定置信水平的概率分布置信域, 并基于该 ϕ-divergence 置信域构造了分布不确定集 (ambiguity set) 与分布鲁棒优化问题, 并探讨了其数学性质. Mohajerin Esfahani 和 Kuhn (2018) 以及 Zhao 和 Guan (2018) 将 Wasserstein 距离引入数据驱动鲁棒优化, 并分别从期望值目标与风险目标研究了基于 Wasserstein 距离的分布鲁棒优化问题, 并证明了对于单阶段问题, 这类 Wasserstein 数据驱动分布鲁棒优化模型可以等价地转化为多项式可解的凸优化问题. Chen 等 (2020) 提出了一类事件依赖 (event-wise) 分布鲁棒优化建模框架, 能够将上述基于统计距离的分布鲁棒优化模型进行统一表示. 最近, Qi 等 (2022) 以及 Zhang 等 (2023) 基于 Wasserstein 分布鲁棒优化方法分别研究了分位数回归 (quantile regression) 以及考虑商品需求特征的报童问题 (feature-based newsvendor problem) 等单维度统计及运营决策问题. Hu 等 (2022) 基于时间序列预测模型, 开发了预测驱动型多阶段分布鲁棒优化建模框架, 并基于此框架研究了多阶段枢纽设施选址问题. 此外, 分布鲁棒优化方法还可与统计机器学习方法相结合, 进一步提升模型的数据驱动性 (Blanchet et al., 2019; Kuhn et al., 2019; Hao et al., 2020; Liu and Li, 2021). 例如, 基于多元回归树的数据驱动分布鲁棒优化模型在出租车调度问题中的应用, 以及基于马氏距离 (Mahalanobis distance) K-均值 (K-means) 的数据驱动分布鲁棒优化模型在能源回收系统选址问题中的应用 (Hao et al., 2020; Liu and Li, 2021).

最近, Liu 等 (2023) 针对分布鲁棒优化模型的不足, 将分布不确定性 (ambiguity) 与模型错误 (missspecification) 同时融入分布鲁棒优化模型中, 进而提出全局分布鲁棒优化 (globalized distributionally robust counterpart) 模型. 具体而言, 全局分布鲁棒优化模型一方面要求鲁棒性约束对分布不确定集中的概率分布严格成立; 另一方面对于不在分布不确定集中的概率分布 (即发生模型错误), 则允许鲁棒性约束在一定程度上被违反, 而违反的程度由模型错误容忍度水平控制. 同时, 通过选择合理的模型输入参数 (分布不确定集和模型错误容忍度水平), 全局分布鲁棒优化模型能够退化到现有文献中各类不确定优化模型, 包括经典的鲁棒优化、全局鲁棒优化、分布鲁棒优化、随机规划以及目标鲁棒性优化模型.

1.4 本书的结构

本书重点介绍不确定环境下设施选址问题的数据驱动优化方法, 按模型复杂度, 从初级基于有限概率分布信息的优化模型, 到高阶基于时间序列预测信息的优化模型进行介绍. 主要内容包括: 基于矩信息和统计距离的分布鲁棒优化设施选址模型, 基于时间序列预测集的多阶段分布鲁棒优化设施选址模型, 以及不确定环境下的结构化鲁棒设施选址模型. 本书前三章主要介绍选址优化问题的背景与不确定最优化方法. 其中, 第 1 章为引言, 介绍设施选址问题的背景, 不确定环境下设施选址问题的研究概述, 以及数据驱动型设施选址模型的相关研究. 第 2 章介绍不确定最优化方法, 包括鲁棒优化、分布鲁棒优化、两阶段随机规划, 以及两阶段分布鲁棒优化方法.

第 3 章和第 4 章介绍基于有限概率分布信息的数据驱动分布鲁棒选址优化方法. 其中, 第 3 章以需求和成本不确定下的枢纽设施选址问题为背景, 基于绝对离差与交叉离差等概率分布信息构建并分析分布鲁棒枢纽选址模型. 第 4 章介绍状态依赖 (state-wise) 需求不确定下的设施选址问题, 基于状态信息构建一般锥结构分布不确定集, 并建立状态依赖数据驱动型分布鲁棒设施选址模型且设计精确求解算法.

第 5 章和第 6 章介绍资源回收系统设施选址问题的结构化鲁棒优化模型. 第 5 章在原料产量 (feedstock volume) 与原料组成成分 (feedstock composition) 具有不确定性的情况下, 结合决策相关 (decision-dependent) 动态鲁棒优化以及逆优化思想, 开发一种系统鲁棒性指标 (robustness index), 作为资源回收系统的决策优化指标. 进一步, 在该指标体系下, 介绍一套针对资源回收系统运营环境 (operating environment) 的多目标鲁棒性分析框架. 第 6 章在公私合作模式 (public-private partnership, PPP) 的框架下, 考虑带有地方政府与资源回收系统运营商的资源回收系统联合运营优化问题. 通过构造一个决策依赖型分布不确定集 (decision-dependent ambiguity set) 来建模政府干预对不确定参数概率分布的影响, 并在分布鲁棒优化框架下将上述问题抽象为具有不确定分布的 Stackelberg 博弈模型. 通过此模型, 对不确定环境下地方政府决策-能源运营商决策进行协同决策分析.

第 7 章和第 8 章介绍基于时间序列预测集的多阶段 (分布) 鲁棒选址优化方法. 第 7 章考虑一个基于预测不确定集的资源回收设施选址问题. 基于原料增长量预测模型 (feedstock growth forecasting model) 和决策者对预测误差的容忍水平 (user-specified levels of forecasting errors), 构建原料产量预测不确定集 (predictive uncertainty set) 来建模多阶段原料产量不确定性, 并给出基于此不确定集的多阶段鲁棒资源回收设施选址模型. 第 8 章考虑具有商品需求分布不确定性的多阶段枢纽选址问题 (multi-period hub location problem). 基于时间序列 (time series) 预测模型, 构建基于 Wasserstein 距离的嵌套型分布不确定集 (nested ambiguity set), 并建立一个预算驱动型 (budget-driven) 多阶段枢纽选址模型.

最后, 附录 A 包括本书所有引理、命题与定理的证明过程. 附录 B 为本书所用到的凸优化基础理论.

1.5 本书的使用方法

本书的知识结构由易到难, 从基础的不确定优化建模理论到高阶的基于预测模型的数据驱动型优化建模方法. 本书可作为高年级本科生或研究生的 “数据驱动优化建模” 课程的教学用书, 还可作为 “不确定环境下数据驱动优化建模” 领域的科研工具书. 对于初学者, 可先学习前三章关于优化建模的基础知识, 进一步学习第 3 章和第 4 章中基础的数据驱动型分布鲁棒优化建模方法, 随后从第 5 章开始学习高阶的数据驱动型分布鲁棒优化建模方法. 对于有一定不确定优化建模理论基础的读者, 可直接从第 3 章开始学习.

1.6 相关数学符号说明

本节对本书中通用的数学符号进行定义和说明. 我们用 $\mathbb{R}^N$ 来表示 N 维实数空间, 用加黑的小写字母表示向量, 用加黑的大写字母表示矩阵, 例如, $\boldsymbol{x} \in \mathbb{R}^N$ 和 $\boldsymbol{A} \in \mathbb{R}^{M\times N}$ 分别表示 $\boldsymbol{x}$ 是一个 N 维的实向量, $\boldsymbol{A}$ 是一个 $M \times N$ 的实矩阵. 此外, 我们用 $\mathcal{P}(\Xi)$ 来表示以集合 Ξ 为支撑集的所有概率分布的集合, 用波浪号加向量的形式 $\tilde{\boldsymbol{z}}$ 表示随机变量, 用 $\tilde{\boldsymbol{z}} \sim \mathbb{P}, \mathbb{P} \in \mathcal{P}(\mathbb{R}^I)$ 将 $\tilde{\boldsymbol{z}}$ 定义为遵循分布 $\mathbb{P}$ 的 I 维随机变量, 其中 $\mathbb{P}[\cdot]$ 和 $\mathbb{E}_{\mathbb{P}}[\cdot]$ 分别表示相应的概率和数学期望. 记 $[N] = \{1, 2, \cdots, N\}$ 为元素从 1 到 N 的整数集合. 给定两个整数 a 和 b, 且 $b > a$, 定义整数集合 $[a : b] := \{a, a+1, \cdots, b\}$. 最后, 本书中所有数值实验结果都是在内存为 64GB 且 CPU 为 Intel Core™ 3.6 GHz 的计算机上使用 Gurobi 优化求解器计算获得.

第 2 章　鲁棒与随机优化模型概要

本章主要介绍与本书内容相关的鲁棒与随机优化建模方法. 针对单阶段 (single stage) 问题, 介绍鲁棒优化模型的一般形式及其鲁棒对等 (robust counterpart) 问题、不同种类不确定集 (uncertainty set) 和对应的线性鲁棒对等问题的高效求解形式 (tractable reformulation), 以及一般情况下鲁棒对等问题的转换. 接下来, 介绍分布鲁棒优化建模方法, 并分别给出了基于矩分布不确定集 (moment-based ambiguity set) 与基于 Wasserstein 距离分布不确定集 (Wasserstein-distance based ambiguity set) 的对偶转化理论. 针对两阶段 (two stage) 问题, 本章介绍考虑补偿 (recourse) 的两阶段随机优化模型以及两阶段分布鲁棒优化模型的相关研究成果. 本章的主要结构如下: 2.1 节主要介绍鲁棒优化模型的构建以及鲁棒对等问题的转化方式. 2.2 节主要介绍分布鲁棒优化建模方法及其高效计算形式的转化. 2.3 节主要介绍两阶段随机规划模型以及求解此模型的 L-shaped 精确算法. 2.4 节主要介绍两阶段分布鲁棒优化模型及其相关研究成果.

2.1　鲁 棒 优 化

在本节中, 我们首先介绍如下具有一般形式的线性优化问题:

$$\begin{aligned}\min_{\boldsymbol{x}} \quad & \boldsymbol{c}^\top \boldsymbol{x} \\ \text{s.t.} \quad & \boldsymbol{A}\boldsymbol{x} \geqslant \boldsymbol{b},\end{aligned}$$

其中, $\boldsymbol{x} \in \mathbb{R}^N$ 为决策变量, 矩阵 $\boldsymbol{A} \in \mathbb{R}^{M\times N}$, 向量 $\boldsymbol{b} \in \mathbb{R}^M$ 和 $\boldsymbol{c} \in \mathbb{R}^N$ 分别为问题约束的系数矩阵、约束的右端向量以及目标函数的系数向量. 上述系数向量和矩阵一般由历史数据估计得来. 因此, 优化问题中的系数 (或参数) 往往具有不确定性. 因为准确地估计这些实际应用问题中的参数一般来说是不可能的, 所以我们将这些参数视为不确定的输入参数.

现有文献表明, 在线性规划问题中, 相对于约束右端系数 $\boldsymbol{b}$ 的扰动, 约束系数矩阵 $\boldsymbol{A}$ 的小幅扰动会导致约束严格不可行 (strictly infeasible). 同时, 目标函数中系数 $\boldsymbol{c}$ 扰动也会导致模型产生较大的偏差. 那么, 当模型参数 (或模型输入) 具有不确定性时, 此类问题的处理成为一大挑战. 鲁棒优化作为处理上述参数不确定性问题的方法, 被广泛应用于供应链管理、库存管理、金融以及收益管理

等实际问题中 (Ben-Tal and Nemirovski, 2002; Bertsimas et al., 2011; Gorissen et al., 2015). 在本节中, 我们将重点介绍鲁棒优化建模方法及其高效求解形式 (tractable reformulation). 由于具有目标函数系数不确定性的问题可以等价地转化为只具有约束参数不确定的问题, 而右端项亦可表达为一种 "系数", 因此我们在本节中只考虑约束中系数的不确定性, 即约束参数矩阵 $\boldsymbol{A}$ 的不确定性.

1973 年, Soyster (1973) 首次利用鲁棒优化思想来解决线性规划问题中的不确定性. Mulvey 等在 1995 年首次提出鲁棒优化的概念. 随后, Ben-Tal、Nemirovski 和 El Ghaoui 的研究又进一步丰富了鲁棒优化建模理论并构建了鲁棒优化建模框架 (Ben-Tal and Nemirovski, 1998; El Ghaoui et al., 1998; Ben-Tal and Nemirovski, 1999, 2000; Ben-Tal et al., 2009). 鲁棒优化方法假定模型的不确定参数在给定的不确定集内变动, 并要求鲁棒约束对于不确定集内的所有参数都要成立. 具体而言, 给定参数不确定集 $\mathcal{U}$, 我们考虑如下鲁棒线性优化问题:

$$\begin{aligned} \min_{\boldsymbol{x}} \quad & \boldsymbol{c}^\top \boldsymbol{x} \\ \text{s.t.} \quad & \boldsymbol{A}\boldsymbol{x} \geqslant \boldsymbol{b},\ \forall \boldsymbol{A} \in \mathcal{U}. \end{aligned}$$

上式一般被称为鲁棒对等 (robust counterpart) 问题. 这里需要说明的是, 上述问题中的鲁棒约束 (robust constraint) $\boldsymbol{A}\boldsymbol{x} \geqslant \boldsymbol{b}, \forall \boldsymbol{A} \in \mathcal{U}$ 指的是对于任意的参数 $\boldsymbol{A} \in \mathcal{U}$, 该约束都必须成立. 可以看出, 正是鲁棒约束使得上述鲁棒优化模型的最优解能够抵抗参数不确定性的影响. 一般来说, 上述鲁棒对等问题是一个半无限规划 (semi-infinite programming), 因此无法直接利用线性规划算法进行求解.

2.1.1　不确定集与鲁棒对等问题

在本节中, 我们列举一些基于常用不确定集的鲁棒对等问题及其高效求解形式. 不失一般性, 我们考虑如下只含有单个约束的鲁棒对等问题:

$$\begin{aligned} \min_{\boldsymbol{x}} \quad & \boldsymbol{c}^\top \boldsymbol{x} \\ \text{s.t.} \quad & \boldsymbol{a}^\top \boldsymbol{x} \geqslant b,\ \forall \boldsymbol{a} \in \mathcal{U}, \end{aligned} \tag{2.1}$$

其中, 向量 $\boldsymbol{a} \in \mathbb{R}^N$ 对于鲁棒约束 $\boldsymbol{a}^\top \boldsymbol{x} \geqslant b,\ \forall \boldsymbol{a} \in \mathcal{U}$, 我们可以理解为约束条件 $\boldsymbol{a}^\top \boldsymbol{x} \geqslant b$ 对于任意的 $\boldsymbol{a} \in \mathcal{U}$ 都要成立, 或者等价地转化为如下约束:

$$\min_{\boldsymbol{a} \in \mathcal{U}} \boldsymbol{a}^\top \boldsymbol{x} \geqslant b. \tag{2.2}$$

接下来, 我们考虑几种常见的不确定集 $\mathcal{U}$.

例 1 (区间不确定集)　考虑如下区间不确定集 $\mathcal{U}_I$:

$$\mathcal{U}_I = [a_1^-, a_1^+] \times [a_2^-, a_2^+] \times \cdots \times [a_N^-, a_N^+].$$

不难证明, 当 $\mathcal{U} = \mathcal{U}_I$ 时, 鲁棒约束 (2.2) 可以被等价地转化为

$$\sum_{i\in[N]} \mu_i x_i - \sigma_i |x_i| \geqslant b,$$

其中, $\mu_i = \dfrac{a_i^- + a_i^+}{2}, \sigma_i = \dfrac{a_i^+ - a_i^-}{2}, \forall i \in [N]$, 且 $[N] = \{1, 2, \cdots, N\}$ 表示包含从 1 到 N 的正整数集合. 进一步, 引入辅助变量 λ_i 将绝对值函数项线性化后, 基于区间不确定集 $\mathcal{U}_I$ 的鲁棒约束 (2.2) 等价于如下约束:

$$\begin{cases} \displaystyle\sum_{i\in[N]} \mu_i x_i - \sigma_i \lambda_i \geqslant b, \\ \lambda_i \geqslant -x_i, \forall i \in [N], \\ \lambda_i \geqslant x_i, \forall i \in [N]. \end{cases}$$

因此, 基于区间不确定集 $\mathcal{U}_I$ 的鲁棒对等问题 (2.1) 可以等价地转化为一个线性规划问题:

$$\begin{aligned} \min_{\boldsymbol{x},\boldsymbol{\lambda}} \quad & \boldsymbol{c}^\top \boldsymbol{x}, \\ \text{s.t.} \quad & \sum_{i\in[N]} \mu_i x_i - \sigma_i \lambda_i \geqslant b, \\ & \lambda_i \geqslant -x_i, \forall i \in [N], \\ & \lambda_i \geqslant x_i, \forall i \in [N]. \end{aligned}$$

例 2 (范数不确定集) 给定任意非奇异矩阵 $\boldsymbol{M}$, 我们定义范数不确定集 (norm-based uncertainty set) 如下:

$$\mathcal{U}_p^\gamma = \{\boldsymbol{a} : \|\boldsymbol{M}(\boldsymbol{a} - \boldsymbol{\mu})\|_p \leqslant \gamma\}, \quad 其中\ p = 1, 2, \infty.$$

以上三种范数不确定集 $\mathcal{U}_p^\gamma$ 都可以看成 $\boldsymbol{a}$ 与 $\boldsymbol{\mu}$ 之间的偏差经过非奇异矩阵 $\boldsymbol{M}$ 变换之后所形成的集合, 其中参数 γ 称为不确定性的预算 (budget of uncertainty) 注意到, 对于任意的 $p = 1, 2, \infty$, 基于范数不确定集的鲁棒约束条件 (2.2) 可以等价地转化为如下约束:

$$-\max_{\boldsymbol{a}:\|\boldsymbol{M}(\boldsymbol{a}-\boldsymbol{\mu})\|_p \leqslant \gamma} -\boldsymbol{a}^\top \boldsymbol{x} \geqslant b.$$

通过变量代换 $\boldsymbol{d} = \boldsymbol{M}(\boldsymbol{a} - \boldsymbol{\mu})/\gamma$, 我们可以将上述约束变为

$$-\max_{\boldsymbol{d}:\|\boldsymbol{d}\|_p \leqslant 1} \boldsymbol{x}^\top (-\gamma \boldsymbol{M}^{-1} \boldsymbol{d} - \boldsymbol{\mu}) \geqslant b,$$

利用对偶范数的定义 (见附录 B.2 中定义 B.9), 我们进一步将鲁棒约束转化为如下形式:

$$\boldsymbol{x}^\top \boldsymbol{\mu} - \gamma\|(\boldsymbol{M}^{-1})^\top \boldsymbol{x}\|_p^* \geqslant b,$$

其中, $\|\cdot\|_p^*$ 为范数 $\|\cdot\|_p$ 的对偶范数. 因此, 基于范数不确定集的鲁棒对等问题 (2.1) 最终等价于如下凸优化问题:

$$\begin{aligned} \min_{\boldsymbol{x}} \quad & \boldsymbol{c}^\top \boldsymbol{x} \\ \text{s.t.} \quad & \boldsymbol{x}^\top \boldsymbol{\mu} - \gamma\|(\boldsymbol{M}^{-1})^\top \boldsymbol{x}\|_p^* \geqslant b. \end{aligned} \tag{2.3}$$

接下来, 我们给出范数的阶数 p 分别取 1, 2 和 ∞ 的情况下, 上述凸优化问题的具体形式.

(1) 若 $p=1$, 注意到 $\|\boldsymbol{y}\|_1^* = \|\boldsymbol{y}\|_\infty = \max_{i\in[N]} |y_i|$, 引入辅助变量 λ 将绝对值函数线性化后, 范数对等问题 (2.3) 转化为如下线性规划问题:

$$\begin{aligned} \min_{\boldsymbol{x},\lambda} \quad & \boldsymbol{c}^\top \boldsymbol{x} \\ \text{s.t.} \quad & \boldsymbol{x}^\top \boldsymbol{\mu} - \gamma\lambda \geqslant b, \\ & \lambda \geqslant (\boldsymbol{M}^{-1})_i^\top \boldsymbol{x},\ \forall i \in [N], \\ & \lambda \geqslant -(\boldsymbol{M}^{-1})_i^\top \boldsymbol{x},\ \forall i \in [N], \end{aligned}$$

其中, $(\boldsymbol{M}^{-1})_i$ 为矩阵 $\boldsymbol{M}^{-1}$ 的第 i 行.

(2) 若 $p=\infty$, 注意到 $\|\boldsymbol{y}\|_\infty^* = \|\boldsymbol{y}\|_1 = \sum_{i\in[N]} |y_i|$, 引入辅助变量 $\boldsymbol{\lambda}$ 将绝对值函数项线性化后, 范数鲁棒对等问题 (2.3) 转化为如下线性规划问题:

$$\begin{aligned} \min_{\boldsymbol{x},\boldsymbol{\lambda}} \quad & \boldsymbol{c}^\top \boldsymbol{x} \\ \text{s.t.} \quad & \boldsymbol{x}^\top \boldsymbol{\mu} - \gamma \sum_{i\in[N]} \lambda_i \geqslant b, \\ & \lambda_i \geqslant (\boldsymbol{M}^{-1})_i^\top \boldsymbol{x},\ \forall i \in [N], \\ & \lambda_i \geqslant -(\boldsymbol{M}^{-1})_i^\top \boldsymbol{x},\ \forall i \in [N]. \end{aligned}$$

(3) 若 $p=2$, 注意到 $\|\boldsymbol{y}\|_2^* = \|\boldsymbol{y}\|_2$, 范数鲁棒对等问题 (2.3) 转化为如下二阶锥规划问题

$$\begin{aligned} \min_{\boldsymbol{x}} \quad & \boldsymbol{c}^\top \boldsymbol{x} \\ \text{s.t.} \quad & \|(\boldsymbol{M}^{-1})^\top \boldsymbol{x}\|_2 \leqslant \left(\boldsymbol{x}^\top \boldsymbol{\mu} - b\right)/\gamma. \end{aligned}$$

至此, 基于区间不确定集与范数不确定集的鲁棒对等问题的高效求解形式如下:

$$\boldsymbol{a}^\top \boldsymbol{x} \geqslant b, \forall \boldsymbol{a} \in \mathcal{U}_I \iff \underbrace{\sum_{i\in[N]} \mu_i x_i}_{\text{均值}} - \underbrace{\sum_{i\in[N]} \sigma_i |x_i|}_{\text{惩罚项}} \geqslant b,$$

$$\boldsymbol{a}^\top \boldsymbol{x} \geqslant b, \forall \boldsymbol{a} \in \mathcal{U}_p^\gamma \iff \underbrace{\boldsymbol{x}^\top \boldsymbol{\mu}}_{\text{均值}} - \underbrace{\gamma \|(\boldsymbol{M}^{-1})^\top \boldsymbol{x}\|_p^*}_{\text{惩罚项}} \geqslant b.$$

我们发现, 区间不确定集与范数不确定集本质上都是通过不确定参数的均值加上扰动项构造得来, 而它们所对应的鲁棒对等问题的高效求解形式都可写成均值项减去惩罚项的结构, 其中范数不确定集的参数 γ 为用来控制惩罚项的系数.

例 3 (多面体不确定集) 给定矩阵 $\boldsymbol{B}$ 与向量 $\boldsymbol{r}$, 可定义多面体不确定集 $\mathcal{U}_{\mathrm{P}} = \{\boldsymbol{a} : \boldsymbol{B}\boldsymbol{a} \geqslant \boldsymbol{r}\}$. 注意到, 基于多面体不确定集 $\mathcal{U}_{\mathrm{P}}$ 的鲁棒约束 (2.2) 可写为

$$\min_{\boldsymbol{a}:\boldsymbol{B}\boldsymbol{a}\geqslant \boldsymbol{r}} \boldsymbol{a}^\top \boldsymbol{x} \geqslant b.$$

利用线性规划对偶理论 (见附录 B.3), 上述约束等价于

$$\max_{\boldsymbol{p}:\boldsymbol{B}^\top \boldsymbol{p}=\boldsymbol{x},\boldsymbol{p}\geqslant \boldsymbol{0}} \boldsymbol{p}^\top \boldsymbol{r} \geqslant b.$$

因此, 基于多面体不确定集 $\mathcal{U}_{\mathrm{P}}$ 的鲁棒对等问题 (2.1) 等价于如下线性规划问题:

$$\begin{aligned}
\min_{\boldsymbol{x},\boldsymbol{p}} \quad & \boldsymbol{c}^\top \boldsymbol{x} \\
\text{s.t.} \quad & \boldsymbol{p}^\top \boldsymbol{r} \geqslant b, \\
& \boldsymbol{B}^\top \boldsymbol{p} = \boldsymbol{x}, \\
& \boldsymbol{p} \geqslant \boldsymbol{0}.
\end{aligned}$$

2.1.2 一般情况下鲁棒对等问题的转化

在本节中, 我们考虑一般情况下的鲁棒对等问题:

$$\begin{aligned}
\min_{\boldsymbol{x}} \quad & \boldsymbol{c}^\top \boldsymbol{x} \\
\text{s.t.} \quad & f(\boldsymbol{a}, \boldsymbol{x}) \leqslant b, \ \forall \boldsymbol{a} \in \mathcal{U}.
\end{aligned}$$

这里我们假设对于任意给定的 $\boldsymbol{x} \in \mathbb{R}^N$, 函数 $f(\boldsymbol{a}, \boldsymbol{x})$ 关于 $\boldsymbol{a}$ 是凹函数, 且不确定集 $\mathcal{U}$ 具有如下表达形式:

$$\mathcal{U} := \{\boldsymbol{a} = \overline{\boldsymbol{a}} + \boldsymbol{A}\boldsymbol{\xi} \mid \boldsymbol{\xi} \in \mathcal{W}\},$$

其中, $\boldsymbol{A} \in \mathbb{R}^{M\times R}$, 且集合 $\mathcal{W} \subseteq \mathbb{R}^R$ 是一个非空的凸紧集. 利用芬切尔 (Fenchel) 对偶定理 (见附录 B.3), 我们可以给出一般情况下鲁棒对等问题的等价转化形式.

定理 2.1 (鲁棒约束的 Fenchel 对偶不等式, (Ben-Tal et al., 2015)) 假设不确定集 $\mathcal{U}$ 具有以下形式:

$$\mathcal{U} = \left\{\boldsymbol{a} = \boldsymbol{a}^0 + \boldsymbol{A}\boldsymbol{\zeta} \mid \boldsymbol{\zeta} \in \mathcal{Z} \subset \mathbb{R}^L\right\},$$

其中, $\boldsymbol{a}^0 \in \mathbb{R}^M$, $\boldsymbol{A} \in \mathbb{R}^{M\times L}$, $\boldsymbol{\zeta}$ 称为扰动项, 集合 $\mathcal{Z}$ 为非空的凸紧集且 $0 \in \text{ri}(\mathcal{Z})$, $\text{ri}(\mathcal{Z})$ 为集合 Z 的相对内部. 那么鲁棒约束 $\{\boldsymbol{x} \in \mathbb{R}^n \mid f(\boldsymbol{a},\boldsymbol{x}) \leqslant b,\ \forall \boldsymbol{a} \in \mathcal{U}\}$ 等价于如下集合:

$$\left\{\boldsymbol{x} \in \mathbb{R}^n \mid \exists \boldsymbol{v} \in \mathbb{R}^M, \boldsymbol{v}^\top \bar{\boldsymbol{a}} + \delta^*(\boldsymbol{A}^\top \boldsymbol{v} \mid \mathcal{W}) + f_*(\boldsymbol{v},\boldsymbol{x}) \leqslant b\right\},$$

其中, $\delta(\cdot|\mathcal{W})$ 为集合 $\mathcal{W}$ 的示性函数, $\delta^*(\cdot|\mathcal{W})$ 为函数 $\delta(\cdot|\mathcal{W})$ 的共轭函数 (conjugate function) 而 $f_*(\cdot,\boldsymbol{x})$ 为函数 $f(\cdot,\boldsymbol{x})$ 的凹共轭函数 (见附录 B.3 的定义 B.12 和例 39).

利用定理 2.1, 我们可以将一些常见的非线性鲁棒对等问题转化为高效求解形式. 具体的例子可以参考 Ben-Tal 等 (2015) 的研究.

2.2 分布鲁棒优化

分布鲁棒优化 (distributionally robust optimization) 作为处理不确定性决策与最优化问题最重要的建模范式之一, 已经被成功地应用于供应链管理、运营管理、交通运输以及健康医疗管理等领域 (Delage and Ye, 2010; Mohajerin Esfahani and Kuhn, 2018; Rahimian and Mehrotra, 2019; Gao and Kleywegt, 2023). 粗略地讲, 分布鲁棒优化是将基于给定概率分布的随机规划模型推广至概率分布不确定情形的最优化建模方法. 因此, 从建模的角度, 分布鲁棒优化方法最大的优势在于不仅利用了已有的概率分布信息, 而且具备对抗分布不确定性的鲁棒性机制 (Delage and Ye, 2010; Wiesemann et al., 2014; Bertsimas et al., 2019).

在分布鲁棒优化模型中, 分布不确定集 (ambiguity set) 是模型的关键组成部分. 一个好的分布不确定集, 一方面其所包含的概率信息量应该足够丰富, 即能够以较高的置信度包含真实的数据生成分布. 另一方面, 应该尽可能精准, 即排除一些不太可能发生的病态分布, 以防止造成模型的决策过于保守. 现有文献中主要关注两类分布不确定集. 一类是矩分布不确定集 (moment-based ambiguity set), 该集合由所有满足某些矩信息约束的概率分布所组成 (Delage and Ye, 2010; Goh and Sim, 2010; Wiesemann et al., 2014; Chen et al., 2019). 随机规划模型的求解首先需要计算一个高维积分, 这被证明是 NP 难问题 (Hanasusanto and Kuhn, 2018), 而基于矩分布不确定集的分布鲁棒优化模型在很多情况下可转化为凸优化问题进行高效求解 (Wiesemann et al., 2014). 另一类分布不确定集是基于统计距离的分布不确定集 (distance-based ambiguity set), 常见的距离函数有 ϕ-散度 (divergence)(Pflug and Wozabal, 2007; Ben-Tal et al., 2013; Hu and Hong, 2013; Jiang and Guan, 2016) 以及 Wasserstein 距离 (Pflug and Wozabal, 2007; Mohajerin Esfahani and Kuhn, 2018; Blanchet and Murthy, 2019; Gao and

Kleywegt, 2023). 基于概率距离函数的分布不确定集由围绕参照分布 (reference distribution) 给定距离以内的所有概率分布组成. 基于这类分布不确定集的分布鲁棒优化模型不但具有良好的高效计算性质 (Kuhn et al., 2019; Shafieezadeh-Abadeh et al., 2019) , 而且具有优良的统计性质 (Gotoh et al., 2018; Gao et al., 2024) , 尤其是有限样本保证 (finite-sample guarantees)(Mohajerin Esfahani and Kuhn, 2018; Gao, 2023).

分布鲁棒优化的中心思想是通过一个分布不确定集来表示参数概率分布的不确定性, 其决策过程可视为决策者与"自然" (nature) 的博弈. 具体而言, 给定分布不确定集 $\mathcal{F}$, 我们考虑如下分布鲁棒优化问题:

$$\min_{\boldsymbol{x}\in\mathcal{X}} \sup_{\mathbb{P}\in\mathcal{F}} \mathbb{E}_{\mathbb{P}}[f(\boldsymbol{x},\tilde{\boldsymbol{\xi}})], \tag{2.4}$$

其中, $f(\cdot,\cdot)$ 为目标函数, $\boldsymbol{x}$ 为决策变量且其可行集为 $\mathcal{X}$, $\tilde{\boldsymbol{\xi}}$ 为随机变量 (或不确定参数). 不难看出, 若分布不确定集 $\mathcal{F}$ 为单点集, 即 $\mathcal{F}=\{\hat{\mathbb{P}}\}$, 上述问题退化到概率分布 $\hat{\mathbb{P}}$ 下的随机规划问题; 若分布不确定集 $\mathcal{F}$ 由一族单点分布所构成, 即 $\mathcal{F}=\{\delta_{\boldsymbol{\xi}} \mid \boldsymbol{\xi}\in\Xi\}$, 那么上述问题退化到经典的鲁棒优化问题. 接下来, 我们分别介绍基于矩分布不确定集与 Wasserstein 分布不确定集的分布鲁棒优化建模方法.

1. 矩分布不确定集

给定函数 $\boldsymbol{g}:\mathbb{R}^L\mapsto\mathbb{R}^I$ 以及正常锥 $\mathcal{K}$ (proper cone, 见附录 B.1 中例 29), 我们考虑如下广义矩分布不确定集:

$$\mathcal{F}=\{\mathbb{P}\in\mathcal{P}(\mathbb{R}^I) \mid \mathbb{P}[\tilde{\boldsymbol{\xi}}\in\Xi]=1,\ \mathbb{E}_{\mathbb{P}}[\tilde{\boldsymbol{\xi}}]=\boldsymbol{\mu},\ \mathbb{E}_{\mathbb{P}}[g(\tilde{\boldsymbol{\xi}})]\preceq_{\mathcal{K}}\boldsymbol{\nu}\},$$

其中, 我们假设支撑集 Ξ 是闭紧集, 且满足 $\boldsymbol{\mu}\in\operatorname{int}\Xi$ 和 $\boldsymbol{g}(\boldsymbol{\mu})\prec_{\mathcal{K}}\boldsymbol{\nu}$, 其中 $\operatorname{int}\Xi$ 为集合 Ξ 的内点集. 上述分布不确定集中, 第一个约束表示支撑集约束, 即随机变量 $\tilde{\boldsymbol{\xi}}$ 的所有取值在集合 Ξ 中, 第二个约束是随机变量 $\tilde{\boldsymbol{\xi}}$ 的概率分布期望值约束, 最后一个约束通过映射 $\boldsymbol{g}$ 来刻画随机变量 $\tilde{\boldsymbol{\xi}}$ 的概率分布特征. 例如, 通过选取不同形式的映射 $\boldsymbol{g}$ 与锥 $\mathcal{K}$, 最后一个约束条件可以刻画平均绝对离差 (absolute dispersion)、(协) 方差以及高阶矩信息. 进一步, 引入辅助变量 $\boldsymbol{\zeta}$, 我们可推出 $\mathcal{F}$ 对应的提升分布不确定集 (lifted ambiguity set) $\mathcal{G}$ 为

$$\mathcal{G}=\{\mathbb{P}\in\mathcal{P}(\mathbb{R}^I\times\mathbb{R}^L) \mid \mathbb{P}[(\tilde{\boldsymbol{\xi}},\tilde{\boldsymbol{\zeta}})\in\bar{\Xi}]=1,\ \mathbb{E}_{\mathbb{P}}[\tilde{\boldsymbol{\xi}}]=\boldsymbol{\mu},\ \mathbb{E}_{\mathbb{P}}[\tilde{\boldsymbol{\zeta}}]\preceq_{\mathcal{K}}\boldsymbol{\nu}\},$$

其中, 扩展支撑集 $\bar{\Xi}=\{(\boldsymbol{\xi},\boldsymbol{\zeta}) \mid \boldsymbol{\xi}\in\Xi,\ \boldsymbol{g}(\boldsymbol{\xi})\preceq_{\mathcal{K}}\boldsymbol{\zeta}\}$. 根据 Wiesemann 等 (2014) 所提出的提升定理 (lifting theorem) 可知, 分布不确定集 $\mathcal{G}$ 中关于随机变量 $\tilde{\boldsymbol{\xi}}$ 的边缘概率分布所构成的集合即为分布不确定集 $\mathcal{F}$, 即 $\mathcal{F}=\Pi_{\boldsymbol{\xi}}\mathcal{G}$. 因此, 对于任意

给定的 $\boldsymbol{x} \in \mathcal{X}$, 我们有

$$\sup_{\mathbb{P} \in \mathcal{F}} \mathbb{E}_{\mathbb{P}}[f(\boldsymbol{x}, \tilde{\boldsymbol{\xi}})] \ = \ \sup_{\mathbb{P} \in \mathcal{G}} \mathbb{E}_{\mathbb{P}}[f(\boldsymbol{x}, \tilde{\boldsymbol{\xi}})].$$

进一步, 利用强对偶定理 ((Hanasusanto et al., 2017) 引理 7), 上述右端问题的对偶形式可等价地转化为

$$\begin{array}{ll} \min & \alpha + \boldsymbol{\beta}^\top \boldsymbol{\mu} + \boldsymbol{\gamma}^\top \boldsymbol{\sigma} \\ \text{s.t.} & \alpha + \boldsymbol{\beta}^\top \boldsymbol{\xi} + \boldsymbol{\gamma}^\top \boldsymbol{\zeta} \geqslant f(\boldsymbol{x}, \boldsymbol{\xi}), \quad \forall (\boldsymbol{\xi}, \boldsymbol{\zeta}) \in \bar{\Xi}, \\ & \alpha \in \mathbb{R}, \ \boldsymbol{\beta} \in \mathbb{R}^I, \ \boldsymbol{\gamma} \in \mathbb{R}_+^L. \end{array}$$

若扩展支撑集 $\bar{\Xi}$ 是锥可表示 (conic representable) 的集合, 我们便可利用锥对偶理论 (见附录 B.3) 将上述问题中的半无限约束 (semi-infinite constraint) 转为有限不等式约束.

例 4 (平均绝对离差) 考虑如下包含随机变量平均绝对离差 (absolute dispersion) 的分布不确定集:

$$\mathcal{F} = \{\mathbb{P} \in \mathcal{P}(\mathbb{R}^I) \mid \mathbb{P}[\tilde{\boldsymbol{\xi}} \in \Xi] = 1, \ \mathbb{E}_{\mathbb{P}}[\tilde{\boldsymbol{\xi}}] = \boldsymbol{\mu}, \ \mathbb{E}_{\mathbb{P}}[\|\tilde{\boldsymbol{\xi}} - \boldsymbol{\mu}\|_1] \leqslant \boldsymbol{\sigma}\},$$

其中, 我们假设支撑集 Ξ 是非空多面体, 即 $\Xi = \{\boldsymbol{\xi} \in \mathbb{R}^I \mid \boldsymbol{A}\boldsymbol{\xi} \leqslant \boldsymbol{b}\}$. 上述分布不确定集包含所有支撑集为 Ξ, 均值为 $\boldsymbol{\mu}$ 且平均绝对偏差上界为 $\boldsymbol{\sigma}$ 的 (离散或连续型) 概率分布. 因此, 对于任意决策 $\boldsymbol{x} \in \mathcal{X}$, 分布鲁棒优化模型 (2.4) 中内层最大化问题可显式地写为如下形式:

$$\begin{array}{ll} \max & \displaystyle\int_{\Xi} f(\boldsymbol{x}, \boldsymbol{\xi}) \mathrm{d}\mathbb{P}(\boldsymbol{\xi}) \\ \text{s.t.} & \displaystyle\int_{\Xi} \mathrm{d}\mathbb{P}(\boldsymbol{\xi}) = 1, \\ & \displaystyle\int_{\Xi} \boldsymbol{\xi} \mathrm{d}\mathbb{P}(\boldsymbol{\xi}) = \boldsymbol{\mu}, \\ & \displaystyle\int_{\Xi} \|\boldsymbol{\xi} - \boldsymbol{\mu}\|_1 \mathrm{d}\mathbb{P}(\boldsymbol{\xi}) \leqslant \boldsymbol{\sigma}, \\ & \mathbb{P} \in \mathcal{P}(\mathbb{R}^I). \end{array}$$

可以看出, 上述问题是一个半无限规划问题. 假设强对偶条件, 即 Slater 条件 (Slater condition) 成立, 即 $\boldsymbol{\mu} \in \text{int}\ \Xi$ 且 $\boldsymbol{\sigma} > \boldsymbol{0}$, 则利用强对偶定理, 上述问题可以转化

为如下鲁棒优化问题:

$$\begin{aligned}\min \quad & \alpha+\boldsymbol{\beta}^{\top}\boldsymbol{\mu}+\boldsymbol{\gamma}^{\top}\boldsymbol{\sigma}\\ \text{s.t.} \quad & \alpha+\boldsymbol{\beta}^{\top}\boldsymbol{\xi}+\boldsymbol{\gamma}^{\top}\|\boldsymbol{\xi}-\boldsymbol{\mu}\|_1 \geqslant f(\boldsymbol{x},\boldsymbol{\xi}), \forall \boldsymbol{\xi}\in\Xi,\\ & \alpha\in\mathbb{R},\ \boldsymbol{\beta}\in\mathbb{R}^I, \boldsymbol{\gamma}\in\mathbb{R}_+^I.\end{aligned}$$

此时, 我们可以利用 2.1.1 节中的鲁棒优化方法对上述问题进行转化.

2. Wasserstein 分布不确定集

给定任意欧氏空间中的范数 $\|\cdot\|$, 我们可定义 Wasserstein 距离——也称作 Kantorovich-Rubinstein 距离——来度量两个概率分布之间的差异. 关于 Wasserstein 距离的更多介绍, 请参见文献 (Villani et al., 2009).

定义 2.1 给定 $p\in[1,\infty)$, 任意两个概率分布 $\mathbb{P}$ 以及 $\hat{\mathbb{P}}$ 之间的 p 型 Wasserstein 距离定义为

$$d_{\mathrm{W}}^{p}(\mathbb{P},\hat{\mathbb{P}})=\inf_{\mathbb{Q}\in\mathcal{Q}(\mathbb{P},\hat{\mathbb{P}})}\left(\mathbb{E}_{\mathbb{Q}}[\|\tilde{\boldsymbol{\xi}}-\tilde{\boldsymbol{\zeta}}\|^{p}]\right)^{\frac{1}{p}},$$

其中 $\tilde{\boldsymbol{\xi}}\sim\mathbb{P}$, $\tilde{\boldsymbol{\zeta}}\sim\hat{\mathbb{P}}$, 且 $\mathcal{Q}(\mathbb{P},\hat{\mathbb{P}})$ 为空间 $\mathbb{R}^I\times\mathbb{R}^I$ 上所有边缘分布为 $\mathbb{P}$ 以及 $\hat{\mathbb{P}}$ 的联合概率分布所构成的集合. 特别地, 若 $p=\infty$, 那么 ∞ 型 Wasserstein 距离定义为

$$d_{\mathrm{W}}^{\infty}(\mathbb{P},\hat{\mathbb{P}})=\inf_{\mathbb{Q}\in\mathcal{Q}(\mathbb{P},\hat{\mathbb{P}})}\mathbb{Q}\text{-}\operatorname*{ess\,sup}_{\Xi\times\Xi}\|\tilde{\boldsymbol{\xi}}-\tilde{\boldsymbol{\zeta}}\|,$$

其中, 联合概率分布 $\mathbb{Q}$ 的本性上确界 (essential supremum) 定义为

$$\mathbb{Q}\text{-}\operatorname*{ess\,sup}_{\Xi\times\Xi}\|\tilde{\boldsymbol{\xi}}-\tilde{\boldsymbol{\zeta}}\| =\inf\left\{\mathrm{M}\mid\mathbb{Q}[\|\tilde{\boldsymbol{\xi}}-\tilde{\boldsymbol{\zeta}}\|>\mathrm{M}]=0\right\}.$$

基于定义 2.1, 给定任意参考分布 $\hat{\mathbb{P}}$ 以及半径参数 $\theta\geqslant 0$, 我们可以定义 Wasserstein 分布不确定集如下:

$$\mathcal{F}_{\mathrm{W}}^{p}(\theta)=\left\{\mathbb{P}\in\mathcal{P}(\Xi)\mid d_{\mathrm{W}}^{p}(\mathbb{P},\hat{\mathbb{P}})\leqslant\theta\right\}.$$

利用 Hölder 不等式, 可以证明对于任意的 $q\geqslant p\geqslant 1$ 以及概率分布 $\mathbb{P}$, $\hat{\mathbb{P}}$, 有 $d_{\mathrm{W}}^{q}(\mathbb{P},\hat{\mathbb{P}})\geqslant d_{\mathrm{W}}^{p}(\mathbb{P},\hat{\mathbb{P}})\geqslant d_{\mathrm{W}}^{1}(\mathbb{P},\hat{\mathbb{P}})$. 这意味着对于任意 $q\geqslant p\geqslant 1$ 以及 $\theta\geqslant 0$, 有 $\mathcal{F}_{\mathrm{W}}^{q}(\theta)\subseteq\mathcal{F}_{\mathrm{W}}^{p}(\theta)$.

接下来, 我们以 1 型 Wasserstein 距离的分布不确定集为例, 介绍 Wasserstein 分布鲁棒优化模型的高效求解形式转化. 假定随机变量 $\tilde{\boldsymbol{\xi}}$ 的支撑集为 Ξ 且参考分

布 $\hat{\mathbb{P}} = \frac{1}{N}\sum_{n\in[N]}\delta_{\hat{\boldsymbol{\xi}}_n}$, 其中 δ_a 为狄拉克 (Dirac) 函数, 即在 a 点函数值为 1, 其余为 0. 则我们可构建如下基于 1 型 Wasserstein 距离的分布鲁棒优化模型:

$$\min_{\boldsymbol{x}\in\mathcal{X}} \sup_{\mathbb{P}\in\mathcal{F}_{\mathrm{W}}^1(\theta)} \mathbb{E}_{\mathbb{P}}[f(\boldsymbol{x},\tilde{\boldsymbol{\xi}})]. \tag{2.5}$$

进一步, 利用强对偶定理 ((Gao and Kleywegt, 2023) 命题 2), 我们可以得到问题 (2.5) 的对偶问题为

$$\begin{aligned} \min\quad & \alpha\theta + \frac{1}{N}\sum_{n\in[N]}\beta_n \\ \text{s.t.}\quad & \beta_s + \alpha\|\boldsymbol{\xi} - \hat{\boldsymbol{\xi}}_n\| \geqslant f(\boldsymbol{x},\boldsymbol{\xi}), \forall\boldsymbol{\xi}\in\Xi,\ n\in[N] \\ & \alpha\geqslant 0,\ \boldsymbol{\beta}\in\mathbb{R}^N,\ \boldsymbol{x}\in\mathcal{X}. \end{aligned} \tag{2.6}$$

最后, 利用 2.1.2 节中提到的一般鲁棒对等问题转化方法, 我们可以把问题 (2.6) 中的半无限约束转为确定性约束, 从而将问题 (2.5) 转化为确定性凸优化问题.

2.3 两阶段随机规划

考虑具有补偿机制的两阶段随机线性规划 (two-stage stochastic linear programming with recourse) 模型:

$$\begin{aligned} \min_{\boldsymbol{x}}\quad & \mathbb{E}_{\mathbb{P}}\left[Q(\boldsymbol{x},\tilde{\boldsymbol{\xi}})\right] + \boldsymbol{c}^\top\boldsymbol{x} \\ \text{s.t.}\quad & \boldsymbol{x}\in\mathcal{X}, \end{aligned} \tag{2.7}$$

其中, 集合 $\mathcal{X}$ 为第一阶段决策变量 $\boldsymbol{x}$ 的可行集, 定义为 $\mathcal{X} := \{\boldsymbol{x} : \boldsymbol{A}\boldsymbol{x} = \boldsymbol{b}, \boldsymbol{x}\geqslant \boldsymbol{0}\}$, $\tilde{\boldsymbol{\xi}}$ 是随机变量, 其支撑集为 Ξ, 而 $Q(\boldsymbol{x},\boldsymbol{\xi})$ 是给定 $\boldsymbol{x}$ 以及在随机变量 $\boldsymbol{\xi}$ 的实现值下第二阶段问题的目标函数, 其定义如下:

$$\begin{aligned} Q(\boldsymbol{x},\boldsymbol{\xi}) := \min_{\boldsymbol{y}}\quad & \boldsymbol{y}^\top\boldsymbol{q}(\boldsymbol{\xi}) \\ \text{s.t.}\quad & \boldsymbol{T}(\boldsymbol{\xi})\boldsymbol{x} + \boldsymbol{W}(\boldsymbol{\xi})\boldsymbol{y} = \boldsymbol{h}(\boldsymbol{\xi}), \\ & \boldsymbol{y}\geqslant 0. \end{aligned} \tag{2.8}$$

这里 $\boldsymbol{T}(\boldsymbol{\xi})$, $\boldsymbol{q}(\boldsymbol{\xi})$, $\boldsymbol{W}(\boldsymbol{\xi})$, $\boldsymbol{h}(\boldsymbol{\xi})$ 都是含有 $\boldsymbol{\xi}$ 的随机变量, 矩阵 $\boldsymbol{W}(\boldsymbol{\xi})$ 被称为补偿矩阵 (recourse matrix), 变量 $\boldsymbol{y}$ 为第二阶段决策变量, 其依赖于上述随机变量的实现值.

如果存在某个 $\boldsymbol{x}$ 以及 $\boldsymbol{\xi} \in \Xi$, 使得第二阶段问题 (2.8) 无解, 则记 $Q(\boldsymbol{x},\boldsymbol{\xi}) = +\infty$; 若第二阶段问题无界, 则记 $Q(\boldsymbol{x},\boldsymbol{\xi}) = -\infty$. 因此, 两阶段随机规划模型 (2.7) 成立的一个关键的条件是

$$\mathbb{E}_{\mathbb{P}}\left[\left|Q(\boldsymbol{x},\tilde{\boldsymbol{\xi}})\right|\right] < \infty, \quad \forall \boldsymbol{x} \in \mathcal{X}.$$

接下来, 我们给出两阶段随机规划模型 (2.7) 的一些特殊形式.

定义 2.2 (固定补偿以及完全补偿) 如果补偿矩阵 $\boldsymbol{W}(\boldsymbol{\xi})$ 不含有随机变量 $\boldsymbol{\xi}$, 即 $\boldsymbol{W}(\boldsymbol{\xi}) \equiv \boldsymbol{W}, \forall \boldsymbol{\xi} \in \Xi$, 那么我们称该问题为固定补偿 (fixed recourse) 问题. 如果对于任意的 $\boldsymbol{\eta}$, 多面体 $\boldsymbol{P}(\boldsymbol{\eta}) := \left\{\boldsymbol{y} : \boldsymbol{W}\boldsymbol{y} = \boldsymbol{\eta}, \boldsymbol{y} \geqslant \boldsymbol{0}\right\}$ 都是非空的, 那么我们称该问题为完全补偿 (complete recourse) 问题.

根据线性规划对偶理论, 在固定补偿的情形下, 两阶段随机规划问题 (2.7) 是完全补偿问题当且仅当第二阶段问题的对偶问题可行域是有界的, 即对于任意的 $\boldsymbol{q}$, $\Pi(\boldsymbol{q}) := \left\{\boldsymbol{p} : \boldsymbol{W}^\top \boldsymbol{p} \leqslant \boldsymbol{q}\right\}$ 是有界的. 此外, 若 $\Pi(\boldsymbol{q})$ 是非空集合, 那么它的回收锥 (recession cone) 一定仅包含零点.

例 5 (简单补偿) 一类更加特殊的问题是简单补偿 (simple recourse) 问题. 在简单补偿问题中, 矩阵 $\boldsymbol{W} = [\boldsymbol{I}; -\boldsymbol{I}]$, 且矩阵 $\boldsymbol{T}(\boldsymbol{\xi})$ 和向量 $\boldsymbol{q}(\boldsymbol{\xi})$ 都是固定值, 不含有随机元素, 即 $\boldsymbol{T}(\boldsymbol{\xi}) \equiv \boldsymbol{T}, \boldsymbol{q}(\boldsymbol{\xi}) \equiv \boldsymbol{q}, \forall \boldsymbol{\xi} \in \Xi$, 且 $\boldsymbol{q}$ 的各个分量均为正数. 可以看出, 简单补偿问题一定是固定完全补偿问题.

例 6 (相对完全补偿) 另一类重要的特殊问题是相对完全补偿 (relative complete recourse) 问题. 在相对完全补偿的假设下, 对于任意 $\boldsymbol{x} \in \mathcal{X}$, 第二阶段问题的可行域是几乎处处非空的, 即

$$\mathbb{P}\left\{Q(\boldsymbol{x},\tilde{\boldsymbol{\xi}}) < \infty\right\} = 1, \quad \forall \boldsymbol{x} \in \mathcal{X}.$$

不难看出, 两阶段随机优化问题 (2.7) 的求解涉及高维积分 $\mathbb{E}_{\mathbb{P}}[Q(\boldsymbol{x},\tilde{\boldsymbol{\xi}})]$ 的计算. 因此, 这类问题通常很难直接求解. 求解两阶段随机规划模型 (2.7) 的一个切实可行的方法是使用样本 $\{\hat{\boldsymbol{\xi}}_i, i \in [N]\}$ 来对期望值 $\mathbb{E}_{\mathbb{P}}[Q(\boldsymbol{x},\tilde{\boldsymbol{\xi}})]$ 进行估计, 即使用样本概率分布

$$\mathbb{P}\{\tilde{\boldsymbol{\xi}} = \hat{\boldsymbol{\xi}}_i\} = \frac{1}{N}, \quad \forall i \in [N]$$

来近似不确定参数 $\tilde{\boldsymbol{\xi}}$ 的真实分布, 并且在上述样本概率分布下构建问题 (2.7) 的近似形式. 于是, 我们可以得到原问题 (2.7) 的样本平均近似 (sample average approximation, SAA) 形式如下:

$$
\begin{aligned}
\min_{\boldsymbol{x},\boldsymbol{y}_i}\quad & \boldsymbol{c}^\top\boldsymbol{x}+\frac{1}{N}\sum_{i=1}^{N}\boldsymbol{q}_i^\top\boldsymbol{y}_i\\
\text{s.t.}\quad & \boldsymbol{T}_i\boldsymbol{x}+\boldsymbol{W}\boldsymbol{y}_i=\boldsymbol{h}_i,\forall i\in[N],\\
& \boldsymbol{y}_i\geqslant 0,\forall i\in[N],\\
& \boldsymbol{x}\in\mathcal{X},
\end{aligned}
\tag{2.9}
$$

其中, $\boldsymbol{q}_i := \boldsymbol{q}(\hat{\boldsymbol{\xi}}_i)$, $\boldsymbol{T}_i := \boldsymbol{T}(\hat{\boldsymbol{\xi}}_i)$ 及 $\boldsymbol{h}_i := \boldsymbol{h}(\hat{\boldsymbol{\xi}}_i)$, $\forall i\in[N]$. 该方法叫做样本平均近似方法. 利用 SAA 方法, 我们可以将两阶段随机规划问题 (2.7) 近似地转化为线性规划问题 (2.9) 来进行求解. 然而, SAA 方法的缺点是近似问题 (2.9) 的规模可能很大. 例如, 假设随机变量 $\boldsymbol{\xi}$ 的各个分量 $\xi_1,\xi_2,\cdots,\xi_M$ 是独立同分布的, 并且每个 $\xi_m, m\in[M]$ 都服从一个两点分布, 那么需要生成的样本数量是指数级的, 即 $N=2^M$.

接下来, 我们介绍一种可以高效求解问题 (2.9) 的优化算法——L-shaped 算法 (Birge and Louveaux, 2011). L-shaped 算法的中心思想是采用分解迭代的方式求解大规模的线性规划问题, 而在每一次迭代中只需要求解一个小规模的线性规划问题. 下面, 我们简要介绍 L-shaped 算法的求解步骤.

给定一组样本 $\hat{\boldsymbol{\xi}}_i, i\in[N]$, 我们考虑如下两阶段随机规划模型的样本平均近似形式:

$$
\begin{aligned}
\min_{\boldsymbol{x}}\quad & G(\boldsymbol{x})+\boldsymbol{c}^\top\boldsymbol{x}\\
\text{s.t.}\quad & \boldsymbol{x}\in\mathcal{X},
\end{aligned}
\tag{2.10}
$$

其中, 我们定义 $G(\boldsymbol{x}) := \dfrac{1}{N}\sum_{i\in[N]}Q(\boldsymbol{x},\hat{\boldsymbol{\xi}}_i)$, $Q(\boldsymbol{x},\hat{\boldsymbol{\xi}}_i)$ 是给定样本 $\hat{\boldsymbol{\xi}}_i$ 下第二阶段问题:

$$
\begin{aligned}
Q(\boldsymbol{x},\hat{\boldsymbol{\xi}}_i) := \min_{\boldsymbol{y}}\quad & \boldsymbol{q}_i^\top\boldsymbol{y}\\
\text{s.t.}\quad & \boldsymbol{T}_i\boldsymbol{x}+\boldsymbol{W}\boldsymbol{y}=\boldsymbol{h}_i,\\
& \boldsymbol{y}\geqslant 0.
\end{aligned}
$$

此外, 我们假设第二阶段问题是固定补偿问题, 并且对于任意 $\boldsymbol{x}\in\mathcal{X}$ 以及样本 $\hat{\boldsymbol{\xi}}_i$, $Q(\boldsymbol{x},\hat{\boldsymbol{\xi}}_i)<\infty$. 由于第二阶段问题 $Q(\boldsymbol{x},\hat{\boldsymbol{\xi}}_i)$ 是线性规划问题, 我们可以利用线性规划的强对偶定理得到其对偶问题如下:

$$
Q(\boldsymbol{x},\hat{\boldsymbol{\xi}}_i)=\max_{\boldsymbol{p}}\left\{\boldsymbol{p}^\top(\boldsymbol{h}_i-\boldsymbol{T}_i\boldsymbol{x}):\boldsymbol{W}^\top\boldsymbol{p}\leqslant\boldsymbol{q}_i\right\}.
$$

通过上式我们可以看出, 对于任意 $\hat{\boldsymbol{\xi}}_i, i\in[N]$, 函数 $Q(\boldsymbol{x},\hat{\boldsymbol{\xi}}_i)$ 是关于 $\boldsymbol{x}$ 的凸

函数. 此外, 对于每一个对偶问题, 给定 $\boldsymbol{x}$ 我们都可以求出其最优解 $\boldsymbol{p}_i^\star$, 即

$$\boldsymbol{p}_i^\star := \underset{\boldsymbol{p}}{\operatorname{argmax}} \left\{ \boldsymbol{p}^\top (\boldsymbol{h}_i - \boldsymbol{T}_i \boldsymbol{x}) : \boldsymbol{W}^\top \boldsymbol{p} \leqslant \boldsymbol{q}_i \right\}.$$

根据上述定义, 我们可以推导出 $-\boldsymbol{T}_i^\top \boldsymbol{p}_i^\star$ 为 $Q(\boldsymbol{x}, \hat{\boldsymbol{\xi}}_i)$ 在点 $\boldsymbol{x}$ 的次梯度, 即 $-\boldsymbol{T}_i^\top \boldsymbol{p}_i^\star \in \partial Q(\boldsymbol{x}, \hat{\boldsymbol{\xi}}_i)$.

基于上述结果, 我们进一步推导出函数 $G(\boldsymbol{x})$ 在给定点 $\hat{\boldsymbol{x}}$ 处的一个次梯度向量 $\boldsymbol{s}$, 其定义为

$$\boldsymbol{s} := -\frac{1}{N} \sum_{i \in [N]} \boldsymbol{T}_i^\top \boldsymbol{p}_i^\star \in \frac{1}{N} \sum_{i \in [N]} \partial Q(\hat{\boldsymbol{x}}, \hat{\boldsymbol{\xi}}_i) = \partial G(\hat{\boldsymbol{x}}).$$

接下来, 利用次梯度的性质, 我们可以用 $G(\boldsymbol{x})$ 在点 $\hat{\boldsymbol{x}}$ 处的一阶近似 $G(\hat{\boldsymbol{x}}) + \boldsymbol{s}^\top (\boldsymbol{x} - \hat{\boldsymbol{x}})$ 替换 $G(\boldsymbol{x})$, 从而得到问题 (2.10) 的线性近似问题:

$$\min_{\boldsymbol{x} \in \mathcal{X}} \left[G(\hat{\boldsymbol{x}}) + \boldsymbol{s}^\top (\boldsymbol{x} - \hat{\boldsymbol{x}}) \right] + \boldsymbol{c}^\top \boldsymbol{x}.$$

不难发现, 上述近似问题的最优值是原问题 (2.10) 最优值的下界, 即

$$\min_{\boldsymbol{x} \in \mathcal{X}} \left[G(\hat{\boldsymbol{x}}) + \boldsymbol{s}^\top (\boldsymbol{x} - \hat{\boldsymbol{x}}) \right] + \boldsymbol{c}^\top \boldsymbol{x} \leqslant \min_{\boldsymbol{x} \in \mathcal{X}} G(\boldsymbol{x}) + \boldsymbol{c}^\top \boldsymbol{x}.$$

给定一组点 $\{\hat{\boldsymbol{x}}_\ell : \hat{\boldsymbol{x}}_l \in \mathcal{X}, \ell \in [L]\}$, 假设我们可以得到函数 $G(\boldsymbol{x})$ 在每个点 $\hat{\boldsymbol{x}}_\ell$ 处的次梯度向量 $\boldsymbol{s}_\ell, \ell \in [L]$, 那么我们可以构造问题 (2.10) 的近似问题如下:

$$\begin{aligned} \min_{\boldsymbol{x}, \theta} \quad & \theta + \boldsymbol{c}^\top \boldsymbol{x} \\ \text{s.t.} \quad & \theta \geqslant G(\hat{\boldsymbol{x}}_\ell) + \boldsymbol{s}_\ell^\top (\boldsymbol{x} - \hat{\boldsymbol{x}}_\ell), \forall \ell \in [L], \\ & \boldsymbol{x} \in \mathcal{X}. \end{aligned} \tag{2.11}$$

上述问题同样是问题 (2.10) 的下界, 并且是一个规模适中的线性规划问题. 通过求解问题 (2.11), 我们可以得到一个新的解 $\boldsymbol{x}_{L+1}$ 与其对应的次梯度向量 $\boldsymbol{s}_{L+1}$. 然后, 我们可以在问题 (2.11) 的约束条件中增加一个割平面 $\theta \geqslant G(\boldsymbol{x}_{L+1}) + \boldsymbol{s}_{L+1}^\top (\boldsymbol{x} - \boldsymbol{x}_{L+1})$, 进而得到一个新的近似问题 (2.11). 重复上述步骤, 问题 (2.11) 的可行域将逐渐缩小使得其最优值增大, 并且向原问题 (2.10) 的最优值收敛.

最后, 我们考虑 L-shaped 算法的停止条件: 一方面, 我们可以通过求解线性近似问题 (2.11) 得到解 $\boldsymbol{x}^\star$, $\theta^\star$ 与问题 (2.10) 的一个下界 $\theta^\star + \boldsymbol{c}^\top \boldsymbol{x}^\star$; 另一方面, 我们将近似问题 (2.11) 的解 $\boldsymbol{x}^\star$ 代入原问题 (2.10) 中, 则可以求出原问题的一个上界 $G(\boldsymbol{x}^\star) + \boldsymbol{c}^\top \boldsymbol{x}^\star$. 那么给定误差限 ϵ, 算法迭代停止条件可设为上界与下界的差值不超过误差限 ϵ.

2.4　两阶段分布鲁棒优化

在本节中, 我们简要介绍两阶段分布鲁棒优化方法. 给定一个分布不确定集 $\mathcal{F}$, 我们可以构建如下两阶段分布鲁棒优化模型:

$$
\begin{aligned}
\min_{\boldsymbol{x}} \quad & \boldsymbol{c}^\top \boldsymbol{x} + \sup_{\mathbb{P}\in\mathcal{F}} \mathbb{E}_{\mathbb{P}}\big[Q(\boldsymbol{x},\tilde{\boldsymbol{\xi}})\big] \\
\text{s.t.} \quad & \boldsymbol{x} \in \mathcal{X},
\end{aligned}
$$

其中, 问题 $Q(\boldsymbol{x},\tilde{\boldsymbol{\xi}})$ 为固定补偿问题.

在给定均值与协方差矩阵的情况下, Bertsimas 等 (2010) 研究了风险厌恶型两阶段分布鲁棒优化模型. 在一定条件下, 他们证明了该模型可转化为半正定规划 (semi-definite programming) 并且可在多项式时间内进行求解. 当分布不确定集 $\mathcal{F}$ 具有二阶锥结构 (second-order conic representable) 且强对偶条件成立时, Bertsimas 等 (2019) 利用提升定理以及锥对偶理论, 将两阶段分布鲁棒优化模型等价地转化为确定性锥规划 (conic programming) 问题. 然而, 该问题的规模可能很大, 且依赖于第二阶段问题的对偶问题可行域的极值点数量. 因此, Bertsimas 等 (2019) 还提出了一种新的线性决策准则 (linear decision rule) 来近似第二阶段优化问题的最优值, 并且与传统线性决策准则相比能够有效降低该方法的保守性. 此外, 我们也可以利用 2.3 节中提到的 L-shaped 算法来对两阶段分布鲁棒优化模型进行求解, 具体细节可参阅第 4 章中的内容以及 Liu 等 (2022) 的研究. 在上述研究的基础上, Zhen 等 (2018) 将两阶段分布鲁棒优化转化为两阶段鲁棒优化, 并且利用 Fourier-Motzkin 消去法, 高效地移除了冗余的约束. 通过一系列数值实验证明, 在小规模的算例上, 该算法能够快速找到最优解; 而在中等规模的算例上, 该算法能够有效改进线性决策准则的逼近质量. 基于 Wasserstein 分布不确定集, Hanasusanto 和 Kuhn (2018) 将两阶段分布鲁棒优化模型等价地转化为了一个锥规划问题. Saif 和 Delage (2021) 研究基于 Wasserstein 分布不确定集的单阶段以及两阶段设施选址问题, 并提出了一种能够进行高效求解的列-和-约束生成算法 (column-and-constraint generation algorithm, CCG). 此外, 还有部分文献研究了基于 ϕ-散度分布不确定集的两阶段分布鲁棒优化模型, 如 (Bayraksan and Love, 2015; Zhao and Guan, 2015; Jiang and Guan, 2018).

2.5　本 章 小 结

本章主要介绍与本书内容相关的优化建模方法, 包括鲁棒优化、分布鲁棒优化、两阶段随机规划以及两阶段分布鲁棒优化. 基于区间、范数以及多面体不确定集, 我们利用线性规划对偶理论将鲁棒优化模型转化为高效求解形式; 针对一般的不确定集, 我们利用 Fenchel 对偶理论进行模型转化. 此外, 基于矩信息以及 Wasserstein 距离的分布不确定集, 我们也介绍了如何利用锥对偶理论将对应的分布鲁棒优化模型转化为鲁棒优化模型. 对于两阶段问题, 我们简单介绍了两阶段随机规划模型以及两阶段分布鲁棒优化模型相关研究.

由于本书篇幅有限, 还有一些常用的不确定优化模型, 我们不在本书中做一一介绍. 如两阶段鲁棒优化, 作为两阶段分布鲁棒优化的特例 (当分布不确定集中仅包含支撑集信息), 也被广泛应用于设施选址、电力系统以及资源调度等诸多领域. 还有机会约束规划模型, 最早由 Charnes 和 Cooper (1959) 提出, 是投资组合与风险管理领域的常用模型之一, 并且和 (条件) 风险价值等风险度量模型有较强的相关性. 与鲁棒优化模型相比, 该模型允许解在一定程度上不满足约束条件. 由于概率机会约束算子是非凸的, 该类模型通常是 NP 难问题. 因此, 机会约束模型求解难度较大, 一般只能通过近似方法进行逼近求解. 关于两阶段鲁棒优化的相关研究, 读者可以参考文献: (Ben-Tal et al., 2004; Bertsimas and Goyal, 2012; Zeng and Zhao, 2013; Gorissen et al., 2015; Bertsimas and de Ruiter, 2016; Yanikoglu et al., 2019). 关于机会约束规划的相关研究, 读者可以参考文献 (Nemirovski and Shapiro, 2006; Calafiore and El Ghaoui, 2006; Zymler et al., 2013; Xie and Ahmed, 2018; Shen and Jiang, 2021; Xie, 2021; Chen et al., 2024; Küçükyavuz and Jiang, 2022; Jiang and Xie, 2022; Chen et al., 2023; Jiang and Xie, 2023).

第 3 章　两阶段分布鲁棒枢纽选址优化

本章主要介绍商品需求和运输成本不确定下两阶段分布鲁棒枢纽选址模型 (two-stage distributionally robust hub location model) 的构建与其高效求解形式 (tractable reformulation) 的转化. 由于问题包含两类不确定参数, 我们首先在不确定参数相互独立假设下, 利用边际离差 (marginal dispersion) 与交叉离差 (cross dispersion) 等概率分布信息构建分布不确定集. 接下来, 基于此分布不确定集, 我们构造两阶段分布鲁棒容量充足型 (uncapacitated) 枢纽选址模型, 并给出模型的高效计算形式. 其次, 在放松不确定参数相互独立假设下, 我们分别介绍两阶段分布鲁棒容量充足型枢纽选址模型与容量限制型 (capacitated) 枢纽选址模型的构建, 以及它们的高效计算形式. 最后, 我们设计数值实验来比较随机枢纽选址模型、鲁棒枢纽选址模型与本章所介绍的分布鲁棒枢纽选址模型处理参数不确定性问题的效果. 本章主要结构如下: 3.1 节主要介绍不确定性环境下枢纽选址问题的背景. 3.2 节主要介绍确定性容量充足型枢纽选址模型. 3.3 节主要介绍不确定参数相互独立假设下分布鲁棒容量充足型枢纽选址模型. 3.4 节主要介绍不确定参数非独立假设下分布鲁棒容量充足型枢纽选址模型. 3.5 节主要介绍分布鲁棒容量限制型枢纽选址模型. 最后, 3.6 节主要介绍如何通过数值实验来验证分布鲁棒枢纽选址模型处理环境不确定性问题的有效性.

3.1　背　　景

由于经济成本的限制和运输技术的制约, 在起始地-目的地 (origin-destination) 之间直接大规模地运输多种商品实际上是不可行的. 为了解决上述问题, 枢纽-辐射式网络 (hub-and-spoke networks) 运输系统被广泛地应用于物流行业 (Bryan and O'kelly, 1999). 枢纽-辐射式网络运输系统将来自不同起始地的商品集中到枢纽并对具有相同目的地的商品进行统一运输配送, 通过产生经济规模效应达到节约成本与提升运输效率的目的 (Pels, 2021). 枢纽选址问题主要研究枢纽-辐射式网络中枢纽节点的选择与配置问题, 其目标一般是最小化枢纽建设的固定成本 (fixed cost) 与商品运输成本之和. 一般来说, 枢纽选址问题是一个两阶段规划问题 (Campbell and O'Kelly, 2012), 其中, 第一阶段问题为确定最优的枢纽选址决策, 此决策是在观测到未来参数实现值 (realization) 之前确定的, 被称为现时决策 (here-and-now decision). 第二阶段问题是在给定选址决策以及观测到未来参数实现值之后选择最优的商品运输决策. 因此, 第二阶段运输决策被称为等候决策 (wait-and-see decision). 枢纽选址问题中选址决策实施的长期性与问题的两阶段结构导致其在优化建模方面有两大挑战: ① 枢纽运行期内参数的不确定性, 例如, 商品需求和运输成本的不确定性; ② 运输决策针对未来参数实现值的适应性 (adaptivity) 或可行性 (Wang et al., 2020). 接下来, 我们对上述两大挑战进行详细分析.

枢纽选址问题的第一个挑战是其参数的不确定性. 枢纽选址决策作为一个长期决策, 受未来商品需求和运输成本不确定性影响较大. 枢纽选址决策一旦被确定, 被建造的枢纽通常会运行几年甚至更长的时间. 在这段时间内, 商品的需求和运输成本受市场环境波动的影响会发生巨大变化, 并且准确地预测未来商品的需求和运输成本几乎是不可能的. 这导致枢纽选址决策被确定时的参数与其被实施时的参数存在巨大差异. 因此, 枢纽选址决策能否对抗未来不确定的市场环境并保证运输企业的效益是决策者关心的重要问题. 有资料显示, 忽视市场环境的不确定性会导致枢纽选址优化模型的解是次优的 (Snyder, 2006; Unnikrishnan et al., 2009).

枢纽选址问题的另一个挑战来自于其运输决策的适应性. 与选址决策不同, 运输决策是等候决策, 其必须适应未来的参数环境, 以保证关于商品需求和枢纽容量的约束成立, 且使得运输成本最小. 因此, 在第二阶段问题中, 我们需要把约束条件中的商品需求和目标函数中的运输成本建模成不确定参数, 这增加了问题的求解难度. 在两阶段分布鲁棒优化问题中, 对于第二阶段子问题的约束右端向量以及目标函数同时含有不确定参数 (即, 不确定商品需求和运输成本) 情形的求解, 已被认为是 NP 难问题 (Bertsimas and Shtern, 2018; Bertsimas et al., 2019). 因此, 运输决策的适应性导致不确定环境下两阶段枢纽选址问题难以被求解.

3.2 确定性容量充足型枢纽选址模型

在本节中, 我们介绍两种确定性容量充足型枢纽选址模型. 首先, 我们考虑有 N 个备选枢纽地址和 K 种商品的枢纽选址问题, 其中, 每种商品的需求量记为 u_k, 其通过一个或两个枢纽从起始地 o_k 运送到目的地 ς_k. 此外, 第 k 种商品的运输路径用 (o_k, i, j, ς_k) 来表示, 其中 i 和 j 分别表示第 i 个和第 j 个枢纽, 若 $i=j$ 则表示商品在运输过程中只通过一个枢纽. 我们用 c_i 表示枢纽 i 的固定成本, 并用向量 $\boldsymbol{V}=(V_{ijk})_{i,j\in[N],k\in[K]}$ 表示运输成本, 其中 V_{ijk} 表示第 k 种商品通过路径 (o_k, i, j, ς_k) 运输的单位运输成本. 对于问题的决策变量, 我们用向量 $\boldsymbol{x}=(x_i)_{i\in[N]}$ 来表示 0-1 整数选址变量, 其中当且仅当枢纽 i 被选择使用时, $x_i=1$. 我们定义向量 $\boldsymbol{z}=(z_{ijk})_{i,j\in[N],k\in[K]}$ 为第二阶段问题的运输决策变量, 其中 z_{ijk} 表示通过

路径 (o_k, i, j, ς_k) 运输第 k 种商品量.

当需求量和成本 $(\boldsymbol{u}, \boldsymbol{V})$ 给定时, 确定性容量充足型枢纽选址模型 (deterministic uncapacitated hub location model) 可以由如下总量模型 (quantity model) 来表示:

$$
\begin{aligned}
\min \quad & \boldsymbol{c}^\top \boldsymbol{x} + \sum_{k\in[K]} \sum_{i,j\in[N]} V_{ijk} z_{ijk} \\
\text{s.t.} \quad & \sum_{i,j\in[N]} z_{ijk} = u_k, \forall k \in [K], \\
& \sum_{j\in[N]} z_{ijk} + \sum_{j\in[N]\setminus\{i\}} z_{jik} \leqslant x_i u_k,\ \forall i \in [N],\ k \in [K], \\
& z_{ijk} \geqslant 0, \forall i, j \in [N],\ k \in [K], \\
& \boldsymbol{x} \in \{0,1\}^N,
\end{aligned}
\tag{3.1}
$$

其中, 目标函数的第一项为总固定成本, 第二项为总运输成本. 此外, 第一组约束保证每一种商品的需求都会得到满足. 第二组约束限制商品只能通过已被选定的枢纽来运输.

接下来, 我们重新定义运输变量为 $y_{ijk}, i \in [N], j \in [N], k \in [K]$, 其表示通过路径 (o_k, i, j, ς_k) 运输第 k 种商品的量占此商品总量的比例. 此外, 对于每种商品 k, 我们还定义向量 $\boldsymbol{y}_k := (y_{ijk})_{i,j\in[N]}$ 与 $\boldsymbol{y} := (\boldsymbol{y}_k)_{k\in[K]}$. 基于上述变量, 确定性容量充足型枢纽选址总量模型 (3.1) 可以等价地转化为如下确定性容量充足型枢纽选址分数模型 (fraction model):

$$
\begin{aligned}
\min \quad & \boldsymbol{c}^\top \boldsymbol{x} + \sum_{k\in[K]} u_k \sum_{i,j\in[N]} V_{ijk} y_{ijk} \\
\text{s.t.} \quad & \sum_{i,j\in[N]} y_{ijk} = 1, \forall k \in [K], \\
& \sum_{j\in[N]} y_{ijk} + \sum_{j\in[N]\setminus\{i\}} y_{jik} \leqslant x_i, \forall i \in [N],\ k \in [K], \\
& y_{ijk} \geqslant 0, \forall i, j \in [N],\ k \in [K], \\
& \boldsymbol{x} \in \{0,1\}^N.
\end{aligned}
\tag{3.2}
$$

给定分数模型 (3.2) 的最优解 $y^\star_{ijk}$, 我们可以通过等式 $z_{ijk} = y_{ijk} u_k$ 反推出总量模型 (3.1) 的最优解 $z^\star_{ijk}$. 此外, 我们可以看出两个模型的运输变量 z_{ijk} 和 y_{ijk} 关于不同商品种类 K 是可分的, 也就是说总运输成本最小化问题可以被拆成 K 个子问题, 即每种商品运输成本最小化问题. 通过对比上述两个模型, 我们还观察到模型的约束右端向量或不确定需求参数 $\boldsymbol{u}$ 可以转移到目标函数中. 利用上述

两个特征, 我们将分数模型 (3.2) 转化为如下形式:

$$\min_{\boldsymbol{x}\in\{0,1\}^N}\left\{\boldsymbol{c}^\top\boldsymbol{x}+\sum_{k\in[K]}u_k\left[\min_{\boldsymbol{y}_k\in\mathcal{Y}(\boldsymbol{x})}\sum_{i,j\in[N]}V_{ijk}y_{ijk}\right]\right\}, \tag{3.3}$$

其中, 给定选址决策 $\boldsymbol{x}$ 的情况下, 第 k 种商品的运输决策变量 $\boldsymbol{y}_k$ 的可行集 $\mathcal{Y}(\boldsymbol{x})$ 有如下形式:

$$\mathcal{Y}(\boldsymbol{x})=\left\{\boldsymbol{y}\in\mathbb{R}_+^{N^2}\ \middle|\ \begin{array}{l}\displaystyle\sum_{i,j\in[N]}y_{ij}=1,\\ \displaystyle\sum_{j\in[N]}y_{ij}+\sum_{j\in[N]\backslash\{i\}}y_{ji}\leqslant x_i,\quad\forall i\in[N]\end{array}\right\}.$$

接下来, 基于上述确定性枢纽选址模型 (3.3), 我们构建一个可以处理商品需求和运输成本不确定性问题的两阶段分布鲁棒枢纽选址模型, 并推导其高效计算形式.

3.3 不确定参数相互独立下分布鲁棒容量充足型枢纽选址模型

在本节中, 我们主要介绍在不确定商品需求和运输成本相互独立假设下, 两阶段分布鲁棒容量充足型枢纽选址模型的构建, 并给出其与自适应鲁棒枢纽选址模型 (adaptive robust hub location model) 之间的理论联系. 不确定参数相互独立假设主要包括两个方面: 一方面, 商品需求和运输成本之间相互独立; 另一方面, 不同种类商品的运输成本相互独立.

3.3.1 分布不确定集的构建

首先, 我们构造不确定商品需求和运输成本的分布不确定集. 针对需求的不确定性, 我们定义包含不确定需求 $\tilde{\boldsymbol{u}}$ 部分概率分布信息的分布不确定集如下:

$$\mathcal{F}_0=\left\{\mathbb{P}\in\mathcal{P}(\mathbb{R}^K)\ \middle|\ \begin{array}{l}\tilde{\boldsymbol{u}}\sim\mathbb{P},\\ \mathbb{E}_{\mathbb{P}}[\tilde{\boldsymbol{u}}]\in[\boldsymbol{u}^-,\boldsymbol{u}^+],\\ \mathbb{E}_{\mathbb{P}}[\|\boldsymbol{D}_\ell(\tilde{\boldsymbol{u}}-\bar{\boldsymbol{u}})\|_1]\leqslant\sigma_\ell,\quad\forall\ell\in[L_0],\\ \mathbb{P}[\tilde{\boldsymbol{u}}\in\mathcal{U}]=1\end{array}\right\}. \tag{3.4}$$

其中, $\boldsymbol{u}^-$, $\boldsymbol{u}^+$, $\bar{\boldsymbol{u}}$ 和 $(\sigma_\ell)_{\ell\in[L_0]}$ 是由商品历史需求数据估计得到的参数. 我们记 $\|\cdot\|_1$ 为 1 范数. 在分布不确定集 (3.4) 中, 第一个约束表示不确定需求 $\tilde{\boldsymbol{u}}$ 的期

望值是有界的且属于区间 $[\boldsymbol{u}^-, \boldsymbol{u}^+]$. 第二组约束是 $\tilde{\boldsymbol{u}}$ 的期望离差约束, 其共包括 L_0 个约束. 我们利用矩阵 $\boldsymbol{D}_\ell$ 来灵活地表示多种实用的离差, 例如, 边际离差和交叉离差. 最后一个约束是 $\tilde{\boldsymbol{u}}$ 的支撑集约束, 它限制 $\tilde{\boldsymbol{u}}$ 的所有取值在一个多面体 (polyhedron) 集合 $\mathcal{U}$ 中, 其中 $\mathcal{U} = \left\{\boldsymbol{u} \in \mathbb{R}_+^K \mid \boldsymbol{B}_0\boldsymbol{u} \leqslant \boldsymbol{b}_0\right\}$, $\boldsymbol{B}_0 \in \mathbb{R}^{M_0\times K}$ 且 $\boldsymbol{b}_0 \in \mathbb{R}^{M_0}$, 并满足条件 $[\boldsymbol{u}^-, \boldsymbol{u}^+] \subseteq \mathcal{U}$. 此外, 一个常用的支撑集构建方法是将一组历史需求样本点 $\{\boldsymbol{u}^s\}_{s\in[S]}$ 的凸包作为支撑集 $\mathcal{U}$, 即 $\mathcal{U} = \operatorname{conv}(\{\boldsymbol{u}^s\}_{s\in[S]})$ (凸包的定义见附录 B.1 的定义 B.2). 接下来, 我们给出两个期望离差约束的例子.

例 7 (边际离差) 假设对于一些 $\ell \in [L_0]$ 和 $k \in [K]$, 我们有 $\boldsymbol{D}_\ell = \boldsymbol{e}_k$, 其中 $\boldsymbol{e}$ 为第 k 个元素为 1 其余元素为 0 的向量. 基于上述矩阵 $\boldsymbol{D}_\ell$, 我们可以构造如下不确定需求 $\tilde{u}_k$ 的期望边际离差约束:

$$\mathbb{E}_{\mathbb{P}}[\|\boldsymbol{D}_\ell(\tilde{\boldsymbol{u}} - \bar{\boldsymbol{u}})\|_1] = \mathbb{E}_{\mathbb{P}}[\|\boldsymbol{e}_k^\top(\tilde{\boldsymbol{u}} - \bar{\boldsymbol{u}})\|_1] = \mathbb{E}_{\mathbb{P}}[\|\tilde{u}_k - \bar{u}_k\|_1] \leqslant \sigma_\ell,$$

其中, σ_ℓ 是期望边际离差的上界. 上述约束限制了 $\tilde{u}_k$ 围绕其均值 $\bar{u}_k$ 的离散程度.

例 8 (交叉离差) 给定一组不确定需求 $\{\tilde{u}_k\}_{k\in\mathcal{K}\subseteq[K]}$, 我们令 $\boldsymbol{D}_\ell = \sum_{k\in\mathcal{K}} \boldsymbol{e}_k^\top$ 来构造下面的期望交叉离差约束:

$$\mathbb{E}_{\mathbb{P}}\left[\left\|\sum_{k\in\mathcal{K}}(\tilde{u}_k - \bar{u}_k)\right\|_1\right] = \mathbb{E}_{\mathbb{P}}\left[\left\|\sum_{k\in\mathcal{K}}\boldsymbol{e}_k^\top(\tilde{\boldsymbol{u}} - \bar{\boldsymbol{u}})\right\|_1\right] \leqslant \sigma_\ell,$$

其表示每个组中商品的绝对离差之和的期望值有一个上界 σ_ℓ 以刻画组内商品需求的相关性.

对于运输成本, 我们将沿路径 (o_k, i, j, ς_k) 运输第 k 种商品的单位运输成本 (unit shipping cost) 记作 V_{ijk}, 并且假设 V_{ijk} 由单位收集成本 (unit collection cost)、单位运送成本 (unit transportation cost) 和单位配送成本 (unit distribution cost) 三个部分组成. 因此, 我们将单位运输成本表示为如下形式:

$$V_{ijk} = d_{o_{ki}}v_{k1} + d_{ij}v_{k2} + d_{j\varsigma_k}v_{k3} = \boldsymbol{d}_{ijk}^\top\boldsymbol{v}_k,$$

其中, 我们定义向量 $\boldsymbol{v}_k = (v_{k1}, v_{k2}, v_{k3})$ 和 $\boldsymbol{d}_{ijk} = (d_{o_{ki}}, d_{ij}, d_{j\varsigma_k})$, 并且将 v_{k1}, v_{k2} 和 v_{k3} 分别记为单位收集成本、单位运送成本和单位配送成本. 类似地, 我们将 $d_{o_{ki}}$, d_{ij} 和 $d_{j\varsigma_k}$ 分别记为从第 k 种商品的起始地到第 i 个枢纽的距离, 从第 i 个枢纽到第 j 个枢纽的距离以及从第 j 个枢纽到第 k 种商品目的地之间的距离. 我们将单位运输成本定义为上述形式的原因主要有两点: 首先, 在实际情况中, 运输成本与运输距离正相关. 因此, 每条路径的单位运输成本可以通过路径长度与单位成本的乘积来估计; 其次, 除了与路径长度有关, 单位运输成本也可能取决于商品的种类. 因为不同种类商品可能使用不同的运输方式来运输, 所以即使在同一个路径上, 不同种类商品的单位运输成本也可能不同. 例如, 电商会根据客户对运输时间的要求对运输方式进行分类标价, 运输时间为 1 至 2 天的航空运输价格最高, 而运输时间为 4 至 5 天的美国邮政 (USPS) 的价格则相对较低 (Acimovic and Graves, 2014).

接下来, 我们构造第 k 种商品不确定运输成本 $\tilde{\boldsymbol{v}}_k$ 的分布不确定集如下:

$$\mathcal{F}_k = \left\{ \mathbb{P} \in \mathcal{P}(\mathbb{R}^3) \left| \begin{array}{l} \tilde{\boldsymbol{v}}_k \sim \mathbb{P}, \\ \mathbb{E}_{\mathbb{P}}[\tilde{\boldsymbol{v}_k}] \in [\boldsymbol{v}_k^-, \boldsymbol{v}_k^+], \\ \mathbb{E}_{\mathbb{P}}[\|\boldsymbol{C}_{k\ell}(\tilde{\boldsymbol{v}}_k - \bar{\boldsymbol{v}}_k)\|_1] \leqslant \varepsilon_{k\ell}, \quad \forall \ell \in [L_k], \\ \mathbb{P}[\tilde{\boldsymbol{v}}_k \in \mathcal{V}_k] = 1 \end{array} \right. \right\},$$

其中, $\bar{\boldsymbol{v}}_k, \boldsymbol{v}_k^-, \boldsymbol{v}_k^+$ 和 $(\varepsilon_{kl})_{l\in[L_k]}$ 是由历史运输成本数据估计得到的参数. 此外, 不确定运输成本的支撑集为 $\mathcal{V}_k = \left\{\boldsymbol{v} \in \mathbb{R}_+^3 \mid \boldsymbol{B}_k\boldsymbol{v}_k \leqslant \boldsymbol{b}_k\right\}$ 且满足条件 $[\boldsymbol{u}_k^-, \boldsymbol{u}_k^+] \subseteq \mathcal{V}_k$, 其中 $\boldsymbol{B}_k \in \mathbb{R}^{M_k\times 3}$ 和 $\boldsymbol{b}_k \in \mathbb{R}^{M_k}$. 与分布不确定集 $\mathcal{F}_0$ 类似, 分布不确定集 $\mathcal{F}_k$ 可以通过矩阵 $\boldsymbol{C}_{k\ell} \in \mathbb{R}^{N_{k\ell}\times K}$, $\ell \in [L_k]$ 来表示收集、运输和配送成本之间的相关性.

最后, 我们假设不确定需求 $\tilde{\boldsymbol{u}}$ 和不确定成本 $\tilde{\boldsymbol{v}} = (\tilde{\boldsymbol{v}}_k)_{k\in[K]}$ 是相互独立的, 且不同种类商品的运输成本 $\boldsymbol{v}_1, \cdots, \boldsymbol{v}_K$ 也是相互独立的. 基于上述假设, 我们可以构造不确定参数 $(\tilde{\boldsymbol{u}}, \tilde{\boldsymbol{v}})$ 联合分布不确定集如下:

$$\mathcal{F} = \left\{ \mathbb{P} \in \mathcal{P}(\mathbb{R}^K \times \mathbb{R}^{3K}) \;\middle|\; \mathbb{P} = \bigotimes_{k\in[K]\cup\{0\}} \mathbb{P}_k,\ \mathbb{P}_k \in \mathcal{F}_k\ \forall k \in [K] \cup \{0\} \right\}, \tag{3.5}$$

其中, "$\bigotimes$" 是概率分布的直乘算子. 此外, 我们将在 3.4 节和 3.5 节中介绍放松不确定参数相互独立假设下分布不确定集的构建.

在不确定参数相互独立假设下, 我们利用直乘算子构造不确定参数 $(\tilde{\boldsymbol{u}}, \tilde{\boldsymbol{v}})$ 的联合概率分布不确定集 $\mathcal{F}$ 以整合分布不确定集 $\mathcal{F}_0$ 和 $\mathcal{F}_k$ 的分布信息. 接下来, 我们介绍两阶段分布鲁棒容量充足型枢纽选址模型.

3.3.2 两阶段分布鲁棒容量充足型枢纽选址模型

基于分布不确定集 (3.5), 我们首先构建如下两阶段分布鲁棒容量充足型枢纽选址总量模型:

$$\min_{\boldsymbol{x}\in\{0,1\}^N} \left\{ \boldsymbol{c}^\top \boldsymbol{x} + \sup_{\mathbb{P}\in\mathcal{F}} \mathbb{E}_{\mathbb{P}}[\mathrm{Q_u}(\boldsymbol{x}, \tilde{\boldsymbol{u}}, \tilde{\boldsymbol{v}})] \right\}. \tag{3.6}$$

给定选址决策 $\boldsymbol{x}$ 和 $(\boldsymbol{u}, \boldsymbol{v})$, 上述模型的第二阶段问题 $\mathrm{Q_u}(\boldsymbol{x}, \boldsymbol{u}, \boldsymbol{v})$ 可以表示为如下形式:

$$\mathrm{Q_u}(\boldsymbol{x},\boldsymbol{u},\boldsymbol{v}) = \min_{\boldsymbol{z}\in\mathcal{Z}(\boldsymbol{x},\boldsymbol{u})} \sum_{k\in[K]} \sum_{i,j\in[N]} \boldsymbol{d}_{ijk}^{\top}\boldsymbol{v}_k z_{ijk},$$

其中, 运输决策 $\boldsymbol{z}$ 的可行集为

$$\mathcal{Z}(\boldsymbol{x},\boldsymbol{u}) = \left\{ \boldsymbol{z}\in\mathbb{R}_+^{KN^2} \;\middle|\; \begin{array}{l} \displaystyle\sum_{i,j\in[N]} z_{ijk} = u_k,\ \forall k\in[K], \\ \displaystyle\sum_{j\in[N]} z_{ijk} + \sum_{j\in[N]\setminus\{i\}} z_{jik} \leqslant x_i u_k,\ \forall i\in[N],\ k\in[K] \end{array} \right\}.$$

模型 (3.6) 的目标是寻找一个最优枢纽选址决策, 使得固定成本和极值概率分布下期望运输成本 (worst-case expected shipping cost) 之和最小. 我们记最优枢纽选址决策为 $\boldsymbol{x}^{\star}$, 最优商品运输决策为 $\boldsymbol{z}^{\star}(\cdot)$, 其中, 第二阶段最优商品运输量决策 $\boldsymbol{z}^{\star}(\cdot)$ 是关于不确定参数 $(\boldsymbol{u},\boldsymbol{v})$ 的一个映射, 记为 $z_{ijk}^{\star}(\boldsymbol{u},\boldsymbol{v})$, 即通过路径 (o_k,i,j,ς_k) 运输第 k 种商品的总量.

基于确定性枢纽选址分数模型 (3.2), 我们构造如下两阶段分布鲁棒容量充足型枢纽选址分数模型:

$$\min_{\boldsymbol{x}\in\{0,1\}^N} \left\{ \boldsymbol{c}^{\top}\boldsymbol{x} + \sup_{\mathbb{P}\in\mathcal{F}} \mathbb{E}_{\mathbb{P}}[\mathrm{F_u}(\boldsymbol{x},\tilde{\boldsymbol{u}},\tilde{\boldsymbol{v}})] \right\}. \tag{3.7}$$

根据分数模型 (3.2) 的结构特点, 我们可以将上述模型第二阶段问题表示为如下形式:

$$\mathrm{F_u}(\boldsymbol{x},\boldsymbol{u},\boldsymbol{v}) = \sum_{k\in[K]} u_k \left[\min_{\boldsymbol{y}_k\in\mathcal{Y}(\boldsymbol{x})} \sum_{i,j\in[N]} \boldsymbol{d}_{ijk}^{\top}\boldsymbol{v}_k y_{ijk} \right],$$

其中, 运输决策 y_{ijk} 是指第 k 种商品经过枢纽 i 和 j 的运输量占其总需求量的比例. 同样地, 两阶段分布鲁棒容量充足型枢纽选址分数模型 (3.7) 包括最优枢纽选址决策 $\boldsymbol{x}^{\star}$ 和最优商品需求量运输比例决策 $\boldsymbol{y}^{\star}(\cdot)$, 该决策是 $(\boldsymbol{u},\boldsymbol{v})$ 到 $y_{ijk}^{\star}(\boldsymbol{u},\boldsymbol{v})$ 的一个映射, 表示每种商品通过路径 (o_k,i,j,ς_k) 运输的量占其需求总量的最优比例.

给定选址决策 $\boldsymbol{x}$ 和 $(\boldsymbol{u},\boldsymbol{v})$ 的值, 我们可以看出上述两个模型的第二阶段运输问题 $\mathrm{Q_u}(\boldsymbol{x},\boldsymbol{u},\boldsymbol{v})$ 和 $\mathrm{F_u}(\boldsymbol{x},\boldsymbol{u},\boldsymbol{v})$ 的最优运输决策是等价的. 因此, 在任意分布不确定集 $\mathcal{F}$ 下, 总量模型 (3.6) 和分数模型 (3.7) 都是等价的.

引理 3.1 给定任意分布不确定集 $\mathcal{F}$, 两阶段分布鲁棒容量充足型枢纽选址总量模型 (3.6) 等价于两阶段分布鲁棒容量充足型枢纽选址分数模型 (3.7).

一般情况下, 当分布不确定集只含有不确定参数的支撑集约束时, 分布鲁棒优化模型会退化为鲁棒优化模型 (Wiesemann et al., 2014; Chen et al., 2019). 在

不确定参数相互独立假设下, 我们证明在分布不确定集包含支撑集约束以及其他约束 (例如, 期望值和期望离差约束) 的情形下, 两阶段分布鲁棒容量充足型枢纽选址模型等价于两阶段鲁棒容量充足型枢纽选址模型.

定理 3.1 两阶段分布鲁棒容量充足型枢纽选址分数模型 (3.7) 等价于如下两阶段鲁棒容量充足型枢纽选址模型:

$$\min_{\boldsymbol{x}\in\{0,1\}^N}\left\{\boldsymbol{c}^\top\boldsymbol{x}+\max_{(\boldsymbol{u},\boldsymbol{v})\in\mathcal{W}}\mathrm{F_u}(\boldsymbol{x},\boldsymbol{u},\boldsymbol{v})\right\},\tag{3.8}$$

其中, 基于分布不确定集 $\mathcal{F}$ (3.5), 我们定义如下不确定集:

$$\bar{\mathcal{U}}\triangleq\left\{\boldsymbol{u}\in\mathbb{R}_+^K\mid\boldsymbol{B}_0\boldsymbol{u}\leqslant\boldsymbol{b}_0,\boldsymbol{u}\in[\boldsymbol{u}^-,\boldsymbol{u}^+],\|\boldsymbol{D}_\ell(\boldsymbol{u}-\bar{\boldsymbol{u}})\|_1\leqslant\sigma_\ell\ \forall\ell\in[L_0]\right\},$$

$$\begin{aligned}\bar{\mathcal{V}}_k\triangleq\big\{\boldsymbol{v}_k\in\mathbb{R}_+^3\mid\boldsymbol{B}_k\boldsymbol{v}_k\leqslant\boldsymbol{b}_k,\boldsymbol{v}_k\in[\boldsymbol{v}_k^-,\boldsymbol{v}_k^+],\|\boldsymbol{C}_{k\ell}(\boldsymbol{v}_k-\bar{\boldsymbol{v}}_k)\|_1\\ \leqslant\varepsilon_{k\ell}\ \forall\ell\in[L_k]\big\},\quad\forall k\in[K],\end{aligned}$$

以及关于 $(\boldsymbol{u},\boldsymbol{v})$ 的联合不确定集:

$$\mathcal{W}\triangleq\{(\boldsymbol{u},\boldsymbol{v})\in\mathbb{R}^K\times\mathbb{R}^{3K}\mid\boldsymbol{u}\in\bar{\mathcal{U}},\boldsymbol{v}=(\boldsymbol{v}_1,\cdots,\boldsymbol{v}_K):\boldsymbol{v}_k\in\bar{\mathcal{V}}_k\ \forall k\in[K]\}.\tag{3.9}$$

通过定理 3.1, 我们可以得到在不确定参数相互独立假设下两阶段分布鲁棒容量充足型枢纽选址模型 (3.7) 与两阶段鲁棒容量充足型枢纽选址模型 (3.8) 之间的理论联系: 首先, 我们可以得出两个模型的最优值相等; 其次, 不确定集 $\mathcal{W}$ 使用与分布不确定集 $\mathcal{F}$ 中期望约束形式相似的约束, 来整合分布不确定集 $\mathcal{F}$ 中的期望值和期望离差等概率分布信息. 例如, 假设我们在分布鲁棒优化模型 (3.7) 的分布不确定集 $\mathcal{F}$ 中加入一个期望值约束 $\mathbb{E}_\mathbb{P}[\tilde{\boldsymbol{u}}]\in[\boldsymbol{u}^-,\boldsymbol{u}^+]$. 根据定理 3.1, 我们可得上述操作与在鲁棒优化模型 (3.8) 的不确定集 $\mathcal{W}$ 中添加约束 $\boldsymbol{u}\in[\boldsymbol{u}^-,\boldsymbol{u}^+]$ 的建模效果相同. 此外, 我们可以从随机规划的角度来解释定理 3.1. 假设不确定参数 $(\tilde{\boldsymbol{u}},\tilde{\boldsymbol{v}})$ 的期望值服从在其取值处概率为 1 的 Dirac 分布, 即 $\mathbb{G}=\delta_{(\mathbb{E}_\mathbb{P}(\tilde{\boldsymbol{u}},\tilde{\boldsymbol{v}}))}$, 并且 $\mathbb{G}\in\mathcal{F}$. 则两阶段分布鲁棒容量充足型枢纽选址模型 (3.7) 可以被看作是基于概率分布 $\mathbb{G}$ 的随机枢纽选址问题的鲁棒对等形式 (robust counterpart), 其中不确定参数 $(\tilde{\boldsymbol{u}},\tilde{\boldsymbol{v}})$ 的期望值 $\mathbb{E}_\mathbb{P}(\tilde{\boldsymbol{u}},\tilde{\boldsymbol{v}})$ 是不确定变量, 且属于 $\mathcal{W}$.

3.3.3 最优静态运输决策

利用两阶段鲁棒容量充足型枢纽选址模型 (3.8), 我们现在讨论两阶段分布鲁棒容量充足型枢纽选址模型最优运输决策的结构特征. 因为在不确定参数相互独立假设下, 集合 $\mathcal{W}$ 满足条件 $\mathcal{W}=\bar{\mathcal{U}}\otimes\bar{\mathcal{V}}_1\otimes\cdots\otimes\bar{\mathcal{V}}_K$, 所以模型 (3.8) 的不确定集是约束依赖的 (constraint-wise) (Ben-Tal and Nemirovski, 1999). 根据这个性质,我们证明两阶段鲁棒容量充足型枢纽选址模型 (3.8) 的最优运输决策具有非适应 (nonadaptive) 结构.

定理 3.2 给定选址决策 $\boldsymbol{x}$, 对于任意商品 $k \in [K]$, 其对应的最优运输决策 $\boldsymbol{y}_k^\star$ 是非自适应的, 并且可以通过求解如下鲁棒优化问题来获得

$$\min_{\boldsymbol{y}_k \in \mathcal{Y}(\boldsymbol{x})} \max_{\boldsymbol{v}_k \in \bar{\mathcal{V}}_k} \sum_{i,j \in [N]} \boldsymbol{d}_{ijk}^\top \boldsymbol{v}_k y_{ijk}. \tag{3.10}$$

此外, 两阶段鲁棒枢纽选址模型 (3.8) 等价于如下非自适应鲁棒优化问题 (non-adaptive robust optimization problem):

$$\begin{aligned} \min_{\boldsymbol{x} \in \{0,1\}^N} & \left\{ \boldsymbol{c}^\top \boldsymbol{x} + \max_{\boldsymbol{u} \in \bar{\mathcal{U}}} \sum_{k \in [K]} u_k \lambda_k \right\} \\ \text{s.t.} \quad & \lambda_k \geqslant \sum_{i,j \in [N]} \boldsymbol{d}_{ijk}^\top \boldsymbol{v}_k y_{ijk}, \ \forall \boldsymbol{v}_k \in \bar{\mathcal{V}}_k, \ k \in [K], \\ & \boldsymbol{y}_k \in \mathcal{Y}(\boldsymbol{x}), \ \lambda_k \in \mathbb{R}, \ \forall k \in [K], \end{aligned} \tag{3.11}$$

其中 λ_k, $k \in [K]$ 是辅助变量.

通过定理 3.1 和定理 3.2, 我们可以得到如下结论: 首先, 在不确定参数相互独立假设下, 模型 (3.7) 可以等价地转化为一个非自适应鲁棒优化问题 (3.11); 其次, 我们证明模型 (3.7) 的最优运输决策是静态的 (static). 对于任意 $(\boldsymbol{u}, \boldsymbol{v}) \in \mathcal{W}$, 最优静态运输决策 $\boldsymbol{y}^\star$ 满足 $\boldsymbol{y}^\star(\boldsymbol{u}, \boldsymbol{v}) = \boldsymbol{y}^\star$. 此外, 通过求解一系列线性规划问题 (3.10), 我们可以得到最优静态运输决策 $\boldsymbol{y}^\star$; 最后, 根据分布鲁棒优化理论, 我们可知存在一个关于 $(\tilde{\boldsymbol{u}}, \tilde{\boldsymbol{v}})$ 的极值概率分布 $\mathbb{P}^\star \in \mathcal{F}$, 并且最优运输决策 $\boldsymbol{y}^\star(\cdot)$ "静态" 适应 ("statically" adapt) $\mathbb{P}^\star$. 基于定理 3.1, 我们可知这个极值概率分布是一个 Dirac 分布 $\mathbb{P}^\star = \delta_{(\boldsymbol{u}^\star, \boldsymbol{v}^\star)}$, 其中 $(\boldsymbol{u}^\star, \boldsymbol{v}^\star) \in \mathcal{W}$.

3.3.4 高效计算形式与极值概率分布

接下来, 我们利用非自适应鲁棒枢纽选址问题 (3.11) 来推导分布鲁棒模型 (3.7) 的高效求解形式 (tractable reformulation). 此外, 给定最优选址决策 $\boldsymbol{x}^\star$, 我们给出分布不确定集 (3.5) 中极值概率分布 $\mathbb{P}^\star$ 与最优运输决策 $\boldsymbol{y}^\star(\cdot)$ 的反推方法. 在实际应用中, 决策者可以利用极值概率分布通过压力测试来检验枢纽选址决策的鲁棒性.

利用强对偶理论, 我们可以将鲁棒优化问题 (3.10) 等价地转化为线性规划问题, 以及将鲁棒优化问题 (3.11) 等价地转化为混合整数线性规划问题, 具体结论如下.

定理 3.3 给定分布不确定集 $\mathcal{F}$ (3.5) 和枢纽选址决策 $\boldsymbol{x}$, 对于任意 $k \in [K]$,

第 k 种商品的最优运输决策 $\boldsymbol{y}_k^\star$ 可以通过求解如下线性规划问题来获得,

$$
\begin{aligned}
\min\quad & \overline{\boldsymbol{\phi}}_k^\top \boldsymbol{v}_k^+ - \underline{\boldsymbol{\phi}}_k^\top \boldsymbol{v}_k^- + \boldsymbol{\gamma}_k^\top \boldsymbol{b}_k + \sum_{\ell\in[L_k]} (\eta_{k\ell}\varepsilon_{k\ell} - \boldsymbol{\psi}_k^\top \boldsymbol{C}_{k\ell}\bar{\boldsymbol{v}}_k)\\
\text{s.t.}\quad & \overline{\boldsymbol{\phi}}_k - \underline{\boldsymbol{\phi}}_k + \boldsymbol{B}_k^\top \boldsymbol{\gamma}_k - \sum_{\ell\in[L_k]} \boldsymbol{C}_{k\ell}^\top \boldsymbol{\psi}_{k\ell} - \sum_{i,j\in[N]} \boldsymbol{d}_{ijk} y_{ijk} \geqslant \boldsymbol{0},\\
& \eta_{k\ell}\in\mathbb{R},\ \boldsymbol{\psi}_{k\ell}\in\mathbb{R}_+^{N_{k\ell}},\ \eta_{k\ell}\geqslant \|\boldsymbol{\psi}_{k\ell}\|_\infty, \forall \ell\in[L_k],\\
& \overline{\boldsymbol{\phi}}_k, \underline{\boldsymbol{\phi}}_k \in \mathbb{R}_+^3,\ \boldsymbol{\gamma}_k\in\mathbb{R}_+^{M_k},\ \boldsymbol{y}_k\in\mathcal{Y}(\boldsymbol{x}),
\end{aligned}
\tag{3.12}
$$

其中, $\overline{\boldsymbol{\phi}}_k, \underline{\boldsymbol{\phi}}_k, \boldsymbol{\gamma}_k, \eta_{k\ell}, \boldsymbol{\psi}_{k\ell}$ 是辅助决策变量. 最优枢纽选址决策 $\boldsymbol{x}^\star$ 和静态运输策略 $\boldsymbol{y}^\star$ 可以通过求解如下混合整数线性规划问题来获得,

$$
\begin{aligned}
\min_{\boldsymbol{x}\in\{0,1\}^N}\quad & \boldsymbol{c}^\top \boldsymbol{x} + \overline{\boldsymbol{\phi}}_0^\top \boldsymbol{u}^+ - \underline{\boldsymbol{\phi}}_0^\top \boldsymbol{u}^- + \boldsymbol{\gamma}_0^\top \boldsymbol{b}_0 + \sum_{\ell\in[L_0]} (\eta_\ell\sigma_\ell - \boldsymbol{\psi}_\ell^\top \boldsymbol{D}_\ell \bar{\boldsymbol{u}})\\
\text{s.t.}\quad & \overline{\boldsymbol{\phi}}_0 - \underline{\boldsymbol{\phi}}_0 + \boldsymbol{B}_0^\top \boldsymbol{\gamma}_0 - \sum_{\ell\in[L_0]} \boldsymbol{D}_\ell^\top \boldsymbol{\psi}_\ell \geqslant \boldsymbol{\lambda},\\
& \lambda_k \geqslant \overline{\boldsymbol{\phi}}_k^\top \boldsymbol{v}_k^+ - \underline{\boldsymbol{\phi}}_k^\top \boldsymbol{v}_k^- + \boldsymbol{\gamma}_k^\top \boldsymbol{b}_k + \sum_{\ell\in[L_k]} (\eta_{k\ell}\varepsilon_{k\ell} - \boldsymbol{\psi}_{k\ell}^\top \boldsymbol{C}_{k\ell}\bar{\boldsymbol{v}}_k), \forall k\in[K],\\
& \overline{\boldsymbol{\phi}}_k - \underline{\boldsymbol{\phi}}_k + \boldsymbol{B}_k^\top \boldsymbol{\gamma}_k - \sum_{\ell\in[L_k]} \boldsymbol{C}_{k\ell}^\top \boldsymbol{\psi}_{k\ell} - \sum_{i,j\in[N]} \boldsymbol{d}_{ijk} y_{ijk} \geqslant \boldsymbol{0}, \forall k\in[K],\\
& \eta_\ell\in\mathbb{R},\ \boldsymbol{\psi}_\ell\in\mathbb{R}_+^{N_{0\ell}},\ \eta_\ell\geqslant \|\boldsymbol{\psi}_\ell\|_\infty, \forall \ell\in[L_0],\\
& \overline{\boldsymbol{\phi}}_0, \underline{\boldsymbol{\phi}}_0 \in \mathbb{R}_+^K,\ \boldsymbol{\gamma}_0\in\mathbb{R}_+^{M_0},\ \boldsymbol{\lambda}\in\mathbb{R}^K,\\
& \eta_{k\ell}\in\mathbb{R},\ \boldsymbol{\psi}_{k\ell}\in\mathbb{R}_+^{N_{k\ell}},\ \eta_{k\ell}\geqslant \|\boldsymbol{\psi}_{k\ell}\|_\infty, \forall k\in[K],\ \ell\in[L_k],\\
& \overline{\boldsymbol{\phi}}_k, \underline{\boldsymbol{\phi}}_k \in \mathbb{R}_+^3,\ \boldsymbol{\gamma}_k\in\mathbb{R}_+^{M_k},\ \boldsymbol{y}_k\in\mathcal{Y}(\boldsymbol{x}), \forall k\in[K],
\end{aligned}
\tag{3.13}
$$

其中, $\boldsymbol{\lambda}, \overline{\boldsymbol{\phi}}_0, \underline{\boldsymbol{\phi}}_0, \boldsymbol{\gamma}_0, \eta_\ell, \boldsymbol{\psi}_\ell, \overline{\boldsymbol{\phi}}_k, \underline{\boldsymbol{\phi}}_k, \boldsymbol{\gamma}_k, \eta_{k\ell}, \boldsymbol{\psi}_{k\ell}$ 是辅助决策变量.

给定枢纽选址决策 $\boldsymbol{x}$, 基于定理 3.2, 可以反推出不确定参数 $(\tilde{\boldsymbol{u}}, \tilde{\boldsymbol{v}})$ 的极值概率分布 $\mathbb{P}^\star$, 具体步骤如下.

算法 1　极值概率分布的反推算法

输入: 枢纽选址决策 $\boldsymbol{x}$.

1. 对于任意 $k\in[K]$, 求解鲁棒优化问题

$$
\lambda_k^\star = \min_{\boldsymbol{y}_k\in\mathcal{Y}(\boldsymbol{x})} \max_{\boldsymbol{v}_k\in\bar{\mathcal{V}}_k} \sum_{i,j\in[N]} \boldsymbol{d}_{ijk}^\top \boldsymbol{v}_k y_{ijk},
$$

获得最优运输决策 $\boldsymbol{y}_k^\star$ 和最优辅助变量 $\lambda_k^\star$.

---

2. 对于任意 $k \in [K]$, 固定变量 $\boldsymbol{y}_k$ 的值为 $\boldsymbol{y}_k^\star$ 并求解线性规划问题

$$\boldsymbol{v}_k^\star \in \arg\max_{\boldsymbol{v}_k \in \bar{\mathcal{V}}_k} \sum_{i,j\in[N]} \boldsymbol{d}_{ijk}^\top \boldsymbol{v}_k y_{ijk}^\star,$$

获得极值概率分布下第 k 种商品运输成本为 $\boldsymbol{v}_k^\star$.

3. 固定变量 λ_k 的值为 $\lambda_k^\star$, 求解线性规划问题

$$\boldsymbol{u}^\star \in \arg\max_{\boldsymbol{v}_k \in \bar{\mathcal{V}}_k} \sum_{k\in[K]} u_k \lambda_k^\star,$$

获得极值概率分布下第 k 种商品需求量为 $\boldsymbol{u}^\star$.

输出: 极值概率分布 $\mathbb{P}^\star = \delta_{(\boldsymbol{u}^\star, \boldsymbol{v}^\star)}$.

3.4 不确定参数非独立下分布鲁棒容量充足型枢纽选址模型

在本节中, 我们主要介绍在放松不确定参数相互独立假设下, 两阶段分布鲁棒容量充足型枢纽选址模型的构建. 一般来说, 在模型的两种不确定参数不相互独立时, 我们很难获得模型的高效计算形式. 然而, 通过对不确定参数的支撑集做出合理的假设, 我们可以证明两阶段分布鲁棒容量充足型枢纽选址模型等价于一个混合整数线性规划问题. 首先, 我们考虑如下分布不确定集:

$$\mathcal{F}^\dagger = \left\{ \mathbb{P} \in \mathcal{P}(\mathbb{R}^K \times \mathbb{R}^{3K}) \;\middle|\; \begin{array}{l} (\tilde{\boldsymbol{u}}, \tilde{\boldsymbol{v}}) \sim \mathbb{P}, \\ \mathbb{E}_{\mathbb{P}}[\tilde{\boldsymbol{u}}] \in [\boldsymbol{u}^-, \boldsymbol{u}^+], \\ \mathbb{E}_{\mathbb{P}}[\tilde{\boldsymbol{v}}_k] \in [\boldsymbol{v}_k^-, \boldsymbol{v}_k^+],\ \forall k \in [K], \\ \mathbb{E}_{\mathbb{P}}[\|\boldsymbol{C}_l(\tilde{\boldsymbol{v}} - \bar{\boldsymbol{v}})\|_1] \leqslant \varepsilon_l,\ \forall l \in [L], \\ \mathbb{P}[\tilde{\boldsymbol{u}} \in \mathcal{U}_{\mathrm{CH}}, \tilde{\boldsymbol{v}} \in \mathcal{V}] = 1 \end{array} \right\}, \tag{3.14}$$

其中, 我们定义集合 $\mathcal{U}_{\mathrm{CH}}$ 是一组参数样本 $\{\boldsymbol{u}^s\}_{s\in[S]}$ 的凸包, 即 $\mathcal{U}_{\mathrm{CH}} = \mathrm{conv}(\{\boldsymbol{u}^s\}_{s\in[S]})$, 和集合 $\mathcal{V} = \{\boldsymbol{v} \in \mathbb{R}_+^{3K} \mid \boldsymbol{B}\boldsymbol{v} \leqslant \boldsymbol{b}\}$. 我们假设支撑集 $\mathcal{V}$ 中的矩阵 $\boldsymbol{B} \in \mathbb{R}^{M\times 3K}$ 可以表示为列向量的形式 $\boldsymbol{B} = (\boldsymbol{B}_1\ \boldsymbol{B}_2\ \cdots\ \boldsymbol{B}_K)$ 且对于任意 $k \in [K]$, $\boldsymbol{B}_k \in \mathbb{R}^{M\times 3}$ 满足 $\boldsymbol{B}\boldsymbol{v} = \sum_{k\in[K]} \boldsymbol{B}_k \boldsymbol{v}_k$. 同样地, 对于任意 $l \in [L]$, 矩阵 $\boldsymbol{C}_l \in \mathbb{R}^{N_l \times 3K}$ 也可以用列向量来表示, 即 $\boldsymbol{C}_l = (\boldsymbol{C}_{l1}\ \boldsymbol{C}_{l2}\ \cdots\ \boldsymbol{C}_{lK})$, 且对于任意 $k \in [K]$, $\boldsymbol{C}_{lk} \in \mathbb{R}^{N\times 3}$ 满足 $\boldsymbol{C}_l \boldsymbol{v} = \sum_{k\in[K]} \boldsymbol{C}_{lk} \boldsymbol{v}_k$.

接下来, 我们讨论分布不确定集 $\mathcal{F}^\dagger$ (3.14) 和 $\mathcal{F}$ (3.5) 在结构上的区别: 首先, 在构造分布不确定集 $\mathcal{F}^\dagger$ 的过程中, 我们没有假设不确定商品需求 $\tilde{\boldsymbol{u}}$ 和运输成本 $\tilde{\boldsymbol{v}}$ 是相互独立的. 因此, $\tilde{\boldsymbol{u}}$ 和 $\tilde{\boldsymbol{v}}$ 在极值概率分布情况下可能是相关的; 其次, 不同种类商品的运输成本 $\tilde{\boldsymbol{v}}_k$ 不一定相互独立. 因为在分布不确定集 $\mathcal{F}^\dagger$ 中, 所有种类

商品的运输成本都包含于一个支撑集 $\mathcal{V}$ 中. 然而, 在分布不确定集 $\mathcal{F}$ 中, 我们针对每种商品 $k \in [K]$ 分别构建其运输成本的支撑集 $\mathcal{V}_k$. 此外, 为了将分布鲁棒模型转化为高效计算形式, 我们在分布不确定集 $\mathcal{F}^\dagger$ 中仅仅考虑了运输成本 $\tilde{\boldsymbol{v}}$ 的期望交叉离差约束. 最后, 分布不确定集 $\mathcal{F}^\dagger$ 的商品需求支撑集是一个凸包 $\mathcal{U}_{\mathrm{CH}}$ 而不是 $\mathcal{F}$ 中的多面体支撑集 $\mathcal{U}$. 实际上, 商品需求支撑集 $\mathcal{U}_{\mathrm{CH}}$ 是凸包的假设是合理的, 主要原因有以下三点. 首先, 在实际应用中, 我们可以得到商品的历史需求数据. 因此, 我们可以利用预测模型来预测未来商品需求, 并且未来真实的商品需求有很大的概率包含于需求预测值的凸包中. 其次, 我们可以用多种估计方法来获得 $\{\boldsymbol{u}^s\}_{s\in[S]}$, 例如, 专家估计与订单分析估计. 因此, 支撑集 $\mathcal{U}_{\mathrm{CH}}$ 可以看作是对不同方法估计值的加权. 最后, 对于常见的多面体支持集 (polyhedron support set), 例如, 单纯形不确定集 (simplex uncertainty set)、箱式不确定集 (box uncertainty set) 以及 Bertsimas 和 Sim (2004) 提出的预算不确定集 (budget uncertainty set), 我们都可以有效地识别出它们的极值点. 因此, 样本集 $\{\boldsymbol{u}^s\}_{s\in[S]}$ 可以是这些集合的极值点.

定理 3.4 给定分布不确定集 $\mathcal{F}^\dagger$ (3.14), 两阶段分布鲁棒容量充足型枢纽选址模型 (3.7) 等价于如下混合整数线性规划问题:

$$
\begin{aligned}
\min\quad & \boldsymbol{c}^\top \boldsymbol{x} + \alpha + \overline{\boldsymbol{\beta}}^\top \boldsymbol{u}^+ - \underline{\boldsymbol{\beta}}^\top \boldsymbol{u}^- + \sum_{k\in[K]} (\overline{\boldsymbol{\gamma}}_k^\top \boldsymbol{v}_k^+ - \underline{\boldsymbol{\gamma}}_k^\top \boldsymbol{v}_k^-) + \sum_{l\in[L]} \rho_l \varepsilon_l \\
\text{s.t.}\quad & \alpha + (\overline{\boldsymbol{\beta}} - \underline{\boldsymbol{\beta}})^\top \boldsymbol{u}^s - \boldsymbol{b}^\top \boldsymbol{\zeta}^s + \bar{\boldsymbol{v}}^\top \sum_{l\in[L]} \boldsymbol{C}_l^\top \boldsymbol{\xi}_l^s \geqslant 0, \forall s \in [S], \\
& \boldsymbol{B}_k^\top \boldsymbol{\zeta}^s - \sum_{l\in[L]} \boldsymbol{C}_{lk}^\top \boldsymbol{\xi}_l^s + \overline{\boldsymbol{\gamma}}_k - \underline{\boldsymbol{\gamma}}_k \geqslant \sum_{i,j\in[N]} \boldsymbol{d}_{ijk} u_k^s y_{ijk}^s, \forall s \in [S],\ k \in [K], \\
& \rho_l \geqslant \|\boldsymbol{\xi}_l^s\|_\infty,\ \boldsymbol{\xi}_l^s \in \mathbb{R}^{N_l}, \forall s \in [S],\ l \in [L], \\
& \boldsymbol{\zeta}^s \in \mathbb{R}_+^M,\ \overline{\boldsymbol{\gamma}}_k, \underline{\boldsymbol{\gamma}}_k \in \mathbb{R}_+^3,\ \boldsymbol{y}_k^s \in \mathcal{Y}(\boldsymbol{x}), \forall s \in [S],\ k \in [K], \\
& \alpha \in \mathbb{R},\ \overline{\boldsymbol{\beta}}, \underline{\boldsymbol{\beta}} \in \mathbb{R}_+^K,\ \boldsymbol{\rho} \in \mathbb{R}_+^L,\ \boldsymbol{x} \in \{0,1\}^N,
\end{aligned}
\tag{3.15}
$$

其中, $\alpha, \overline{\boldsymbol{\beta}}, \underline{\boldsymbol{\beta}}, \overline{\boldsymbol{\gamma}}_k, \underline{\boldsymbol{\gamma}}_k, \boldsymbol{\rho}, \boldsymbol{\zeta}^s, \boldsymbol{\xi}_l^s$ 是辅助决策变量.

定理 3.4 表明支撑集 $\mathcal{U}_{\mathrm{CH}}$ 的凸包结构可以帮助我们将模型 (3.7) 转化为混合整数线性规划问题. 当我们假设 $\tilde{\boldsymbol{v}}$ 的支撑集也是一组样本的凸包 $\mathcal{V}_{\mathrm{CH}}$, 且分布不确定集 $\mathcal{F}^\dagger$ (3.14) 只包含 $\tilde{\boldsymbol{u}}$ 的期望离差约束时, 我们同样可以推导出与定理 3.4 类似的结果. 针对两种不确定参数 $\tilde{\boldsymbol{u}}$ 和 $\tilde{\boldsymbol{v}}$, 模型高效计算形式存在的主要原因是容量充足型枢纽选址模型具有对称结构 (symmetric structure). 特别地, 容量充足型枢纽选址分数模型 (3.2) 可以将约束右端项向量 $\tilde{\boldsymbol{u}}$ 等价地转移到目标函数中, 这使得两种不确定变量 $\tilde{\boldsymbol{u}}$ 和 $\tilde{\boldsymbol{v}}$ 在模型结构上具有良好的对称性. 然而, 在容量限制型枢纽选址问题中, 上述不确定参数的对称性不成立, 我们将在 3.5 节具体分析.

此外, 定理 3.4 还表明当我们放松不确定参数相互独立假设时, 最优运输决策 $\boldsymbol{y}^{\star}(\cdot)$ 有效地适用于不确定商品需求和运输成本的实现值. 特别地, 当支撑集为 $\mathcal{U}_{\mathrm{CH}}$ 时, 最优运输决策 $\boldsymbol{y}^{\star}(\cdot)$ 是一个场景依赖 (scenario-wise) 决策, 即对于任意 $s \in [S]$ 都满足 $\boldsymbol{y}^{\star}(\boldsymbol{u}^s, \boldsymbol{v}^s) = \boldsymbol{y}^{s\star}$, 其中 $\boldsymbol{y}^{s\star} = (\boldsymbol{y}_k^{s\star})_{k\in[K]}$ 是问题 (3.15) 的最优运输决策. 接下来, 我们给出极值概率分布 $\mathbb{P}_v^{\dagger} \in \mathcal{F}^{\dagger}$.

命题 3.1 假设 $(\boldsymbol{y}^{s\star})_{s\in[S]}, (\overline{\gamma}_k^{\star}, \underline{\gamma}_k^{\star})_{k\in[K]}, (\rho_l^{\star})_{l\in[L]}$ 是问题 (3.15) 的最优解. 则基于最优枢纽选址决策 $\boldsymbol{x}^{\star}$ 的极值概率分布 $\mathbb{P}_v^{\dagger} \in \mathcal{F}^{\dagger}$ 有如下形式:

$$\mathbb{P}_v^{\dagger}[(\tilde{\boldsymbol{u}}, \tilde{\boldsymbol{v}}) = (\boldsymbol{u}^s, \boldsymbol{v}^s)] = p^s, \quad \forall s \in [S],$$

其中, 对于任意 $s \in [S]$, 极值概率分布下运输成本 $\boldsymbol{v}^s$ 可通过求解如下线性规划问题得到

$$\max_{\boldsymbol{v}\in\mathcal{V}} \left\{ \sum_{i,j\in[N]} \sum_{k\in[K]} \boldsymbol{d}_{ijk}^{\top} \boldsymbol{v}_k u_k^s y_{ijk}^{s\star} - \sum_{k\in[K]} (\overline{\gamma}_k^{\star} - \underline{\gamma}_k^{\star})^{\top} \boldsymbol{v}_k - \sum_{l\in[L]} \rho_l^{\star} \|\boldsymbol{C}_l(\boldsymbol{v} - \bar{\boldsymbol{v}})\|_1 \right\}, \tag{3.16}$$

并且极值概率 $(p^s)_{s\in[S]} = \boldsymbol{p}^{\star}$ 可以通过求解如下给定 $(\boldsymbol{v}^s)_{s\in[S]}$ 的线性规划问题获得,

$$\begin{aligned} \Omega^{\star} = \max_{\boldsymbol{p}\geqslant \boldsymbol{0}} \quad & \sum_{s\in[S]} p^s \sum_{i,j\in[N]} \sum_{k\in[K]} \boldsymbol{d}_{ijk}^{\top} \boldsymbol{v}_k^s u_k^s y_{ijk}^{s\star} \\ \text{s.t.} \quad & \sum_{s\in[S]} p^s = 1, \\ & \sum_{s\in[S]} p^s \boldsymbol{u}^s \in [\boldsymbol{u}^-, \boldsymbol{u}^+], \\ & \sum_{s\in[S]} p^s \boldsymbol{v}_k^s \in [\boldsymbol{v}_k^-, \boldsymbol{v}_k^+], \forall k \in [K], \\ & \sum_{s\in[S]} p^s \|\boldsymbol{C}_l(\boldsymbol{v}^s - \bar{\boldsymbol{v}})\|_1 \leqslant \varepsilon_l, \forall l \in [L]. \end{aligned} \tag{3.17}$$

事实上, 我们可以将分布不确定集 $\mathcal{F}^{\dagger}$ 的支撑集约束替换为 $\mathbb{P}[\tilde{\boldsymbol{u}} \in \mathcal{U}, \tilde{\boldsymbol{v}} \in \mathcal{V}_{\mathrm{CH}}] = 1$, 其中 $\mathcal{U}$ 是一个多面体集, 且支撑集是给定样本 $\{\boldsymbol{v}^s\}_{s\in[S]}$ 的凸包, 即 $\mathcal{V}_{\mathrm{CH}} = \mathrm{conv}(\{\boldsymbol{v}^s\}_{s\in[S]})$. 同时, 将分布不确定集 $\mathcal{F}^{\dagger}$ 中关于不确定运输成本 $\tilde{\boldsymbol{v}}$ 的期望离差约束 $\mathbb{E}_{\mathbb{P}}[\|\boldsymbol{C}_l(\tilde{\boldsymbol{v}} - \bar{\boldsymbol{v}})\|_1] \leqslant \varepsilon_l$ 替换为不确定商品需求 $\tilde{\boldsymbol{u}}$ 的类似约束 $\mathbb{E}_{\mathbb{P}}[\|\boldsymbol{D}_{\ell}(\tilde{\boldsymbol{u}} - \bar{\boldsymbol{u}})\|_1] \leqslant \sigma_{\ell}$. 基于上述分布不确定集, 我们可以得到与定理 3.4 和命题 3.1 类似的理论结果. 然而, 当分布不确定集 $\mathcal{F}^{\dagger}$ 同时包含商品需求和运输成本的期望离差约束时, 定理 3.4 的结果则不成立.

3.5 分布鲁棒容量限制型枢纽选址模型

在本节中, 我们介绍两阶段分布鲁棒容量限制型枢纽选址模型 (two-stage distributionally robust capacitated hub location model). 容量限制型枢纽是指通过第 $i \in [N]$ 个枢纽运输的商品总量不能超过枢纽 i 的容量 q_i. 特别地, 根据实际问题背景, 我们假设每种商品可以从起始地通过多条路径运输到目的地. 首先, 我们构建如下确定性容量限制型枢纽选址模型 (deterministic capacitated hub location model):

$$
\begin{aligned}
\min \quad & \sum_{i\in[N]} \boldsymbol{c}^\top \boldsymbol{x} + \sum_{k\in[K]} \sum_{i,j\in[N]} \boldsymbol{d}_{ijk}^\top \boldsymbol{v}_k z_{ijk} + \sum_{k\in[K]} h_k \left(u_k - \sum_{i,j\in[N]} z_{ijk} \right) \\
\text{s.t.} \quad & \sum_{i,j\in[N]} z_{ijk} = u_k, \forall k \in [K], \\
& \sum_{k\in[K]} \left(\sum_{j\in[N]} z_{ijk} + \sum_{j\in[N]\setminus\{i\}} z_{jik} \right) \leqslant x_i q_i, \forall i \in [N], \\
& z_{ijk} \geqslant 0, \forall i,j \in [N],\ k \in [K], \\
& \boldsymbol{x} \in \{0,1\}^N,
\end{aligned}
\tag{3.18}
$$

其中, 目标函数的最后一项为未满足商品需求的惩罚成本, 且参数 h_k 是第 k 种商品需求未满足时的单位惩罚成本. 第二组约束是保证商品通过被选中的枢纽运输, 且通过枢纽 i 运输的商品总量不超过其容量 q_i. 其余约束与容量充足型枢纽选址模型 (3.1) 的约束相同.

与容量充足型枢纽选址模型不同, 模型 (3.18) 含有容量约束. 因此, 模型 (3.18) 的运输决策 z_{ijk} 不能根据商品种类 $k \in [K]$ 进行划分. 虽然容量限制型枢纽选址模型 (3.18) 可以被转化为分数模型的形式 (Meraklı and Yaman, 2017), 但是容量约束使得需求参数 $\boldsymbol{u}$ 不能从约束中转化到目标函数中. 因此, 容量限制型枢纽选址模型不具有关于 $\boldsymbol{u}$ 和 $\boldsymbol{v}$ 的对称结构, 并且我们也无法假设不确定参数相互独立. 尽管如此, 我们仍然可以推导出两阶段分布鲁棒容量限制型枢纽选址模型的高效计算形式.

接下来, 我们基于分布不确定集 $\mathcal{F}^\dagger$ (3.14) 构造两阶段分布鲁棒容量限制型枢纽选址模型:

$$
\min_{\boldsymbol{x}\in\{0,1\}^N} \left\{ \boldsymbol{c}^\top \boldsymbol{x} + \sup_{\mathbb{P}\in\mathcal{F}^\dagger} \mathbb{E}_{\mathbb{P}}[\mathrm{Q_c}(\boldsymbol{x}, \tilde{\boldsymbol{u}}, \tilde{\boldsymbol{v}})] \right\}, \tag{3.19}
$$

其中, 给定枢纽选址决策 $\boldsymbol{x}$ 和 $(\boldsymbol{u}, \boldsymbol{v})$, 我们可以将第二阶段运输问题表示为如下

形式:

$$\mathrm{Q_c}(\boldsymbol{x},\boldsymbol{u},\boldsymbol{v}) = \min_{\boldsymbol{z}\in\mathcal{Z}_c(\boldsymbol{x},\boldsymbol{u})} \left\{ \sum_{k\in[K]} \sum_{i,j\in[N]} \left(\boldsymbol{d}_{ijk}^{\top}\boldsymbol{v}_k - h_k\right) z_{ijk} + \boldsymbol{h}^{\top}\tilde{\boldsymbol{u}} \right\},$$

同时, 运输决策 $\boldsymbol{z}$ 的可行集定义如下:

$$\mathcal{Z}_c(\boldsymbol{x},\boldsymbol{u}) := \left\{ \boldsymbol{z}\in\mathbb{R}_+^{KN^2} \;\middle|\; \begin{array}{l} \displaystyle\sum_{i,j\in[N]} z_{ijk} = u_k, \forall k\in[K], \\ \displaystyle\sum_{k\in[K]} \left(\sum_{j\in[N]} z_{ijk} + \sum_{j\in[N]\setminus\{i\}} z_{jik} \right) \leqslant x_i q_i, \forall i\in[N] \end{array} \right\}.$$

接下来, 证明两阶段分布鲁棒容量限制型枢纽选址模型 (3.19) 等价于一个混合整数线性规划问题.

定理 3.5　给定分布不确定集 $\mathcal{F}^{\dagger}$ (3.14), 两阶段分布鲁棒容量限制型枢纽选址模型 (3.19) 等价于如下混合整数线性规划问题:

$$\begin{aligned}
\min\quad & \boldsymbol{c}^{\top}\boldsymbol{x} + \alpha + \overline{\boldsymbol{\beta}}^{\top}\boldsymbol{u}^{+} - \underline{\boldsymbol{\beta}}^{\top}\boldsymbol{u}^{-} + \sum_{k\in[K]} \left(\overline{\boldsymbol{\gamma}}_k^{\top}\boldsymbol{v}_k^{+} - \underline{\boldsymbol{\gamma}}_k^{\top}\boldsymbol{v}_k^{-}\right) + \sum_{l\in[L]} \rho_l \varepsilon_l \\
\text{s.t.}\quad & \alpha + (\overline{\boldsymbol{\beta}} - \underline{\boldsymbol{\beta}} - \boldsymbol{h})^{\top}\boldsymbol{u}^{s} + \sum_{k\in[K]}\sum_{i,j\in[N]} h_k z_{ijk}^{s} - \boldsymbol{b}^{\top}\boldsymbol{\zeta}^{s} \\
& +\bar{\boldsymbol{v}}^{\top}\sum_{l\in[L]} \boldsymbol{C}_l^{\top}\boldsymbol{\xi}_l^{s} \geqslant 0, \forall s\in[S], \\
& \boldsymbol{B}_k^{\top}\boldsymbol{\zeta}^{s} - \sum_{l\in[L]} \boldsymbol{C}_{lk}^{\top}\boldsymbol{\xi}_l^{s} + \overline{\boldsymbol{\gamma}}_k - \underline{\boldsymbol{\gamma}}_k \geqslant \sum_{i,j\in[N]} \boldsymbol{d}_{ijk} z_{ijk}^{s}, \forall s\in[S],\ k\in[K], \\
& \rho_l \geqslant \|\boldsymbol{\xi}_l^{s}\|_{\infty}, \boldsymbol{\xi}_l^{s}\in\mathbb{R}^{N_l}, \forall s\in[S],\ l\in[L], \\
& \boldsymbol{\zeta}^{s}\in\mathbb{R}_+^{M}, \overline{\boldsymbol{\gamma}}_k, \underline{\boldsymbol{\gamma}}_k\in\mathbb{R}_+^{3},\ \boldsymbol{z}^{s}\in\mathcal{Z}_c(\boldsymbol{x},\boldsymbol{u}^{s}), \forall s\in[S],\ k\in[K], \\
& \alpha\in\mathbb{R},\ \overline{\boldsymbol{\beta}}, \underline{\boldsymbol{\beta}}\in\mathbb{R}_+^{K},\ \boldsymbol{\rho}\in\mathbb{R}_+^{L},\ \boldsymbol{x}\in\{0,1\}^{N},
\end{aligned} \tag{3.20}$$

其中, $\alpha, \overline{\boldsymbol{\beta}}, \underline{\boldsymbol{\beta}}, \overline{\boldsymbol{\gamma}}_k, \underline{\boldsymbol{\gamma}}_k, \boldsymbol{\rho}, \boldsymbol{\zeta}^{s}, \boldsymbol{\xi}_l^{s}$ 是辅助决策变量.

利用定理 3.5, 我们得到两阶段分布鲁棒容量限制型枢纽选址模型 (3.19) 的混合整数线性规划问题形式. 需要说明的是, 在 $\tilde{\boldsymbol{u}}$ 的支撑集 $\mathcal{U}_{\mathrm{CH}}$ 是给定样本 $\{\boldsymbol{u}^{s}\}_{s\in[S]}$ 凸包的假设下, 定理 3.5 成立, 然而在 $\tilde{\boldsymbol{v}}$ 的支撑集是凸包的假设下, 定理 3.5 并不成立. 上述结果产生的主要原因是模型 (3.19) 不具有关于 $\boldsymbol{u}$ 和 $\boldsymbol{v}$ 的对称结构.

此外, 模型 (3.19) 第二阶段问题的可行集 $\mathcal{Z}_c(\boldsymbol{x},\boldsymbol{u})$ 依赖于枢纽选址决策 $\boldsymbol{x}$ 和商品需求变量 $\boldsymbol{u}$. 尽管如此, 在给定 $\mathcal{F}^\dagger$ 和最优选址决策 $\boldsymbol{x}^\star$ 的情况下, 我们仍然可以得出分布鲁棒容量限制型枢纽选址模型的最优运输决策 $\boldsymbol{z}^\star(\cdot)$ 具有适应性结构, 这与分布鲁棒容量充足型枢纽选址模型最优运输决策 $\boldsymbol{y}^\star(\cdot)$ 的适应性结构相一致. 接下来, 反推不确定参数 $(\tilde{\boldsymbol{u}},\tilde{\boldsymbol{v}})$ 极值概率分布 $\mathbb{P}_v^\dagger\in\mathcal{F}^\dagger$.

命题 3.2 假设 $(\boldsymbol{z}^{s\star})_{s\in[S]},(\overline{\boldsymbol{\gamma}}_k^\star,\underline{\boldsymbol{\gamma}}_k^\star)_{k\in[K]},(\rho_l^\star)_{l\in[L]}$ 是问题 (3.20) 的最优解. 则基于最优枢纽选址决策 $\boldsymbol{x}^\star$ 的极值概率分布有如下形式:

$$\mathbb{P}_v^\dagger[(\tilde{\boldsymbol{u}},\tilde{\boldsymbol{v}})=(\boldsymbol{u}^s,\boldsymbol{v}^s)]=p^s,\quad \forall s\in[S],$$

其中, 对于任意 $s\in[S]$, 极值概率分布下运输成本 $\boldsymbol{v}^s$ 可以通过求解如下线性优化问题得到,

$$\max_{\boldsymbol{v}\in\mathcal{V}}\left\{\sum_{i,j\in[N]}\sum_{k\in[K]}\boldsymbol{d}_{ijk}^\top\boldsymbol{v}_k z_{ijk}^{s\star}-\sum_{k\in[K]}(\overline{\boldsymbol{\gamma}}_k^\star-\underline{\boldsymbol{\gamma}}_k^\star)^\top\boldsymbol{v}_k-\sum_{l\in[L]}\rho_l^\star\|\boldsymbol{C}_l(\boldsymbol{v}-\bar{\boldsymbol{v}})\|_1\right\},$$

并且概率 $(p^s)_{s\in[S]}=\boldsymbol{p}^\star$ 可以通过求解给定 $(\boldsymbol{v}^s)_{s\in[S]}$ 的线性规划问题来获得,

$$\begin{aligned}\max_{\boldsymbol{p}\geqslant\boldsymbol{0}}\quad&\sum_{s\in[S]}p^s\sum_{i,j\in[N]}\sum_{k\in[K]}\boldsymbol{d}_{ijk}^\top\boldsymbol{v}_k^s z_{ijk}^{s*}\\ \text{s.t.}\quad&\sum_{s\in[S]}p^s=1,\\&\sum_{s\in[S]}p^s\boldsymbol{u}^s\in[\boldsymbol{u}^-,\boldsymbol{u}^+],\\&\sum_{s\in[S]}p^s\boldsymbol{v}_k^s\in[\boldsymbol{v}_k^-,\boldsymbol{v}_k^+],\forall k\in[K],\\&\sum_{s\in[S]}p^s\|\boldsymbol{C}_l(\boldsymbol{v}^s-\bar{\boldsymbol{v}})\|_1\leqslant\varepsilon_l,\forall l\in[L].\end{aligned}$$

3.6 数值实验

本节使用美国民用航空委员会 (US Civil Aeronautics Board, CAB) 的货运数据集设计数值实验, 来验证提出的分布鲁棒枢纽选址模型解决参数不确定性问题的有效性①. 此数据集包含美国每对城市之间的距离和多种商品需求量. 在实验中, 我们选择数据集中 25 个城市以及城市间运输的 $K=625$ 种商品, 并将其余 $N=20$ 个城市作为备选枢纽地址②. 本节的实验主要验证以下五个方面:

① 参考美国民用航空委员会数据: O'KELLY M E, 2014. Cab100 mok[2024-5-27].https://www. researchgate.

② 数据集中不包含枢纽固定成本数据, 枢纽固定成本的计算方法可参考文献 (de Camargo et al., 2008).

① 期望边际离差与交叉离差信息的价值; ② 分布不确定集参数对最优枢纽选址决策和运输决策的影响; ③ 分布鲁棒枢纽选址模型的价值; ④ 容量值对最优枢纽选址决策的影响.

3.6.1 期望边际离差与交叉离差信息的价值

在本节中, 我们设计如下实验验证分布鲁棒容量充足型枢纽选址模型的分布不确定集 $\mathcal{F}$ (3.5) 包含期望边际离差与交叉离差信息的价值. 在 $\tilde{\boldsymbol{u}}$ 和 $\tilde{\boldsymbol{v}}$ 相互独立假设下, 对于任意商品种类 $k \in [K]$, 我们假设不确定需求 $\tilde{\boldsymbol{u}}$ 的经验分布 (empirical distribution) 服从一个正态分布 $\tilde{u}_k \sim \mathbb{N}(w_k, 0.09w_k^2)$, 其中 w_k 是 CAB 数据集中第 k 种商品的需求量. 此外, 不确定运输成本的经验分布服从一个均匀分布 $\tilde{\boldsymbol{v}}_k \sim \mathbb{U}[\boldsymbol{v}_k^-, \boldsymbol{v}_k^+]$, 其中对于所有 $k \in [K]$, 运输成本的上界 $\boldsymbol{v}_k^- = (1,1,1)$ 且下界 $\boldsymbol{v}_k^+ = (4,2,3)$. 利用上述经验分布, 我们随机生成 1000 个样本来构造样本分布 $\hat{\mathbb{P}}$. 实验主要比较三种模型: 鲁棒枢纽选址模型 (RO)、基于边际离差信息的分布鲁棒枢纽选址模型 (DRO-M) 以及基于边际离差与交叉离差信息的分布鲁棒枢纽选址模型 (DRO-C).

表 3.1 给出了三种模型的最优选址决策、最优值和计算时间. 结果表明: DRO-C 模型的最优选址决策为只建设一个枢纽, 且总成本最小. 因此 DRO-C 模型的最优选址决策最具经济性. 此外, RO 模型的最优选址决策是最保守的, 其选择三个枢纽并且总成本在三个模型中最高, 甚至是 DRO-C 模型总成本的两倍. 最后, 三个模型都可以在几秒内被精确求解.

表 3.1 DRO-C、DRO-M 和 RO 模型的最优选址决策、最优值和计算时间

模型	最优选址决策	最优值	计算时间/秒
DRO-C	$\{4\}$	2.95×10^8	4.56
DRO-M	$\{4, 18\}$	3.83×10^8	5.18
RO	$\{13, 17, 19\}$	6.43×10^8	4.96

此外, 我们还设计如下压力测试实验验证 DRO 模型相比于 RO 模型具有额外的边际离差和交叉离差概率分布信息的价值, 以及 DRO-C 模型相比于 DRO-M 模型具有额外的期望交叉离差信息的价值. 首先, 利用算法 1 计算出 DRO-C 模型的极值概率分布 $\mathbb{P}^\star \in \mathcal{F}$. 为了保证测试的公平性, 我们用参数 δ 对极值概率分布 $\mathbb{P}^\star$ 进行扰动获得一个参数化的均匀分布 $[(1-\delta)\boldsymbol{u}^\star, (1+\delta)\boldsymbol{u}^\star] \otimes [(1-\delta)\boldsymbol{v}^\star, (1+\delta)\boldsymbol{v}^\star]$, 其中参数 δ 的值越大表示扰动程度 (variation level) 越大. 然后, 我们利用上述均匀分布随机生成 1000 个样本进行压力测试验证 DRO-C 模型、DRO-M 模型与 RO 模型的最优选址决策的样本外表现 (out-of-sample performance). 图 3.1 给出了不同压力水平下对应三个模型最优选址决策的样本外 (out-of-sample) 总成本

分布. 此外, 表 3.2 给出了在不同压力水平下每个模型样本外总成本分布的统计量信息, 其中包括均值、标准差、在险价值 (value-at-risk, V@R) 和条件在险价值 (conditional value-at-risk, CV@R).

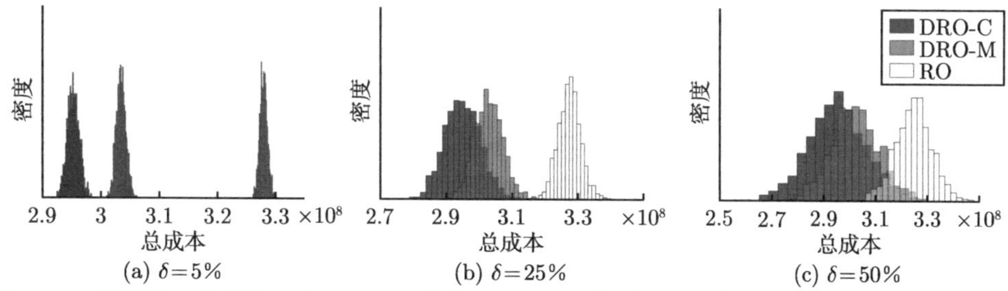

图 3.1 不同压力水平下对应 DRO-C 模型、DRO-M 模型和 RO 模型最优选址决策的样本外总成本分布

表 3.2 不同压力水平下 DRO-C 模型、DRO-M 模型与 RO 模型的样本外总成本分布的统计信息 ($\times 10^8$)

δ	模型	均值	标准差	90%V@R	95%V@R	99%V@R	90%CV@R	95%CV@R	99%CV@R
5%	DRO-C	2.95	0.01	2.97	2.97	2.98	2.97	2.97	2.98
	DRO-M	3.03	0.01	3.05	3.05	3.05	3.05	3.05	3.06
	RO	3.28	0.01	3.28	3.29	3.30	3.29	3.29	3.30
25%	DRO-C	2.95	0.05	3.02	3.04	3.08	3.05	3.06	3.10
	DRO-M	3.03	0.04	3.08	3.10	3.14	3.11	3.12	3.15
	RO	3.28	0.03	3.32	3.33	3.36	3.34	3.35	3.37
50%	DRO-C	2.96	0.11	3.11	3.15	3.22	3.16	3.20	3.25
	DRO-M	3.02	0.09	3.14	3.16	3.23	3.17	3.20	3.25
	RO	3.25	0.07	3.34	3.37	3.41	3.37	3.40	3.44

图 3.1 和表 3.2 的结果表明: ① 在所有扰动水平下 ($\delta = 5\%, 25\%$ 和 50%), 虽然 RO 模型的样本外总成本的标准差在三个模型中最低, 但是 DRO-C 模型和 DRO-M 模型在均值、V@R 和 CV@R 三个统计指标下压力测试表现都比 RO 模型好, 这验证了出在分布不确定集中引入额外分布信息的价值. ② 在扰动水平低的情况下 ($\delta = 5\%$), 不确定参数 $(\tilde{\boldsymbol{u}}, \tilde{\boldsymbol{v}})$ 的真实概率分布可以被分布不确定 $\mathcal{F}$ 准确描述, 此时 DRO 模型的样本外成本远低于 DRO 的样本外成本, 进一步说明考虑分布不确定性的 DRO 模型相比于 RO 模型有明显地处理不确定性问题的优势. ③ 在扰动水平 $\delta = 5\%$ 和 $\delta = 25\%$ 的情况下, DRO-C 模型对应的样本外总成本的均值、V@R 和 CV@R 比 DRO-M 模型的低, 这是由于 DRO-C 模型额外包含了反映商品需求相关性的交叉离差信息. 但是, 在高扰动水平下, 样本的相关性减

弱. 此时 DRO-C 模型的样本外测试表现仍好于 DRO-M 模型, 这说明在 DRO 模型中同时引入边际离散和交叉离散概率分布信息能够提高选址决策质量.

3.6.2 分布不确定集参数对最优枢纽选址决策和运输决策的影响

在本节中, 我们主要基于 DRO-C 模型设计实验来观察分布不确定集参数的变化对最优枢纽选址决策和运输决策的影响. 首先, 我们利用参数 κ 按照如下四种规则分别扰动分布不确定集的参数:

(1) 扰动规则一: 将商品单位运输成本之和的上界 Γ 增加至 $(1+\kappa)\Gamma$.

(2) 扰动规则二: 将商品需求的期望值区间 $[\boldsymbol{u}^-,\boldsymbol{u}^+]$ 扩大到 $[(1-\kappa)\boldsymbol{u}^-,(1+\kappa)\boldsymbol{u}^+]$, 且同时将所有商品运输成本的期望值区间 $[\boldsymbol{v}_k^-,\boldsymbol{v}_k^+]$ 扩大到 $[(1-\kappa)\boldsymbol{v}_k^-,(1+\kappa)\boldsymbol{v}_k^+]$.

(3) 扰动规则三: 将所有商品需求的期望边际离差上界 τ_k 增加至 $(1+\kappa)\tau_k$.

(4) 扰动规则四: 将所有商品需求分组的期望交叉离差上界 δ_ℓ 增加至 $(1+\kappa)\delta_\ell$.

表 3.3 给出了不同扰动水平下 DRO-C 模型的最优选址决策. 结果表明: 分布不确定集参数的扰动会造成 DRO-C 模型最优选址决策的变化. 随着商品需求期望值和运输成本期望值区间的扩大, 或者商品需求的期望边际离差和期望交叉离差的增大, DRO-C 模型将会选择更多的枢纽来保证极值概率分布下商品的运输. 此外, 最优选址决策受商品需求期望离差扰动的影响大于其余参数扰动的影响, 说明在分布不确定集中引入期望交叉离差的重要性.

表 3.3 分布不确定集参数不同扰动水平下 DRO-C 模型的最优选址决策

κ	扰动规则一	扰动规则二	扰动规则三	扰动规则四
0	{4}	{4}	{4}	{4}
0.05	{4}	{4,18}	{4}	{4,18}
0.25	{4}	{4,18}	{4,18}	{4,18}
0.5	{4,18}	{4,18}	{4,18}	{13,17,18}

接下来, 我们设计如下实验来观察测试样本分布的变化对第二阶段运输决策的影响. 首先, 基于上述实验中最优选址决策为 $\{13,17,18\}$ 的 DRO-C 模型, 利用 3.6.1 节中提出的方法构造参数化均匀分布 $[(1-\delta)\boldsymbol{u}^\star,(1+\delta)\boldsymbol{u}^\star]\otimes[(1-\delta)\boldsymbol{v}^\star,(1+\delta)\boldsymbol{v}^\star]$, 并利用此分布在不同的扰动水平 δ 下分别生成 10000 个测试样本. 在每个扰动水平下, 固定选址决策为 $\{13,17,18\}$ 求解第二阶段运输问题获得最优运输决策, 并计算最优运输决策为经过一个或两个枢纽运输商品的测试样本比例. 图 3.2 和图 3.3 分别给出了不同扰动水平下经过一个和两个枢纽运输商品的测试样本占总测试样本的比例, 结果表明: 随着扰动水平的增加, 更多种类的商品将会只通过一个枢纽运输, 而通过两个枢纽运输的商品种类明显减少.

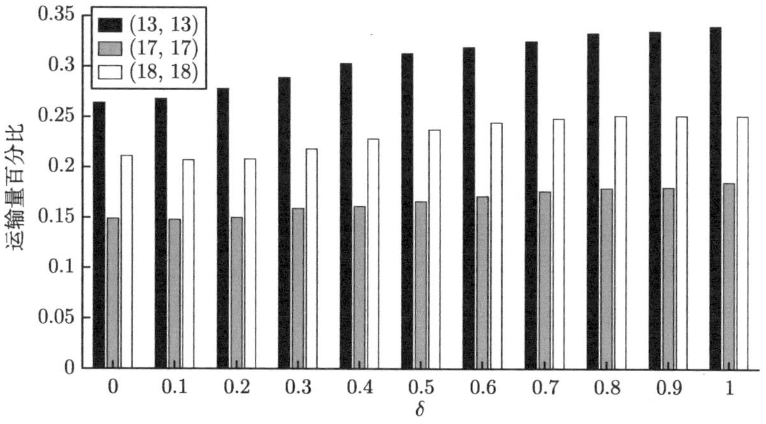

图 3.2　经过单一枢纽运输商品的测试样本占总测试样本的比例

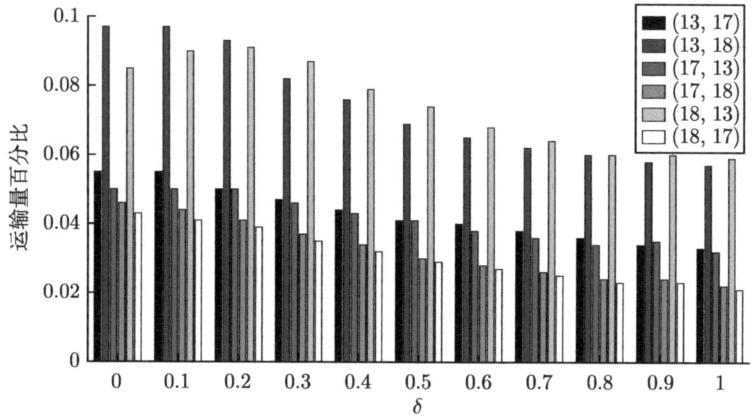

图 3.3　经过两个枢纽运输商品的测试样本占总测试样本的比例

3.6.3　分布鲁棒枢纽选址模型的价值

为了评估分布鲁棒枢纽选址模型的价值, 我们利用压力测试来比较 DRO-C 模型与随机枢纽选址模型 (stochastic hub location model) 处理不确定性问题的效果, 其中随机枢纽选址模型与 Contreras 等 (2011) 研究中的模型一致. 我们记随机枢纽选址模型为 “SAA” 模型. 由于 SAA 模型的规模与样本数量和商品种类有关, 为了高效求解 SAA 模型并保证实验的公平性, 我们将商品种类减少至 100 种. 同时, 我们将 100 种商品平均分为 25 个互不相交的组, 用于估计 DRO-C 模型期望离差. 最后, DRO-C 模型中其余分布不确定集参数的估计方法与 3.6.1 节的方法相同, 并且我们利用 3.6.1 节给出的经验分布生成 100 个样本构建 SAA 模型.

表 3.4 给出了两个模型的最优选址决策、最优值和计算时间. 结果表明: DRO-C 模型的最优选址个数比 SAA 模型的多一个, 这是由于 SAA 模型的最优选址决策是样本内最优的, 而 DRO-C 模型的最优选址决策是在极值概率分布下最优的. 这从侧面反映了 SAA 模型的优化偏差 (optimistic bias). 接下来, 我们将进一步用压力测试验证两个模型最优选址决策的质量. 为了保证实验的公平性, 在不同的扰动水平下, 我们分别利用经验分布和参数化均匀分布 $[(1-\delta)\boldsymbol{u}^\star,(1+\delta)\boldsymbol{u}^\star]\otimes[(1-\delta)\boldsymbol{v}^\star,(1+\delta)\boldsymbol{v}^\star]$ 生成 5000 个样本, 并基于这 10000 个样本, 在不同扰动水平下, 计算两个模型的最优选址决策对应的总成本分布及其统计量.

表 3.4 DRO-C 模型和 SAA 模型的最优选址决策、最优值和计算时间

模型	最优选址决策	最优值	计算时间/秒
DRO-C	$\{3,17,18\}$	1.27×10^8	0.45
SAA	$\{3,16\}$	0.82×10^8	49.09

图 3.4 和表 3.5 分别给出了不同压力水平下 DRO-C 模型与 SAA 模型的最优枢纽选址对应的样本外总成本分布及其统计信息, 结果表明: ① 在所有扰动水平下, DRO-C 模型的最优选址决策对应的样本外总成本及其统计量都低于 SAA 模型最优选址决策对应的, 这说明 DRO-C 模型不仅节约成本, 而且与 SAA 模型相比具有更强的鲁棒性. ② 随着扰动水平的增加, DRO-C 模型与 SAA 模型之间

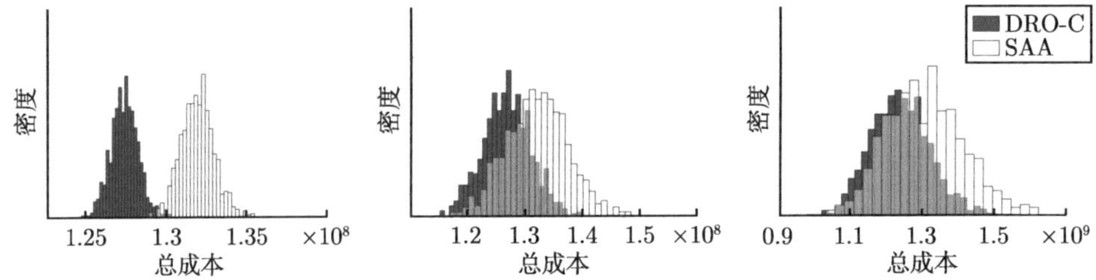

(a) $\delta=5\%$ (b) $\delta=25\%$ (c) $\delta=50\%$

图 3.4 不同压力水平下 DRO-C 模型和随机优化模型的样本外总成本分布

表 3.5 不同压力水平下 DRO-C 模型和 SAA 模型的样本外总成本分布的统计信息

$(\times10^8)$

δ	模型	均值	标准差	90%V@R	95%V@R	99%V@R	90%CV@R	95%CV@R	99%CV@R
5%	DRO-C	1.27	0.01	1.28	1.29	1.29	1.29	1.29	1.30
	SAA	1.32	0.01	1.33	1.34	1.35	1.34	1.34	1.35
25%	DRO-C	1.27	0.04	1.33	1.34	1.37	1.34	1.36	1.38
	SAA	1.32	0.05	1.39	1.41	1.45	1.42	1.43	1.47
50%	DRO-C	1.24	0.08	1.35	1.38	1.46	1.40	1.43	1.49
	SAA	1.31	0.11	1.45	1.50	1.59	1.51	1.55	1.62

样本外实验效果的差距逐渐缩小. 也就是说, 当分布不确定集不能准确地刻画真实分布信息时, DRO-C 模型的建模优势会随之消失, 这说明分布不确定集能够准确刻画真实分布信息的重要性.

3.6.4 容量值对最优枢纽选址决策的影响

在本节中, 我们设计如下实验观察容量值对最优枢纽选址决策的影响. 我们主要考虑基于分布不确定集 $\mathcal{F}^{\dagger}$ 的分布鲁棒容量限制型枢纽选址模型 (3.14). 首先, 利用经验分布 $\mathbb{N}(w_k, 0.09w_k^2)$ 生成商品需求样本 $\{\boldsymbol{u}^s\}_{s\in[S]}$. 然后, 我们用向量 $(q_i)_{i\in[N]} = \boldsymbol{q}$ 表示基础容量水平. 每个枢纽的容量值 q_i 从均匀分布 $\mathbb{U}[1.2q_0, 1.5q_0]$ 中随机产生, 其中 $q_0 = \sum_{s\in S} \boldsymbol{e}^\top \boldsymbol{u}^s/NS$. 为了计算简便, 我们假设模型的分布不确定集只包含一个期望交叉离差约束 $\mathbb{E}_{\mathbb{P}}[\|\boldsymbol{e}^\top(\tilde{\boldsymbol{v}} - \bar{\boldsymbol{v}})\|_1] \leqslant \varepsilon$. 分布不确定集中其余参数的估计方法与前四节实验中的方法一样. 此外, 我们用 $h_k = 2(v_{k1} + v_{k2} + v_{k3})d_{o_k\varsigma_k}$ 来估计未满足商品的单位惩罚成本, 其中 $\boldsymbol{v}_k$ 是单位运输成本, $d_{o_k\varsigma_k}$ 是商品运输起点到目的地的距离. 我们考虑两种枢纽: 第一种是当容量值从 $\boldsymbol{q}$ 增加至 $(1+\omega)\boldsymbol{q}$ 时, 固定成本也随着从 $\boldsymbol{c}$ 增加至 $(1+\omega)\boldsymbol{c}$ 的枢纽, 记为容量敏感型枢纽, 其中 $\omega \geqslant 0$; 第二种是当容量值增加时, 固定成本不会随之增加的枢纽, 记为非容量敏感型枢纽.

表 3.6 给出了不同容量增长比例下容量敏感型枢纽与非容量敏感型枢纽的最优枢纽选址和最优总成本. 结果表明: 两种枢纽对应的最优枢纽选址数量都随着容量提高而减少, 这是非常直观的结果. 此外, 相比于非容量敏感枢纽, 容量敏感型枢纽选址模型的最优选址数量受容量值变化的影响更加明显, 这是容量敏感型枢纽的特性所导致的.

表 3.6 不同容量增长比例下容量敏感型枢纽 (左表) 与非容量敏感型枢纽 (右表) 的最优枢纽选址和最优总成本

ω	最优枢纽选址	最优总成本
0.5	{16,17,18,20}	1.53×10^8
1	{16,17,20}	1.57×10^8
2	{16,17}	1.58×10^8
4	{16}	1.68×10^8

ω	最优枢纽选址	最优总成本
0.5	{16,17,18,20}	1.40×10^8
1	{16,17,18,20}	1.32×10^8
2	{16,17,20}	1.25×10^8
4	{17,18,20}	1.17×10^8

3.7 本 章 小 结

本章主要介绍在不确定商品需求和运输成本下两阶段分布鲁棒枢纽选址模型的构建及其高效计算形式转化. 我们介绍了如何利用边际离差和交叉离差等数据信息构建分布不确定集, 以建模商品需求和运输成本的不确定性. 在不确定参数相互独立假设下, 我们证明了两阶段分布鲁棒容量充足型枢纽选址模型等价于两阶段鲁棒容量充足型枢纽选址模型, 并得出第二阶段最优运输决策是最优静态运输决策的结论. 根据上述结果, 我们进一步给出了极端概率分布的反推算法. 在放松不确定参数相互独立假设下, 我们介绍了两阶段分布鲁棒容量充足型枢纽选址模型与容量限制型枢纽选址模型的构建.

不确定环境下的枢纽选址问题一直是供应链管理与交通运输领域的热点研究问题. 读者可以参考其他相关文献: Shahabi 和 Unnikrishnan (2014) 研究了不确定商品需求下具有单一和多重分配机制的容量充足型枢纽设施选址问题. Ghaffari-Nasab 等 (2015) 研究了不确定商品需求下容量限制型鲁棒枢纽选址模型的构建及求解方法; 他们使用 Bertsimas 和 Sim (2004) 提出的预算不确定集 (budget uncertainty set) 来建模需求不确定性. Boukani 等 (2016) 在 Ghaffari-Nasab 等 (2015) 的研究基础上考虑了枢纽容量和固定成本不确定性的枢纽选址问题, 并提出了最小最大后悔准则枢纽选址模型 (minimax regret hub location model). Meraklı 和 Yaman (2016) 研究了不确定商品需求下具有多重配给机制的容量充足型 p-枢纽中位问题 (p-hub median problem), 并开发了两种不确定优化模型: 管式不确定模型 (hose uncertainty model) 和混合不确定模型 (hybrid uncertainty model). Zetina 等 (2017) 利用基于预算不确定集的鲁棒优化方法研究了容量充足型枢纽选址问题. 针对需求不确定、成本不确定以及需求和成本联合不确定三种情况, 他们分别构建鲁棒枢纽选址模型, 并设计割平面算法来求解模型. 关于不确定环境下枢纽选址问题更全面的综述, 读者可以参考文献 (Campbell and O'Kelly, 2012; Laporte et al., 2019).

第 4 章　两阶段状态依赖分布鲁棒选址优化

第 3 章介绍了容量充足型 (uncapacitated) 选址模型, 本章主要介绍状态依赖需求不确定性下 (state-wise demand uncertainty) 的两阶段产能限制型 (capacitated) 选址问题. 首先, 我们利用状态依赖分布不确定集 (state-wise ambiguity set) 建模具有一般锥结构不确定性 (general conic representable uncertainty) 的不确定需求, 并构建两阶段分布鲁棒工厂选址模型. 接下来, 针对分布不确定集的参数扰动以及工厂选址决策变动, 分别进行敏感性分析 (sensitivity analysis), 并给出针对选址韧性分析的一系列有效边界. 进一步, 基于极值概率分布下第二阶段期望成本函数的有限次梯度表达 (finite subgradient representation), 设计一类针对带有一般锥结构不确定性的嵌套次梯度 Benders 分解算法 (subgradient-based nested Benders decomposition algorithm). 最后, 我们设计实验验证模型处理多状态需求不确定性问题的效果. 本章的主要结构如下: 4.1 节介绍问题背景. 4.2 节介绍确定性选址模型, 以及状态依赖分布不确定集和两阶段分布鲁棒工厂选址模型的构建. 4.3 节介绍极值概率分布下期望成本函数针对分布不确定集参数扰动和选址决策变动的敏感性分析. 4.4 节介绍一种基于次梯度的嵌套 Benders 分解精确算法. 4.5 节介绍实验及结果分析.

4.1　背　　景

选址优化问题作为离散优化与选址科学中的一类重要问题, 被广泛应用于物流、交通运输和电信通信等领域 (Laporte et al., 2019). 一般来说, 工厂选址问题主要研究在一组工厂备选地址中, 如何选择最优的建厂地址并制定从工厂到客户的产品运输决策, 以满足客户需求且最小化工厂建设成本与货物运输成本. 此外, 工厂的产能都是有限制的, 这会造成部分客户需求无法被满足, 因此, 产能限制型工厂选址问题的目标函数还包括未满足客户需求的惩罚成本, 也可以看作是外包成本 (Liu et al., 2022). 与第 3 章的枢纽选址问题类似, 不确定性环境下的工厂选址问题是一个两阶段选址问题, 其中, 第一阶段决策为工厂选址决策, 第二阶段决策是在观测到不确定需求实现值后所作出的货物运输决策 (Laporte et al., 2019). 在第 3 章中, 我们重点关注容量充足型选址优化模型, 分析了运输决策的适应性. 在本章中, 我们将重点关注容量限制型选址优化, 并重点讨论状态依赖需求不确定性的建模.

由于选址优化是一个长期规划问题, 未来的客户需求会受到季节和社会经济状况等因素的影响, 例如, 游泳用品在夏季的销量较高, 而滑雪装备在冬季的需求量最大. 此外, 商品需求也会受到 GDP、消费者收入、人口数量、政府政策和紧急事件的影响 (Lee and Wilhelm, 2010). 我们称上述现象为需求的状态依赖效应 (state-dependent effects). 图 4.1 给出了 2009 年 ~ 2019 年某市某商品的季度零售额分布. 从图 4.1 中, 我们可以看出不同季度的零售额分布不同, 这从侧面表明产品的销量或需求与季度高度相关. 因此, 在本章我们主要考虑状态依赖需求不确定性.

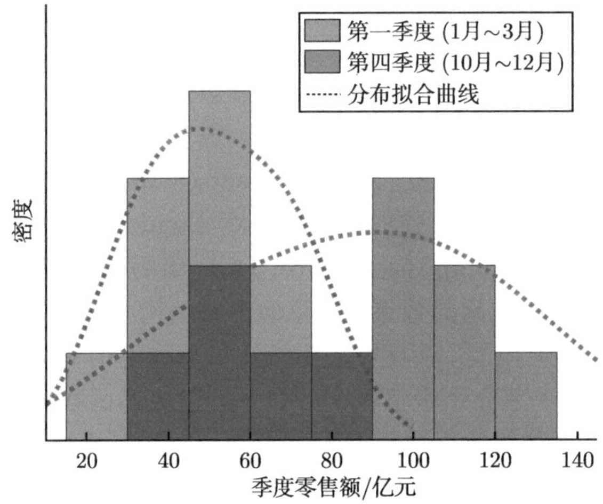

图 4.1 2009 年 ~ 2019 年某市某商品季度零售额分布

4.2 状态依赖需求不确定性下两阶段产能限制型工厂选址模型

4.2.1 确定性产能限制型工厂选址模型

我们首先介绍确定性工厂 (设施) 选址模型 (deterministic facility location model). 考虑有 I 个备选工厂地址以及 J 个客户的工厂选址问题. 同时, 我们记工厂 i 的产能为 q_i, 且客户 j 的产品需求量为 d_j. 此外, 我们记 f_i 为工厂 $i \in [I]$ 的固定成本, c_{ij} 是从工厂 $i \in [I]$ 向客户 $j \in [J]$ 运输单位产品的运输成本, r_j 表示客户 j 未被满足需求的单位惩罚成本. 该问题主要包含两个决策变量: y_i 为 0-1 整数选址决策变量, 当且仅当在第 i 个备选地址处建设工厂时, $y_i = 1$, 否则, $y_i = 0$; x_{ij} 是运输决策变量, 表示从工厂 $i \in [I]$ 向客户 $j \in [J]$ 运输的产品数量. 当产品需求 $\boldsymbol{d}$ 给定时, 我们有如下确定性工厂选址模型:

$$
\begin{aligned}
\min \quad & \boldsymbol{f}^\top \boldsymbol{y} + \sum_{i\in[I]}\sum_{j\in[J]} c_{ij}x_{ij} + \sum_{j\in[J]} r_j\left(d_j - \sum_{i\in[I]} x_{ij}\right), \\
\text{s. t.} \quad & \sum_{i\in[I]} x_{ij} \leqslant d_j, \forall j \in [J], \\
& \sum_{j\in[J]} x_{ij} \leqslant q_i y_i, \forall i \in [I], \\
& x_{ij} \geqslant 0, \forall i \in [I],\ j \in [J], \\
& \boldsymbol{y} \in \mathcal{Y},
\end{aligned}
\tag{4.1}
$$

其中, $\boldsymbol{y} = (y_i)_{i\in[I]}$, 且 $\boldsymbol{f} = (f_i)_{i\in[I]}$. 集合 $\mathcal{Y}$ 是选址变量 $\boldsymbol{y}$ 的可行集, 例如, 其可以表示建设工厂总数量的限制. 一般情况下, 我们定义 $\mathcal{Y} := \left\{\boldsymbol{y} \in \{0,1\}^I \mid \mathbf{G}\boldsymbol{y} \geqslant \boldsymbol{h}\right\}$.

上述模型 (4.1) 的目标函数由三部分组成, 分别是工厂的总固定成本或建设成本、产品的总运输成本以及未满足产品需求的惩罚成本. 此外, 模型的第一组约束条件保证工厂运送给客户 j 的产品总量不超过客户的需求量 d_j. 第二组约束条件限制每个工厂 i, $i \in [I]$ 的产品供应量不超过其产能 q_i. 最后, 我们需要保证运输决策 $x_{ij}, i \in [I], j \in [J]$ 是非负的. 接下来, 我们介绍如何建模状态依赖型不确定性需求.

4.2.2 状态依赖分布不确定集

为了更准确地建模需求的不确定性, 我们将需求的状态依赖性引入分布不确定集, 构建如下状态依赖需求分布不确定集 (state-wise ambiguity set):

$$
\mathcal{F} = \left\{ \mathbb{P} \in \mathcal{P}(\mathbb{R}^J \times [K]) \;\middle|\; \begin{array}{l}
(\tilde{\boldsymbol{d}}, \tilde{k}) \sim \mathbb{P}, \\
\mathbb{E}_{\mathbb{P}}\left[\tilde{\boldsymbol{d}} \middle| \tilde{k} = k\right] \in [\boldsymbol{d}_k^-, \boldsymbol{d}_k^+], \forall k \in [K], \\
\mathbb{E}_{\mathbb{P}}\left[\boldsymbol{g}_k(\tilde{\boldsymbol{d}}) \middle| \tilde{k} = k\right] \leqslant \boldsymbol{\sigma}_k,\ \forall k \in [K], \\
\mathbb{P}\left[\tilde{\boldsymbol{d}} \in \mathcal{D}_k \middle| \tilde{k} = k\right] = 1, \forall k \in [K], \\
\mathbb{P}\left[\tilde{k} = k\right] = p_k, \forall k \in [K]
\end{array} \right\}. \tag{4.2}
$$

在上述分布不确定集 $\mathcal{F}$ 中, $\mathcal{P}$ 表示以集合 $\mathbb{R}^J \times [K]$ 为支撑集的概率分布集合, 其中 K 是状态的数量. 第一组约束表示每个状态 k 下, 不确定需求的条件期望值有限且属于区间 $[\boldsymbol{d}_k^-,\ \boldsymbol{d}_k^+]$. 第二组约束通过广义正常凸映射 (general proper convex mapping) $\boldsymbol{g}_k(\cdot):\ \mathbb{R}^J \mapsto \mathbb{R}^{n_k}$ 和参数 $\boldsymbol{\sigma}_k \in \mathbb{R}^{n_k}$ 来建模每个状态 k 下, 不确定需求的条件期望离差 (dispersion) 与相关性度量, 例如, 绝对离

差 (absolute dispersion) 和 Mahalanobis 距离. 第四组约束是每个状态 k 下, 需求 $\tilde{\boldsymbol{d}}$ 的支撑集约束, 其限制在第 k 种状态下 $\tilde{\boldsymbol{d}}$ 的所有取值在一个有界多面体 (bounded polyhedron) 集合 $\mathcal{D}_k$ 中, 其具体形式为 $\mathcal{D}_k := \left\{\boldsymbol{d} \in \mathbb{R}_+^J \mid \boldsymbol{A}_k\boldsymbol{d} \leqslant \boldsymbol{b}_k\right\}$, 其中 $\boldsymbol{A}_k = [\boldsymbol{a}_{1k}, \boldsymbol{a}_{2k}, \cdots, \boldsymbol{a}_{Jk}] \in \mathbb{R}^{m_k \times J}$, 并且满足 $[\boldsymbol{d}_k^-, \boldsymbol{d}_k^+] \subseteq \mathcal{D}_k$. 最后一组约束表示每个状态发生的概率为 p_k. 最后, 我们假设上述分布不确定集满足 Slater 条件①.

特别地, 上述分布不确定集中第二组约束的广义正常凸映射 $\boldsymbol{g}_k(\cdot)$ 可以灵活地建模各种概率分布信息, 例如, 不确定需求之间的相关性信息. 我们给出如下几个关于 $\boldsymbol{g}_k(\cdot)$ 的具体例子.

例 9 (绝对离差) 假设分布不确定集 $\mathcal{F}$ 中的映射 $\boldsymbol{g}_k(\boldsymbol{d}): \mathbb{R}^J \mapsto \mathbb{R}^{n_k}$ 定义为

$$\boldsymbol{g}_k(\boldsymbol{d}) := \left(\left\|\boldsymbol{U}_{lk}\boldsymbol{d} - \boldsymbol{\theta}_{lk}\right\|_1\right)_{l \in [N_k]}, \quad \forall k \in [K], \tag{4.3}$$

其中, $[N_k] := \{1, \cdots, n_k\}$, $\|\cdot\|_1$ 为 ℓ_1 范数, $\boldsymbol{U}_{lk} \in \mathbb{R}^{L_l \times J}$, $k \in [K]$, $l \in [N_k]$ 并且每个 $\boldsymbol{\theta}_{lk} \in \mathbb{R}^{L_l}$ 都是通过历史产品需求估计得到的参数. 则分布不确定集中的第二组约束可以表示为

$$\mathbb{E}_{\mathbb{P}}\left[\left\|\boldsymbol{U}_{lk}\tilde{\boldsymbol{d}} - \boldsymbol{\theta}_{lk}\right\|_1 \middle| \tilde{k} = k\right] \leqslant \sigma_{lk}, \quad \forall l \in [N_k],\, k \in [K].$$

特别地, 对于某些 l 和 k, 如果我们令 $\boldsymbol{U}_{lk} := \dfrac{1}{J}\boldsymbol{e}^\top \in \mathbb{R}^{1 \times J}$ 且 $\boldsymbol{\theta}_{lk} := \bar{d}_{lk} \in \mathbb{R}$, 其中 $\bar{d}_{lk}$ 表示需求的名义值 (nominal value), 则我们可以得到

$$\mathbb{E}_{\mathbb{P}}\left[\left\|\boldsymbol{U}_{lk}\tilde{\boldsymbol{d}} - \boldsymbol{\theta}_{lk}\right\|_1 \middle| \tilde{k} = k\right] = \mathbb{E}_{\mathbb{P}}\left[\left|\frac{1}{J}\sum_{j \in [J]} \tilde{d}_j - \bar{d}_{lk}\right| \middle| \tilde{k} = k\right] \leqslant \sigma_{lk}.$$

上述约束为一阶交叉中心矩度量 (first-order cross central-moment metric) 约束. 对于每个状态 $k \in [K]$, 上述约束刻画了不同客户需求之间的相关性信息.

例 10 (Mahalanobis 距离) Mahalanobis 距离 (de Maesschalck et al., 2000) 可用于度量每个状态 k 下, 不确定需求 $\boldsymbol{d}$ 相对于其估计均值 $\bar{\boldsymbol{d}}_k$ 偏离程度, 其中偏离程度通过协方差矩阵 $\boldsymbol{\Sigma}_k$ 进行归一化. 特别地, 我们取 $n_k = 1, \forall k \in [K]$ 并将映射 $g_k(\boldsymbol{d})$ 定义为

$$g_k(\boldsymbol{d}) := \sqrt{\left(\boldsymbol{d} - \bar{\boldsymbol{d}}_k\right)^\top \boldsymbol{\Sigma}_k^{-1}\left(\boldsymbol{d} - \bar{\boldsymbol{d}}_k\right)}, \tag{4.4}$$

① 当 $\boldsymbol{d}_k^- = \boldsymbol{d}_k^+ = \bar{\boldsymbol{d}}_k$ 时, Slater 条件可以简化为 $\boldsymbol{g}_k(\bar{\boldsymbol{d}}_k) < \boldsymbol{\sigma}_k$ 以保证强对偶条件成立, 即存在 $\boldsymbol{d}_k^\dagger \in (\boldsymbol{d}_k^-, \boldsymbol{d}_k^+)$ 满足 $\boldsymbol{g}_k(\boldsymbol{d}_k^\dagger) < \boldsymbol{\sigma}_k$.

其中, 我们假设对于所有的 $k \in [K]$, $\boldsymbol{\Sigma}_k$ 是正定矩阵. 则可以将分布不确定集 $\mathcal{F}$ 中的第二组约束表示为

$$\mathbb{E}_{\mathbb{P}}\left[\sqrt{\left(\tilde{\boldsymbol{d}}-\bar{\boldsymbol{d}}_k\right)^{\top}\boldsymbol{\Sigma}_k^{-1}\left(\tilde{\boldsymbol{d}}-\bar{\boldsymbol{d}}_k\right)}\,\middle|\,\tilde{k}=k\right]\leqslant\sigma_k,\quad\forall k\in[K].$$

上述约束描述了不同客户需求间的离差信息.

此外, 也可直接使用需求预测值 $\{\hat{\boldsymbol{d}}_k, k\in[K]\}$ 来构造分布不确定集 $\mathcal{F}$. 在下述例子中, 我们将每个需求预测值 $\hat{\boldsymbol{d}}_k$ 看作一个 "状态". 在此条件下, 映射 $\boldsymbol{g}_k(\cdot)$ 表示为真实需求分布 (true distribution of demand) 与需求预测值 $\hat{\boldsymbol{d}}_k$ 构成的经验分布 (empirical distribution) 之间的统计距离 (statistical distance).

例 11 (Type-∞ Wasserstein 球) 考虑随机向量 $\tilde{\boldsymbol{d}}\sim\mathbb{P}$ 和一个经验分布 $\tilde{\boldsymbol{\zeta}}\sim\widehat{\mathbb{P}}_K$, 其中 $\widehat{\mathbb{P}}_K=\sum_{k\in[K]}p_k\delta_{\hat{\boldsymbol{d}}_k}$, $\delta_{\hat{\boldsymbol{d}}_k}$ 是点 $\hat{\boldsymbol{d}}_k$ 上概率为 1 的 Dirac 分布. 此外, 两个分布的支撑集都是集合 $\mathcal{D}$. 我们将分布 $\mathbb{P}$ 和 $\widehat{\mathbb{P}}_K$ 之间的 type-∞ Wasserstein 距离 (Givens and Shortt, 1984; Xie, 2020; Bertsimas et al., 2023) 定义为

$$\mathrm{W}^{\infty}(\mathbb{P},\widehat{\mathbb{P}}_K):=\inf_{\mathbb{Q}\in\mathcal{P}(\mathcal{D}\times\mathcal{D})}\left\{\mathbb{Q}\text{-ess.sup}\ \left\|\tilde{\boldsymbol{d}}-\tilde{\boldsymbol{\zeta}}\right\|_q\,\middle|\,\begin{array}{l}\mathbb{Q}\text{ 是 }(\tilde{\boldsymbol{d}},\hat{\boldsymbol{\zeta}})\text{ 的联合分布},\\ \text{其边际分布分别是 }\mathbb{P}\text{ 和 }\widehat{\mathbb{P}}_K\end{array}\right\},$$

其中, $\|\cdot\|_q$ 为 $q\in[1,\infty]$ 的 q 范数, $\mathbb{Q}$-ess.sup 为 $\|\cdot\|_q$ 关于概率测度 $\mathbb{Q}$ 的本性上确界 (essential supremum), 即 $\inf\{U:\mathbb{Q}[\|\tilde{\boldsymbol{d}}-\tilde{\boldsymbol{\zeta}}\|_q>U]=0\}$. 那么, 给定半径 $\theta\geqslant 0$, 相应的 type-∞ Wasserstein 球可以定义为 $\mathcal{F}_{\mathrm{W}}^{\infty}(\theta):=\{\mathbb{P}\in\mathcal{P}(\mathcal{D})|W^{\infty}(\mathbb{P},\widehat{\mathbb{P}}_K)\leqslant\theta\}$. Chen 等 (2020) 证明该 type-∞ Wasserstein 球 $\mathcal{F}_{\mathrm{W}}^{\infty}(\theta)$ 与事件依赖分布不确定集具有等价表示形式如下:

$$\mathcal{F}_{\mathrm{W}}^{\infty}(\theta)=\left\{\mathbb{P}\in\mathcal{P}(\mathbb{R}^J\times[K])\,\middle|\,\begin{array}{l}(\tilde{\boldsymbol{d}},\tilde{k})\sim\mathbb{P},\mathbb{P}\left[\tilde{k}=k\right]=p_k,\forall k\in[K],\\ \mathbb{P}\left[\tilde{\boldsymbol{d}}\in\mathcal{D},\|\tilde{\boldsymbol{d}}-\hat{\boldsymbol{d}}_k\|_q\leqslant\theta|\tilde{k}=k\right]=1,\forall k\in[K]\end{array}\right\}.\tag{4.5}$$

在分布不确定集 $\mathcal{F}$ 中, 令对于所有 $k\in[K]$, $\sigma_k=\theta$, 且 $\mathcal{D}_k=\mathcal{D}$, 同时 $[\boldsymbol{d}_k^-,\boldsymbol{d}_k^+]$ 足够大, 并定义

$$g_k(\boldsymbol{d}):=\left\|\boldsymbol{d}-\hat{\boldsymbol{d}}_k\right\|_q,\quad\forall k\in[K],$$

在上述设定下, 分布不确定集 $\mathcal{F}$ (4.2) 实际上等价于上述 type-∞ Wasserstein 球

$\mathcal{F}_{\mathrm{W}}^{\infty}(\theta)$ 的提升形式如下:

$$\mathcal{F}(\theta)=\left\{\mathbb{P}\in\mathcal{P}(\mathbb{R}^J\times\mathbb{R}\times[K])\ \middle|\ \begin{array}{l}(\tilde{\boldsymbol{d}},\tilde{w},\tilde{k})\sim\mathbb{P},\mathbb{P}\left[\tilde{k}=k\right]=p_k,\forall k\in[K],\\ \mathbb{E}_{\mathbb{P}}\left[\tilde{w}\middle|\tilde{k}=k\right]\leqslant\theta,\forall k\in[K],\\ \mathbb{P}\left[\tilde{\boldsymbol{d}}\in\mathcal{D},\|\tilde{\boldsymbol{d}}-\hat{\boldsymbol{d}}_k\|_q\leqslant\tilde{w}\middle|\tilde{k}=k\right]=1,\forall k\in[K]\end{array}\right\}. \tag{4.6}$$

因为 $\mathcal{F}(\theta)$ 在 $(\tilde{\boldsymbol{d}},\tilde{k})$ 空间上的映射等价于 $\mathcal{F}_{\mathrm{W}}^{\infty}(\theta)$, 所以 $\mathcal{F}(\theta)$ 和 $\mathcal{F}_{\mathrm{W}}^{\infty}(\theta)$ 是等价的.

4.2.3 两阶段分布鲁棒产能限制型工厂选址模型

给定状态依赖分布不确定集 $\mathcal{F}$ (4.2), 我们构建两阶段分布鲁棒产能限制型工厂选址模型, 其又被称为鲁棒随机工厂 (设施) 选址问题 (robust stochastic facility location problem, RS-FLP) 模型:

$$\Pi^*:=\min_{\boldsymbol{y}\in\mathcal{Y}}\left\{\boldsymbol{f}^\top\boldsymbol{y}+\sup_{\mathbb{P}\in\mathcal{F}}\mathbb{E}_{\mathbb{P}}\left[\mathrm{Q}(\boldsymbol{y},\tilde{\boldsymbol{d}})\right]\right\}. \tag{4.7}$$

给定选址决策 $\boldsymbol{y}$ 和客户需求 $\boldsymbol{d}$, 第二阶段问题为

$$\mathrm{Q}(\boldsymbol{y},\boldsymbol{d}):=\min_{\boldsymbol{x}\in\mathcal{X}(\boldsymbol{y},\boldsymbol{d})}\sum_{i\in[I]}\sum_{j\in[J]}c_{ij}x_{ij}+\sum_{j\in[J]}r_j\left(d_j-\sum_{i\in[I]}x_{ij}\right), \tag{4.8}$$

其中, $\mathcal{X}(\boldsymbol{y},\boldsymbol{d})$ 是第二阶段问题的可行集:

$$\mathcal{X}(\boldsymbol{y},\boldsymbol{d}):=\left\{\boldsymbol{x}\in\mathbb{R}^{I\times J}\ \middle|\ \begin{array}{l}\sum_{i\in[I]}x_{ij}\leqslant d_j,\forall j\in[J],\\ \sum_{j\in[J]}x_{ij}\leqslant q_iy_i,\forall i\in[I],\\ x_{ij}\geqslant 0,\forall i\in[I],\ j\in[J]\end{array}\right\}. \tag{4.9}$$

上述第二阶段问题 (4.8) 的目标是在给定选址决策 $\boldsymbol{y}$ 和需求 $\boldsymbol{d}$ 的情况下, 寻找最优的运输决策以最小化运输成本与未满足需求的惩罚成本. RS-FLP 模型 (4.7) 的目标是寻找最优的选址决策, 以最小化固定成本与极值概率分布下第二阶段问题期望成本之和.

通过观察状态依赖分布不确定集 $\mathcal{F}$ 的结构, 我们发现: 当 $K=1$ 时, RS-FLP 模型 (4.7) 退化为分布鲁棒优化模型; 其次, 当 RS-FLP 模型 (4.7) 的支撑集 $\mathcal{D}_k$ 为单点集时, 即 $\mathcal{D}_k=\{\hat{\boldsymbol{d}}_k\}$, RS-FLP 模型 (4.7) 则退化为基于经验分布

$\widehat{\mathbb{P}}_K := \sum_{k\in[K]} p_k \delta_{\hat{\boldsymbol{d}}_k}$ 的两阶段随机规划模型. 因此, RS-FLP 模型 (4.7) 可以看作是分布鲁棒工厂选址模型和随机工厂选址模型之间的权衡模型 (trade-off model).

最后, 我们指出当工厂产能充足时, 即当产能满足如下条件时:

$$q_i \geqslant \max_{k\in[K]} \left\{ \sum_{j\in[J]} u_j^k \right\}, \quad \forall i \in [I], \tag{4.10}$$

其中, $u_j^k := \max\{d_j : \boldsymbol{d} \in \mathcal{D}_k\}, \forall j \in [J], k \in [K]$, 问题 (4.7) 转化为一个产能充足型工厂 (设施) 选址问题 (uncapacitated facility location problem, UFLP), 其第二阶段问题的可行集有如下形式:

$$\mathcal{X}_{\mathrm{u}}(\boldsymbol{y}, \boldsymbol{d}) := \left\{ \boldsymbol{x} \in \mathbb{R}^{I\times J} \;\middle|\; \begin{array}{l} \sum_{i\in[I]} x_{ij} \leqslant d_j, \forall j \in [J], \\ x_{ij} \leqslant d_j y_i, \forall i \in [I], j \in [J], \\ x_{ij} \geqslant 0, \forall i \in [I],\ j \in [J] \end{array} \right\}. \tag{4.11}$$

在此情况下, 我们有以下引理成立.

引理 4.1 在工厂产能充足条件 (4.10) 下, 对任意 $\boldsymbol{y} \in \mathcal{Y}$, 我们有

$$\mathbb{P}\Big[\mathcal{X}(\boldsymbol{y}, \tilde{\boldsymbol{d}}) = \mathcal{X}_{\mathrm{u}}(\boldsymbol{y}, \tilde{\boldsymbol{d}})\Big] = 1, \quad \forall \mathbb{P} \in \mathcal{F}. \tag{4.12}$$

引理 4.1 表明, 条件 (4.10) 保证对于任意 $\mathbb{P} \in \mathcal{F}$, $\mathcal{X}(\boldsymbol{y}, \tilde{\boldsymbol{d}})$ 和 $\mathcal{X}_{\mathrm{u}}(\boldsymbol{y}, \tilde{\boldsymbol{d}})$ 对应的第二阶段问题成本 $\mathrm{Q}(\boldsymbol{y}, \tilde{\boldsymbol{d}})$ 几乎处处相等 (coincide almost surely). 因此, 通过引理 4.1, 我们可以将产能充足型鲁棒随机工厂选址问题看作产能限制型鲁棒随机工厂选址问题的特例.

4.3 鲁棒敏感性分析

在本节中, 我们利用产能限制型 RS-FLP 模型 (4.7) 的对偶形式 (dual reformulation), 分析极值概率分布下第二阶段期望成本分别关于分布不确定集参数和选址决策的敏感性.

4.3.1 针对分布不确定集参数的敏感性分析

参数 $(\boldsymbol{d}^+, \boldsymbol{d}^-, \boldsymbol{\sigma}, \boldsymbol{p})$ 是构建分布不确定集 $\mathcal{F}$ 的关键, 其中 $\boldsymbol{d}^+ = (\boldsymbol{d}_k^+)_{k\in[K]}$, $\boldsymbol{d}^- = (\boldsymbol{d}_k^-)_{k\in[K]}, \boldsymbol{\sigma} = (\boldsymbol{\sigma}_k)_{k\in[K]}$ 和 $\boldsymbol{p} = (p_k)_{k\in[K]}$. 这些参数的变动无疑会影响极值概率分布下第二阶段期望成本, 进而影响模型 (4.7) 的最优值. 在本小节中, 我

们在给定选址决策 $\boldsymbol{y}$ 的情况下, 针对参数 $(\boldsymbol{d}^+,\boldsymbol{d}^-,\boldsymbol{\sigma},\boldsymbol{p})$ 的变动进行敏感性分析, 以深入剖析其对第二阶段期望成本的影响.

我们定义基于参数 $\boldsymbol{\Theta}$ 构建的分布不确定集 $\mathcal{F}$ 为 $\mathcal{F}(\boldsymbol{\Theta})$, 其中 $\boldsymbol{\Theta}$ 可以表示参数 $(\boldsymbol{d}^+,\boldsymbol{d}^-,\boldsymbol{\sigma},\boldsymbol{p})$ 中的部分或全部参数. 此外, 给定选址决策 $\boldsymbol{y}$, 我们用

$$\Pi(\boldsymbol{\Theta}|\boldsymbol{y}) := \sup_{\mathbb{P}\in\mathcal{F}(\boldsymbol{\Theta})} \mathbb{E}_{\mathbb{P}}\left[\mathrm{Q}(\boldsymbol{y},\tilde{\boldsymbol{d}})\right] \tag{4.13}$$

表示关于参数 $\boldsymbol{\Theta}$ 的极值概率分布下第二阶段期望成本函数.

接下来, 我们利用 2.2 节提到的分布鲁棒优化方法, 将式 (4.13) 等价地转化为其对偶形式. 利用 Shapiro (2001) 提出的强对偶理论, 我们有如下引理.

引理 4.2　给定选址决策 $\boldsymbol{y}$, 极值概率分布下第二阶段期望成本函数可以表示为

$$\Pi(\boldsymbol{\Theta}|\boldsymbol{y}) = \sum_{k\in[K]} p_k \Pi_k(\boldsymbol{d}_k^+,\boldsymbol{d}_k^-,\boldsymbol{\sigma}_k|\boldsymbol{y}), \tag{4.14}$$

其中,

$$\begin{aligned}&\Pi_k(\boldsymbol{d}_k^+,\boldsymbol{d}_k^-,\boldsymbol{\sigma}_k|\boldsymbol{y})\\ :=&\left[\begin{array}{ll}\min & \alpha_k+\overline{\boldsymbol{\beta}}_k^\top\boldsymbol{d}_k^+-\underline{\boldsymbol{\beta}}_k^\top\boldsymbol{d}_k^-+\boldsymbol{\rho}_k^\top\boldsymbol{\sigma}_k,\\ \text{s.t.} & \alpha_k+(\overline{\boldsymbol{\beta}}_k-\underline{\boldsymbol{\beta}}_k)^\top\boldsymbol{d}+\boldsymbol{\rho}_k^\top\boldsymbol{w}\geqslant \mathrm{Q}(\boldsymbol{y},\boldsymbol{d}),\forall(\boldsymbol{d},\boldsymbol{w})\in\mathcal{W}_k,\\ & \alpha_k\in\mathbb{R},\ \overline{\boldsymbol{\beta}}_k,\underline{\boldsymbol{\beta}}_k\in\mathbb{R}_+^J,\ \boldsymbol{\rho}_k\in\mathbb{R}_+^{n_k}\end{array}\right],\ k\in[K],\end{aligned}$$

并且

$$\mathcal{W}_k := \left\{(\boldsymbol{d},\boldsymbol{w})\in\mathbb{R}^J\times\mathbb{R}^{n_k}\,\middle|\,\boldsymbol{d}\in\mathcal{D}_k,\ \boldsymbol{g}_k(\boldsymbol{d})\leqslant\boldsymbol{w}\right\}. \tag{4.15}$$

同时, $\overline{\boldsymbol{\beta}}_k,\underline{\boldsymbol{\beta}}_k,\boldsymbol{\rho}_k,\alpha_k,k\in[K]$ 是分布不确定集 (4.2) 中约束对应的对偶变量, $\mathrm{Q}(\boldsymbol{y},\boldsymbol{d})$ 为式 (4.8).

接下来, 我们利用问题 (4.14) 的最优解 $(\alpha_k^\star,\overline{\boldsymbol{\beta}}_k^\star,\underline{\boldsymbol{\beta}}_k^\star,\boldsymbol{\rho}_k^\star)_{k\in[K]}$ 分析极值概率分布下期望成本函数 $\Pi(\boldsymbol{\Theta}|\boldsymbol{y})$ 关于参数 $\boldsymbol{\Theta}$ 的敏感性. 首先, 根据问题 (4.14) 的结构特征, 我们可以看出 $\Pi(\boldsymbol{\Theta}|\boldsymbol{y})$ 关于状态概率 p_k 的梯度 $\nabla\Pi(p_k|\boldsymbol{y})$ 恰好是在状态 k 下极值概率分布下的第二阶段期望成本 $(\alpha_k^\star+\overline{\boldsymbol{\beta}}_k^{\star\top}\boldsymbol{d}_k^+-\underline{\boldsymbol{\beta}}_k^{\star\top}\boldsymbol{d}_k^-+\boldsymbol{\rho}_k^{\star\top}\boldsymbol{\sigma}_k)$. 此外, 关于分布不确定集参数 $\boldsymbol{d}^-,\boldsymbol{d}^+$ 和 $\boldsymbol{\sigma}$ 的敏感性分析, 我们有如下命题.

命题 4.1　给定选址决策 $\boldsymbol{y}$, 用 $(\alpha_k^\star,\overline{\boldsymbol{\beta}}_k^\star,\underline{\boldsymbol{\beta}}_k^\star,\boldsymbol{\rho}_k^\star)_{k\in[K]}$ 表示问题 (4.14) 的最优解. 此外, 记 $\Delta_{\boldsymbol{d}^+},\Delta_{\boldsymbol{d}^-}$ 和 $\Delta_{\boldsymbol{\sigma}}$ 分别为参数 $\boldsymbol{d}^+,\boldsymbol{d}^-$ 和 $\boldsymbol{\sigma}$ 的变动量. 用 $\Delta\Pi(\boldsymbol{d}^+|\boldsymbol{y})$,

$\Delta\Pi(\boldsymbol{d}^-|\boldsymbol{y})$ 和 $\Delta(\boldsymbol{\sigma}|\boldsymbol{y})$ 表示参数变动所导致的极值概率分布下第二阶段期望成本变动的变化量. 我们有以下结果:

$$\Delta\Pi(\boldsymbol{d}^+|\boldsymbol{y}) \leqslant \sum_{k\in[K]} p_k \left(\Delta_{\boldsymbol{d}_k^+}^\top \overline{\boldsymbol{\beta}}_k^\star\right), \Delta\Pi(\boldsymbol{d}^-|\boldsymbol{y})$$
$$\leqslant -\sum_{k\in[K]} p_k \left(\Delta_{\boldsymbol{d}_k^-}^\top \underline{\boldsymbol{\beta}}_k^\star\right), \Delta\Pi(\boldsymbol{\sigma}|\boldsymbol{y}) \leqslant \sum_{k\in[K]} p_k \left(\Delta_{\boldsymbol{\sigma}_k}^\top \boldsymbol{\rho}_k^\star\right).$$

命题 4.1 给出了分布不确定集参数变化所导致的极值概率分布下第二阶段期望成本变化的上界. 特别地, 如果参数 $\boldsymbol{d}^+, \boldsymbol{\sigma}$ 分别增加 $\Delta_{\boldsymbol{d}^+}, \Delta_{\boldsymbol{\sigma}}$, $\boldsymbol{d}^-$ 减少 $\Delta_{\boldsymbol{d}^-}$, 则由此造成的极值概率分布下第二阶段期望成本的增加量, 最多为

$$\sum_{k\in[K]} p_k \left(\Delta_{\boldsymbol{d}_k^+}^\top \overline{\boldsymbol{\beta}}_k^\star\right), \quad \sum_{k\in[K]} p_k \left(\Delta_{\boldsymbol{\sigma}_k}^\top \boldsymbol{\rho}_k^\star\right) \text{ 和 } \sum_{k\in[K]} p_k \left(\Delta_{\boldsymbol{d}_k^-}^\top \underline{\boldsymbol{\beta}}_k^\star\right).$$

利用上述结果, 我们可以分析 RS-FLP 模型 (4.7) 的最优值, 即

$$\Pi^\star(\boldsymbol{\Theta}) := \min_{\boldsymbol{y}\in\mathcal{Y}} \left\{\boldsymbol{f}^\top \boldsymbol{y} + \sup_{\mathbb{P}\in\mathcal{F}(\boldsymbol{\Theta})} \mathbb{E}_{\mathbb{P}}\left[\mathrm{Q}(\boldsymbol{y}, \tilde{\boldsymbol{d}})\right]\right\} = \min_{\boldsymbol{y}\in\mathcal{Y}} \left\{\boldsymbol{f}^\top \boldsymbol{y} + \Pi(\boldsymbol{\Theta}|\boldsymbol{y})\right\}, \quad (4.16)$$

关于分布不确定集参数 $\boldsymbol{\Theta}$ 的敏感性, 其中, $\Pi(\boldsymbol{\Theta}|\boldsymbol{y})$ 为给定选址决策 $\boldsymbol{y}$ 下的问题 (4.14). 我们有如下推论.

推论 4.1 令 $\boldsymbol{y}^\star$ 为问题 (4.16) 的最优解, 令 $\left(\alpha_k^\star(\boldsymbol{y}^\star), \overline{\boldsymbol{\beta}}_k^\star(\boldsymbol{y}^\star), \underline{\boldsymbol{\beta}}_k^\star(\boldsymbol{y}^\star), \boldsymbol{\rho}_k^\star(\boldsymbol{y}^\star)\right)_{k\in[K]}$ 为问题 $\Pi(\boldsymbol{\Theta}|\boldsymbol{y}^\star)$ 的最优解. 令 $\Delta_{\boldsymbol{p}}, \Delta_{\boldsymbol{d}^+}, \Delta_{\boldsymbol{d}^-}$ 和 $\Delta_{\boldsymbol{\sigma}}$ 分别为参数 $\boldsymbol{p}, \boldsymbol{d}^+, \boldsymbol{d}^-$ 和 $\boldsymbol{\sigma}$ 的变动量. 此外, 用 $\Delta\Pi^\star(\boldsymbol{p})$, $\Delta\Pi^\star(\boldsymbol{d}^+)$, $\Delta\Pi^\star(\boldsymbol{d}^-)$ 和 $\Delta\Pi^\star(\boldsymbol{\sigma})$ 表示不确定集参数变动所造成的问题 $\Pi(\boldsymbol{\Theta}|\boldsymbol{y}^\star)$ 最优值变动的变化量. 我们有以下结果成立:

(1) 对于任意的 $\Delta_{\boldsymbol{p}}$, 且满足 $\boldsymbol{p} + \Delta_{\boldsymbol{p}} \geqslant \boldsymbol{0}$ 和 $\boldsymbol{e}^\top(\boldsymbol{p} + \Delta_{\boldsymbol{p}}) = 1$, 我们有

$$\Delta\Pi^\star(\boldsymbol{p}) \leqslant \Delta_{\boldsymbol{p}}^\top \left(\alpha_k^\star(\boldsymbol{y}^\star) + \overline{\boldsymbol{\beta}}_k^\star(\boldsymbol{y}^\star)^\top \boldsymbol{d}_k^+ - \underline{\boldsymbol{\beta}}_k^\star(\boldsymbol{y}^\star)^\top \boldsymbol{d}_k^- + \boldsymbol{\rho}_k^\star(\boldsymbol{y}^\star)^\top \boldsymbol{\sigma}_k\right)_{k\in[K]}.$$

(2) 对于任意的 $\Delta_{\boldsymbol{d}^+}, \Delta_{\boldsymbol{d}^-}, \Delta_{\boldsymbol{\sigma}} \geqslant \boldsymbol{0}$, 我们有

$$\Delta\Pi^\star(\boldsymbol{d}^+) \leqslant \sum_{k\in[K]} p_k \Delta_{\boldsymbol{d}_k^+}^\top \overline{\boldsymbol{\beta}}_k^\star(\boldsymbol{y}^\star),$$
$$\Delta\Pi^\star(\boldsymbol{d}^-) \leqslant -\sum_{k\in[K]} p_k \Delta_{\boldsymbol{d}_k^-}^\top \underline{\boldsymbol{\beta}}_k^\star(\boldsymbol{y}^\star), \Delta\Pi^\star(\boldsymbol{\sigma}) \leqslant \sum_{k\in[K]} p_k \Delta_{\boldsymbol{\sigma}_k}^\top \boldsymbol{\rho}_k^\star(\boldsymbol{y}^\star).$$

上述推论说明, 我们可以利用最优选址决策 $\boldsymbol{y}^\star$ 及其对应的最优对偶变量 $\left(\alpha_k^\star(\boldsymbol{y}^\star), \overline{\boldsymbol{\beta}}_k^\star(\boldsymbol{y}^\star), \underline{\boldsymbol{\beta}}_k^\star(\boldsymbol{y}^\star), \boldsymbol{\rho}_k^\star(\boldsymbol{y}^\star)\right)_{k\in[K]}$ 构造 RS-FLP 模型 (4.7) 最优值的敏感性界限 (sensitivity bound).

4.3.2 针对选址决策的敏感性分析

接下来, 我们分析 RS-FLP 模型 (4.7) 针对选址决策变化的敏感性. 在给定选址决策 $\boldsymbol{y} \in \mathcal{Y}$ 的情况下, 我们将极值概率分布下第二阶段期望成本等价地表示为如下形式:

$$\Pi(\boldsymbol{y}) := \sum_{k \in K} p_k \sup_{\mathbb{P} \in \mathcal{F}_k} \mathbb{E}_{\mathbb{P}} \left[\mathrm{Q}(\boldsymbol{y}, \tilde{\boldsymbol{d}}) \right] = \sup_{\mathbb{P} \in \mathcal{F}} \mathbb{E}_{\mathbb{P}} \left[\mathrm{Q}(\boldsymbol{y}, \tilde{\boldsymbol{d}}) \right], \tag{4.17}$$

其中, $\mathcal{F}_k$ 为状态 $k \in [K]$ 下不确定需求条件分布的分布不确定集, 其包含分布不确定集 $\mathcal{F}$ (4.2) 在状态 k 下的分布信息:

$$\mathcal{F}_k = \left\{ \mathbb{P} \in \mathcal{P}(\mathbb{R}^{|J|}) \left| \begin{array}{l} \tilde{\boldsymbol{d}} \sim \mathbb{P}, \\ \mathbb{E}_{\mathbb{P}} \left[\tilde{\boldsymbol{d}} \right] \in [\boldsymbol{d}_k^-, \boldsymbol{d}_k^+], \mathbb{E}_{\mathbb{P}} \left[\boldsymbol{g}_k(\tilde{\boldsymbol{d}}) \right] \leqslant \boldsymbol{\sigma}_k, \mathbb{P} \left[\tilde{\boldsymbol{d}} \in \mathcal{D}_k \right] = 1 \end{array} \right. \right\}. \tag{4.18}$$

给定选址决策 $\boldsymbol{y} \in \mathcal{Y}$, 我们可以通过分别对问题 (4.14) 中的 $\Pi_k(\boldsymbol{d}_k^+, \boldsymbol{d}_k^-, \boldsymbol{\sigma}_k | \boldsymbol{y})$, $k \in [K]$ 做对偶的方式得到问题 (4.17). 特别地, 给定选址决策 $\boldsymbol{y}$ 和需求 $\boldsymbol{d}$, 问题 $\mathrm{Q}(\boldsymbol{y}, \boldsymbol{d})$ 的对偶问题为

$$\begin{aligned} \mathrm{Q}(\boldsymbol{y}, \boldsymbol{d}) &= \boldsymbol{r}^\top \boldsymbol{d} + \min_{\boldsymbol{x} \in \mathcal{X}(\boldsymbol{y}, \boldsymbol{d})} \sum_{i \in [I]} \sum_{j \in [J]} (c_{ij} - r_j) x_{ij} \\ &= \boldsymbol{r}^\top \boldsymbol{d} + \max_{(\boldsymbol{\nu}, \boldsymbol{\lambda}) \in \mathcal{H}} \left\{ -\boldsymbol{\nu}^\top \boldsymbol{d} - \sum_{i \in [I]} q_i y_i \lambda_i \right\}, \end{aligned} \tag{4.19}$$

其中, $\mathcal{H}$ 为对偶变量 $(\boldsymbol{\nu}, \boldsymbol{\lambda})$ 的可行集:

$$\mathcal{H} := \left\{ (\boldsymbol{\nu}, \boldsymbol{\lambda}) \in \mathbb{R}_+^J \times \mathbb{R}_+^I \mid -\nu_j - \lambda_i \leqslant c_{ij} - r_j, \ \forall i \in [I], \ j \in [J] \right\}. \tag{4.20}$$

由于变量 $\boldsymbol{\nu}$ 和 $\boldsymbol{\lambda}$ 存在下界, 而且 $\boldsymbol{d}$ 和 $q_i y_i, \forall i \in [I]$ 都是非负且有界的, 所以问题 (4.19) 总是有界的. 因此, 我们可以在极值点处求得问题 (4.19) 的最优解.

命题 4.2 给定可行集 $\mathcal{H}$ 的极值点集合 $\{(\hat{\boldsymbol{\nu}}^{e_k}, \hat{\boldsymbol{\lambda}}^{e_k})_{e_k \in [E]}\}$ 和一个选址决策 $\boldsymbol{y} \in \mathcal{Y}$. 对于任意 $k \in [K]$, 分布不确定集 $\mathcal{F}_k$ (4.18) 中的极值概率分布 $\mathbb{P}_k^\star$ 可以表示为一个不超过 E 个点的离散分布:

$$\mathbb{P}_k^\star \left[\tilde{\boldsymbol{d}} = \frac{\boldsymbol{\eta}_\star^{e_k}}{\zeta_\star^{e_k}} \right] = \zeta_\star^{e_k}, \quad \forall e_k \in [E], \tag{4.21}$$

其中, $(\boldsymbol{\eta}_{\star}^{e_k}, \zeta_{\star}^{e_k})_{e_k\in[E]}$ 为下面线性锥规划 (conic linear program) 问题的最优解

$$
\begin{aligned}
\mathbb{E}_{\mathbb{P}_k^\star}\left[\mathrm{Q}(\boldsymbol{y},\tilde{\boldsymbol{d}})\right] = \max_{\boldsymbol{\zeta},\boldsymbol{\eta}} \quad & \sum_{e_k\in[E]} (\boldsymbol{r}-\hat{\boldsymbol{\nu}}^{e_k})^\top \boldsymbol{\eta}^{e_k} - \sum_{e_k\in[E]}\sum_{i\in[I]} q_i y_i \hat{\lambda}_i^{e_k}\zeta^{e_k} \\
\text{s.t.} \quad & \sum_{e_k\in[E]} \zeta^{e_k} = 1, \\
& \sum_{e_k\in[E]} \boldsymbol{\eta}^{e_k} \leqslant \boldsymbol{d}_k^+, \\
& \sum_{e_k\in[E]} \boldsymbol{\eta}^{e_k} \geqslant \boldsymbol{d}_k^-, \\
& \sum_{e_k\in[E]} \boldsymbol{\tau}^{e_k} \leqslant \boldsymbol{\sigma}_k, \\
& \boldsymbol{A}_k\boldsymbol{\eta}^{e_k} - \zeta^{e_k}\boldsymbol{b}_k \leqslant \boldsymbol{0}, \quad \forall e_k \in [E], \\
& (\boldsymbol{\eta}^{e_k}, \boldsymbol{\tau}^{e_k}, \zeta^{e_k}) \in \mathcal{K}_k, \quad \forall e_k \in [E], \\
& \zeta^{e_k} \in \mathbb{R}_+, \boldsymbol{\eta}^{e_k} \in \mathbb{R}^J, \boldsymbol{\tau}^{e_k} \in \mathbb{R}^{n_k}, \ \forall e_k \in [E].
\end{aligned} \tag{4.22}
$$

这里, $\mathcal{K}_k, k \in [K]$ 为一个正常锥 (proper cone), 其定义如下:

$$
\mathcal{K}_k := \mathrm{cl}\left\{(\boldsymbol{d},\boldsymbol{w},t) \in \mathbb{R}_+^J \times \mathbb{R}^{n_k} \times \mathbb{R} \mid \boldsymbol{g}_k(\boldsymbol{d}/t) \leqslant \boldsymbol{w}/t,\ t>0\right\},\ \forall k \in [K], \tag{4.23}
$$

其中 $\mathrm{cl}(A)$ 表示集合 A 的闭包 (closure).

例 12 利用正常锥 $\mathcal{K}_k$ (4.23), 我们可以将不确定集 $\mathcal{W}_k$, $k \in [K]$ 等价地转化为如下锥结构形式 (conic representable form):

$$
\mathcal{W}_k = \left\{(\boldsymbol{d},\boldsymbol{w}) \in \mathbb{R}_+^J \times \mathbb{R}^{n_k} \middle| \boldsymbol{A}_k\boldsymbol{d} \leqslant \boldsymbol{b}_k,\ (\boldsymbol{d},\boldsymbol{w},t) \in \mathcal{K}_k, t=1\right\}, \quad \forall k \in [K]. \tag{4.24}
$$

特别地, 如果我们将映射 $\boldsymbol{g}_k(\cdot), \forall k \in [K]$ 定义为例 9 中的绝对离差形式 (4.3), 则不确定集 $\mathcal{W}_k$ 中的约束 $\boldsymbol{g}_k(\boldsymbol{d}) \leqslant \boldsymbol{w}$ 可表示为 ℓ_1 范数锥 (norm cone) 形式:

$$
\left(\|\boldsymbol{U}_{lk}\boldsymbol{d} - \bar{\boldsymbol{\theta}}_{lk}\|_1\right)_{l\in[N_k]} \leqslant \boldsymbol{w}, \quad \forall k \in [K]. \tag{4.25}
$$

在这种情况下, 问题 (4.22) 为一个线性规划问题. 同样地, 如果我们将 $\boldsymbol{g}_k(\cdot)$ 定义为例 10 中的 Mahalanobis 距离 (4.4), 那么不确定集 $\mathcal{W}_k$ 中的约束 $\boldsymbol{g}_k(\boldsymbol{d}) \leqslant \boldsymbol{w}$ 可表示为二阶锥 (second-order cone) 形式:

$$\left\|\boldsymbol{\Sigma}_k^{-1/2}(\boldsymbol{d}-\bar{\boldsymbol{d}}_k)\right\|_2 \leqslant w,\ \forall k \in [K]. \tag{4.26}$$

在这种情况下, 问题 (4.22) 为一个二阶锥规划问题. 最后, 当范数算子 $\|\cdot\|_q$ 取为 $\|\cdot\|_1, \|\cdot\|_\infty$ 或 $\|\cdot\|_2$ 时, 例 11 中的 Type-∞ Wasserstein 球形式会分别令问题 (4.22) 转化为一个线性规划问题或二阶锥规划问题.

接下来, 我们利用命题 4.2 给出的不同状态 $k \in [K]$ 下的极值概率分布 $\mathbb{P}_k^\star$ 探讨选址决策变动对极值概率分布下第二阶段期望成本的影响. 我们有如下命题.

命题 4.3　给定选址决策 $\boldsymbol{y}$, 其中对于部分工厂地址 $i, l \in [I]$,$y_i = 0$ 且 $y_l = 1$. 设 E 为集合 $\mathcal{H}$ 极值点的总个数. 我们将 $c^-(i)$ 定义为选址决策 $\boldsymbol{y}$ 中的其他决策不变的情况下, 由于在地址 i 新开设工厂所导致的第二阶段极值期望成本的减少量; 同样地, 我们将 $c^+(l)$ 定义为选址决策 $\boldsymbol{y}$ 中的其他决策不变的情况下, 由于关闭地址 l 处的工厂所导致的第二阶段极值期望成本的增加量, 则有

$$0 \leqslant c^-(i) \leqslant q_i\left(\sum_{k\in[K]}\sum_{e^k\in[E]}\hat{\lambda}_i^{e_k}\zeta_\star^{e_k}p_k\right),\quad c^+(l) \geqslant q_l\left(\sum_{k\in[K]}\sum_{e^k\in[E]}\hat{\lambda}_l^{e_k}\zeta_\star^{e_k}p_k\right) \geqslant 0, \tag{4.27}$$

其中, $(\zeta_\star^{e_k})_{e_k\in[E]}$ 是由式 (4.21) 给出的极值概率分布.

通过式 (4.27), 我们获得一个重要的管理学启示: 在一个地址新开设工厂 (或关闭一个已开设的工厂) 所造成的第二阶段极值期望成本增加 (或减少) 量的界限, 等于该工厂的产能 q_i (或 q_l) 与极值概率分布下产能约束所对应的期望影子价格 (expected shadow price) 的乘积, 即

$$\sum_{k\in[K]}p_k\left(\sum_{e^k\in[E]}\hat{\lambda}_i^{e_k}\zeta_\star^{e_k}\right)\quad \text{或} \quad \sum_{k\in[K]}p_k\left(\sum_{e^k\in[E]}\hat{\lambda}_l^{e_k}\zeta_\star^{e_k}\right).$$

换言之, 极值概率分布下第二阶段期望成本的增加或减少量, 受极值概率分布下期望影子产能成本 (worst-case expected shadow-capacity cost) 控制.

4.4　嵌套 Benders 分解精确求解算法

在本节中, 基于 Benders 分解算法 (Benders decomposition algorithm), 我们设计一种可以精确求解产能限制型 RS-FLP 模型 (4.7) 的算法.

首先, 利用引理 4.2, 产能限制型 RS-FLP 模型 (4.7) 可以等价地转化为如下形式:

$$
\begin{aligned}
\min \quad & \boldsymbol{f}^\top \boldsymbol{y} + \sum_{k\in[K]} p_k \left(\alpha_k + \overline{\boldsymbol{\beta}}_k^\top \boldsymbol{d}_k^+ - \underline{\boldsymbol{\beta}}_k^\top \boldsymbol{d}_k^- + \boldsymbol{\rho}_k^\top \boldsymbol{\sigma}_k\right) \\
\text{s.t.} \quad & \alpha_k + (\overline{\boldsymbol{\beta}}_k - \underline{\boldsymbol{\beta}}_k)^\top \boldsymbol{d} + \boldsymbol{\rho}_k^\top \boldsymbol{w} \geqslant \mathrm{Q}(\boldsymbol{y}, \boldsymbol{d}),\ \forall k \in [K],\ (\boldsymbol{d}, \boldsymbol{w}) \in \mathcal{W}_k, \\
& \alpha_k \in \mathbb{R},\ \overline{\boldsymbol{\beta}}_k, \underline{\boldsymbol{\beta}}_k \in \mathbb{R}_+^{|J|},\ \boldsymbol{\rho}_k \in \mathbb{R}_+^{n_k},\ \forall k \in [K],\ \boldsymbol{y} \in \mathcal{Y}.
\end{aligned}
\tag{4.28}
$$

特别地, 当不确定集 $\mathcal{W}_k$ 是一个多面体集 (polyhedron set) 时, 我们记其为 $\mathcal{W}_k^{\mathrm{P}}, k \in [K]$. 在上述情况下, 我们可以设计两种不同的方法构建具有有限约束的主问题 (master problem). 第一种方法利用每个集合 $\mathcal{W}_k^{\mathrm{P}}$ 的极值点构建主问题, 其为如下混合整数线性规划问题:

$$
\begin{aligned}
\min_{\boldsymbol{y}\in\mathcal{Y}} \quad & \boldsymbol{f}^\top \boldsymbol{y} + \sum_{k\in[K]} p_k \left(\alpha_k + \overline{\boldsymbol{\beta}}_k^\top \boldsymbol{d}_k^+ - \underline{\boldsymbol{\beta}}_k^\top \boldsymbol{d}^- + \boldsymbol{\rho}_k^\top \boldsymbol{\sigma}_k\right), \\
\text{s.t.} \quad & \alpha_k \geqslant \sum_{i\in[I]} \sum_{j\in[J]} (c_{ij} - r_j) x_{ij}^{e_k} - \boldsymbol{\rho}_k^\top \hat{\boldsymbol{w}}^{e_k} + (\boldsymbol{r} - \overline{\boldsymbol{\beta}}_k + \underline{\boldsymbol{\beta}}_k)^\top \hat{\boldsymbol{d}}^{e_k}, \\
& \boldsymbol{x}^{e_k} \in \mathcal{X}(\boldsymbol{y}, \hat{\boldsymbol{d}}^{e_k}), \forall k \in [K], e_k \in [E_k], \\
& \alpha_k \in \mathbb{R},\ \overline{\boldsymbol{\beta}}_k, \underline{\boldsymbol{\beta}}_k \in \mathbb{R}_+^{J},\ \boldsymbol{\rho}_k \in \mathbb{R}_+^{n_k},\ \forall k \in [K],
\end{aligned}
\tag{4.29}
$$

其中, $\mathcal{X}(\boldsymbol{y}, \boldsymbol{d})$ 是给定 $(\boldsymbol{y}, \boldsymbol{d})$ 下第二阶段问题的可行集 (4.9), 对于每个状态 $k \in [K]$, $\{(\hat{\boldsymbol{d}}^{e_k}, \hat{\boldsymbol{w}}^{e_k}), e_k \in [E_k]\}$ 是多面体不确定集 $\mathcal{W}_k^{\mathrm{P}}$ 的极值点集合. 上述方法与 Zeng 和 Zhao (2013) 以及 Bertsimas 和 Shtern (2018) 的研究中构建主问题的方法类似.

第二种方法利用第二阶段问题 $\mathrm{Q}(\boldsymbol{y}, \boldsymbol{d})$ 的对偶可行集的极值点来构造主问题, 在此情况下, 主问题同样是一个混合整数线性规划问题.

我们注意到, RS-FLP 模型 (4.7) 中每个集合 $\mathcal{W}_k, k \in [K]$ 都具有广义锥结构 (general conic representable formulation), 这使得第一种构建主问题的方法不适用于 RS-FLP 模型. 主要原因有以下两点: 首先, 如果集合 $\mathcal{W}_k$ 不是多面体集, 则集合 $\mathcal{W}_k$ 一般具有无限个极值点, 这导致 RS-FLP 模型的对偶问题 (4.28) 是一个具有无限约束的混合整数线性规划问题; 其次, 假如不确定集 $\mathcal{W}_k$ 是一个多面体集 $\mathcal{W}_k^{\mathrm{P}}$, RS-FLP 模型 (4.7) 共考虑 K 种状态, 且每种状态下集合 $\mathcal{W}_k^{\mathrm{P}}$ 的极值点个数为 E_k, 则总的极值点个数为 $\sum_{k\in[K]} E_k$, 与状态数 K 有关, 这使得混合整数规划问题 (4.29) 是一个大规模优化问题且难以被高效求解. 因此, 在本章中, 我们使用第二种方法构建算法的主问题. 特别地, 我们介绍一种基于次梯度的嵌套分解算法 (subgradient-based nested decomposition approach) 来精确求解模型 (4.7).

4.4.1 松弛主问题及其分解形式

在本小节中, 我们构建算法的松弛主问题 (relaxed master problem). 首先, 将 RS-FLP 模型 (4.7) 转化为如下形式:

$$\Pi^* = \min_{\boldsymbol{y}\in\mathcal{Y}} \left\{ \boldsymbol{f}^\top \boldsymbol{y} + \sum_{k\in[K]} p_k \sup_{\mathbb{P}\in\mathcal{F}_k} \mathbb{E}_{\mathbb{P}} \left[\mathrm{Q}(\boldsymbol{y}, \tilde{\boldsymbol{d}}) \right] \right\}, \tag{4.30}$$

其中, $\mathcal{F}_k$ 是状态 $k \in [K]$ 下不确定需求的分布不确定集 (4.18).

为了构造松弛主问题, 给定状态 $k \in [K]$, 考虑极值概率分布下第二阶段期望成本 $\sup_{\mathbb{P}\in\mathcal{F}_k} \mathbb{E}_{\mathbb{P}}[\mathrm{Q}(\boldsymbol{y}, \tilde{\boldsymbol{d}})]$ 的下界. 对于任意的 $k \in [K]$ 和 $(\boldsymbol{y}, \boldsymbol{d}) \in \mathcal{Y} \times \mathcal{D}_k$, 记 E 为多面体集 $\mathcal{H}$ (4.20) 的极值点总数. 则有如下等式:

$$\mathrm{Q}(\boldsymbol{y}, \boldsymbol{d}) = \mathrm{Q}^E(\boldsymbol{y}, \boldsymbol{d}) := \boldsymbol{r}^\top \boldsymbol{d} + \max_{e\in[E]} \left\{ -\boldsymbol{d}^\top \boldsymbol{\nu}^e - \sum_{i\in[I]} q_i y_i \lambda_i^e \right\}. \tag{4.31}$$

接下来, 考虑集合 $\mathcal{H}$ 的极值点子集 $[L] \subset [E]$, 我们得到问题 $\mathrm{Q}(\boldsymbol{y}, \boldsymbol{d})$ 的下界:

$$\mathrm{Q}(\boldsymbol{y}, \boldsymbol{d}) \geqslant \mathrm{Q}^L(\boldsymbol{y}, \boldsymbol{d}) := \boldsymbol{r}^\top \boldsymbol{d} + \max_{e\in[L]} \left\{ -\boldsymbol{d}^\top \boldsymbol{\nu}^e - \sum_{i\in[I]} q_i y_i \lambda_i^e \right\}, \tag{4.32}$$

利用式 (4.32), 构建如下松弛主问题:

$$\Pi_L^* := \min_{\boldsymbol{y}\in\mathcal{Y}} \left\{ \boldsymbol{f}^\top \boldsymbol{y} + \sum_{k\in[K]} p_k \sup_{\mathbb{P}\in\mathcal{F}_k} \mathbb{E}_{\mathbb{P}} \left[\mathrm{Q}^L(\boldsymbol{y}, \tilde{\boldsymbol{d}}) \right] \right\}. \tag{4.33}$$

特别地, 当 $L = E$ 时, 上述问题 (4.33) 便为 RS-FLP 模型 (4.30) 的主问题. 接下来, 我们将松弛主问题 (4.33) 转化为一个混合整数线性锥规划问题.

命题 4.4 给定集合 $\mathcal{H}$ 的极值点集的一个子集 $\{(\hat{\boldsymbol{\nu}}^{e_k}, \hat{\boldsymbol{\lambda}}^{e_k})_{e_k\in[L]}\}$, 松弛主问题 (4.33) 等价于如下混合整数线性锥规划问题:

$$\begin{aligned}
\Pi_L^* = \min \quad & \boldsymbol{f}^\top \boldsymbol{y} + \sum_{k\in[K]} p_k \left(\alpha_k + \overline{\boldsymbol{\beta}}_k^\top \boldsymbol{d}_k^+ - \underline{\boldsymbol{\beta}}_k^\top \boldsymbol{d}_k^- + \boldsymbol{\rho}_k^\top \boldsymbol{\sigma}_k \right) \\
\text{s.t.} \quad & \alpha_k - \boldsymbol{b}_k^\top \boldsymbol{\xi}^{e_k} - \gamma^{e_k} + \sum_{i\in[I]} q_i y_i \hat{\lambda}_i^{e_k} \geqslant 0, \forall k \in [K],\ e_k \in [L], \\
& \boldsymbol{\psi}^{e_k} - \boldsymbol{A}_k^\top \boldsymbol{\xi}^{e_k} - \overline{\boldsymbol{\beta}}_k + \underline{\boldsymbol{\beta}}_k + \boldsymbol{r} - \hat{\boldsymbol{\nu}}^{e_k} = \boldsymbol{0}, \\
& (\boldsymbol{\psi}^{e_k}, \boldsymbol{\rho}_k, \gamma^{e_k}) \in \mathcal{K}_k^*, \\
& \boldsymbol{\xi}^{e_k} \in \mathbb{R}_+^{m_k}, \boldsymbol{\psi}^{e_k} \in \mathbb{R}^J, \gamma^{e_k} \in \mathbb{R}, \\
& \alpha_k \in \mathbb{R},\ \overline{\boldsymbol{\beta}}_k, \underline{\boldsymbol{\beta}}_k \in \mathbb{R}_+^J,\ \boldsymbol{\rho}_k \in \mathbb{R}_+^{n_k}, \forall k \in [K], \\
& \boldsymbol{y} \in \mathcal{Y},
\end{aligned} \tag{4.34}$$

其中, 对于任意 $k \in [K]$, $\mathcal{K}_k^*$ 为锥 $\mathcal{K}_k$ (4.23) 的对偶锥 (dual cone).

因为松弛主问题 (4.34) 的规模会随着极值点个数的增加而不断增大, 所以直接求解松弛主问题 (4.34) 的计算效率较低. 接下来, 我们介绍一个分解算法来迭代求解松弛主问题. 利用此分解算法, 在每次迭代时, 我们只需要分别求解 K 个小型线性锥规划问题和混合整数线性规划问题.

基于式 (4.32), 我们可以得出如下结论: 对于任意 $k \in [K]$, 松弛主问题 (4.33) 的极值概率分布下第二阶段期望成本

$$\Pi^L(\boldsymbol{y}) := \sup_{\mathbb{P} \in \mathcal{F}_k} \mathbb{E}_{\mathbb{P}} \left[\mathrm{Q}^L(\boldsymbol{y}, \tilde{\boldsymbol{d}}) \right] = \mathbb{E}_{\mathbb{P}_k^\star} \left[\mathrm{Q}^L(\boldsymbol{y}, \tilde{\boldsymbol{d}}) \right] \tag{4.35}$$

是一个关于 $\boldsymbol{y}$ 的凸函数. 接下来, 我们利用给定 $\boldsymbol{y} \in \mathcal{Y}$ 下, 问题 (4.35) 关于 $\boldsymbol{y}$ 的次梯度 (subgradient) 来构建分解算法. 此次梯度可以通过极值概率分布 $\mathbb{P}_k^\star$ 来构造.

推论 4.2 给定集合 $\mathcal{H}$ 的极值点集的一个子集 $\{(\hat{\boldsymbol{\nu}}^{e_k}, \hat{\boldsymbol{\lambda}}^{e_k})_{e_k \in [L]}\}$ 和选址决策 $\boldsymbol{y} \in \mathcal{Y}$. 对于每个 $k \in [K]$, 问题 (4.35) 中的极值概率分布 $\mathbb{P}_k^\star$ 可以表示为一个由 L 个点构成的离散分布:

$$\mathbb{P}_k^\star \left[\tilde{\boldsymbol{d}} = \frac{\boldsymbol{\eta}_\star^{e_k}}{\zeta_\star^{e_k}} \right] = \zeta_\star^{e_k}, \quad \forall e_k \in [L], \tag{4.36}$$

其中, $(\boldsymbol{\eta}_\star^{e_k}, \zeta_\star^{e_k})_{e_k \in [L]}$ 是下面线性锥规划问题的最优解

$$\begin{aligned}
\mathbb{E}_{\mathbb{P}_k^\star} \left[\mathrm{Q}^L(\boldsymbol{y}, \tilde{\boldsymbol{d}}) \right] = \max_{\boldsymbol{\zeta}, \boldsymbol{\eta}} \quad & \sum_{e_k \in [E]} (\boldsymbol{r} - \hat{\boldsymbol{\nu}}^{e_k})^\top \boldsymbol{\eta}^{e_k} - \sum_{e_k \in [E]} \sum_{i \in [I]} q_i y_i \hat{\lambda}_i^{e_k} \zeta^{e_k} \\
\text{s.t.} \quad & \sum_{e_k \in [E]} \zeta^{e_k} = 1, \sum_{e_k \in [E]} \boldsymbol{\eta}^{e_k} \leqslant \boldsymbol{d}_k^+, \\
& \sum_{e_k \in [E]} \boldsymbol{\eta}^{e_k} \geqslant \boldsymbol{d}_k^-, \sum_{e_k \in [E]} \boldsymbol{\tau}^{e_k} \leqslant \boldsymbol{\sigma}_k, \\
& \boldsymbol{A}_k \boldsymbol{\eta}^{e_k} - \zeta^{e_k} \boldsymbol{b}_k \leqslant \boldsymbol{0}, \forall e_k \in [L], \\
& (\boldsymbol{\eta}^{e_k}, \boldsymbol{\tau}^{e_k}, \zeta^{e_k}) \in \mathcal{K}_k, \forall e_k \in [L], \\
& \zeta^{e_k} \in \mathbb{R}_+, \boldsymbol{\eta}^{e_k} \in \mathbb{R}^J, \boldsymbol{\tau}^{e_k} \in \mathbb{R}^{n_k}, \ \forall e_k \in [L],
\end{aligned} \tag{4.37}$$

其中, $\mathcal{K}_k$ (4.23) 是正常锥. 进一步, 我们可得次梯度:

$$\left(-q_i \sum_{e^k \in [L]} \hat{\lambda}_i^e \zeta_\star^{e_k} \right)_{i \in [I]} \in \partial \Pi^L(\boldsymbol{y}). \tag{4.38}$$

通过推论 4.2, 可以得出以下结论: ① 给定选址决策 $\boldsymbol{y}$, 每个子问题 (4.35) 都是一个可被单独求解的小规模线性锥规划问题 (4.37). ② 通过求解问题 (4.37), 可以得到子问题 $\sup_{\mathbb{P}\in\mathcal{F}_k}\mathbb{E}_{\mathbb{P}}[\mathrm{Q}^L(\boldsymbol{y},\tilde{\boldsymbol{d}})]$ 在点 $\boldsymbol{y}$ 处的次梯度. 利用 $\boldsymbol{y}\in\mathcal{Y}$ 的 0-1 整数结构, 可以构造问题 $\sup_{\mathbb{P}\in\mathcal{F}_k}\mathbb{E}_{\mathbb{P}}[\mathrm{Q}^L(\boldsymbol{y},\tilde{\boldsymbol{d}})]$ 的一个有限分段仿射形式 (finite piecewise affine reformulation). 这个形式可以帮助我们构建一个与松弛主问题 (4.33) 等价的线性规划问题. ③ 利用问题 (4.37), 可以证明问题 $\sup_{\mathbb{P}\in\mathcal{F}_k}\mathbb{E}_{\mathbb{P}}[\mathrm{Q}^L(\boldsymbol{y},\tilde{\boldsymbol{d}})]$ 是有界的.

引理 4.3　对于任意的 $\boldsymbol{y}\in\mathcal{Y}$ 和 $[L]\subseteq[E]$, 我们有

$$\left|\sup_{\mathbb{P}\in\mathcal{F}_k}\mathbb{E}_{\mathbb{P}}\left[\mathrm{Q}^L(\boldsymbol{y},\tilde{\boldsymbol{d}})\right]\right|<\infty,\quad \forall k\in[K]. \tag{4.39}$$

基于引理 4.3, 可以构造松弛主问题 (4.33) 的等价问题.

命题 4.5　给定集合 $\mathcal{H}$ 极值点集的一个子集 $[L]$, 则松弛主问题 (4.33) 等价于如下混合整数线性规划问题:

$$\begin{aligned}\min_{\boldsymbol{y},\boldsymbol{\theta}}\quad & \boldsymbol{f}^\top\boldsymbol{y}+\sum_{k\in[K]}p_k\theta_k\\ \text{s.t.}\quad & \theta_k\geqslant\sum_{e_k\in[L]}(\boldsymbol{r}-\hat{\boldsymbol{\nu}}^{e_k})^\top\boldsymbol{\eta}_\star^{e_k}(\boldsymbol{y}^\diamond)\\ & \qquad -\sum_{e_k\in[L]}\sum_{i\in[I]}q_iy_i\hat{\lambda}_i^{e_k}\zeta_\star^{e_k}(\boldsymbol{y}^\diamond),\ \forall\boldsymbol{y}^\diamond\in\mathcal{Y},k\in[K],\\ & \boldsymbol{y}\in\mathcal{Y},\boldsymbol{\theta}\in\mathbb{R}^K,\end{aligned} \tag{4.40}$$

其中, $(\boldsymbol{\eta}_\star^{e_k}(\boldsymbol{y}^\diamond),\zeta_\star^{e_k}(\boldsymbol{y}^\diamond))_{e_k\in[L]}$ 是给定选址决策 $\boldsymbol{y}^\diamond$ 下问题 (4.37) 的最优解.

命题 4.5 中的混合整数线性规划问题 (4.40) 和命题 4.2 中的线性锥规划问题 (4.37) 是设计松弛主问题 (4.33) 分解算法的关键. 特别地, 给定一个选址决策 $\boldsymbol{y}^n\in\mathcal{Y}$, 对于每个 $k\in[K]$, 我们可以分别求解问题 (4.37), 得到问题的最优值 $\sup_{\mathbb{P}\in\mathcal{F}_k}\mathbb{E}_{\mathbb{P}}[\mathrm{Q}^L(\boldsymbol{y}^n,\tilde{\boldsymbol{d}})]$. 进一步, 我们可以构建松弛主问题 (4.33) 的一个上界:

$$\mathrm{UB}_L=\boldsymbol{f}^\top\boldsymbol{y}^n+\sum_{k\in[K]}p_k\sup_{\mathbb{P}\in\mathcal{F}_k}\mathbb{E}_{\mathbb{P}}[\mathrm{Q}^L(\boldsymbol{y}^n,\tilde{\boldsymbol{d}})]. \tag{4.41}$$

根据推论 4.2, 对于每个 $k\in[K]$, 可以利用问题 (4.37) 的最优解计算得到问题 (4.35) 在点 $\boldsymbol{y}^n$ 处的次梯度:

$$\left(-q_i\sum_{e^k\in[L]}\hat{\lambda}_i^{e_k}\zeta_\star^{e_k}(\boldsymbol{y}^n)\right)_{i\in[I]}.$$

利用上述次梯度, 可以构造一个混合整数规划问题 (4.40) 中的线性约束或割平面 (cutting plane), 进一步, 可以得到混合整数规划问题 (4.40) 的一个下界:

$$
\begin{aligned}
\mathrm{LB}_L = \min_{\boldsymbol{y},\boldsymbol{\theta}} \ & \boldsymbol{f}^\top \boldsymbol{y} + \sum_{k\in[K]} p_k\theta_k, \\
\text{s.t.} \ \ & \theta_k \geqslant \sum_{e_k\in[L]} (\boldsymbol{r} - \hat{\boldsymbol{\nu}}^{e_k})^\top \boldsymbol{\eta}_\star^{e_k}(\boldsymbol{y}^n) \\
& \qquad - \sum_{e_k\in[L]} \sum_{i\in[I]} q_i y_i \hat{\lambda}_i^{e_k} \zeta_\star^{e_k}(\boldsymbol{y}^n), \ \forall n\in[N], k\in[K], \\
& \boldsymbol{y}\in\mathcal{Y}, \boldsymbol{\theta}\in\mathbb{R}^K,
\end{aligned}
\tag{4.42}
$$

其中, N 是一个小于等于选址决策可行集 $\mathcal{Y}$ 的元素总数量的整数.

通过求解问题 (4.42), 可以得到一个新的可行解 $\boldsymbol{y}^{n+1}$. 接下来, 再次求解问题 (4.37) 得到点 $\boldsymbol{y}^{n+1}$ 处的次梯度并更新上界 (4.41). 在算法 2 中, 我们详细阐述了求解松弛主问题 (4.33) 的分解算法.

算法 2 松弛主问题 (4.33) 的分解算法

输入: 模型参数, 集合 $\mathcal{H}$ 极值点集的一个子集 $\{(\boldsymbol{\nu}^{e_k}, \boldsymbol{\lambda}^{e_k})_{e_k\in[L]}\}$.

初始化: 初始化选址决策 $\boldsymbol{y}^0\in\mathcal{Y}$.

$$n\leftarrow 0; \quad \mathrm{LB}_L\leftarrow -\infty; \quad \mathrm{UB}_L\leftarrow +\infty.$$

1. 对于任意 $k\in[K]$, 求解线性锥规划问题 (4.37) 并得到次梯度 $\left(-q_i\sum_{e_k\in[L]}\hat{\lambda}_i^{e_k}\zeta_\star^{e_k}(\boldsymbol{y}^n)\right)_{i\in[I]}$:

$$\mathrm{UB}_L\leftarrow \min\left\{\mathrm{UB},\ \boldsymbol{f}^\top\boldsymbol{y}^n + \sum_{k\in[K]} p_k \sup_{\mathbb{P}\in\mathcal{F}_k} \mathbb{E}_{\mathbb{P}}[\mathrm{Q}^L(\boldsymbol{y}^n,\tilde{\boldsymbol{d}})]\right\}; \quad //\text{更新上界}$$

2. 在混合整数线性规划问题 (4.42) 中加入基于 $\left(-q_i\sum_{e_k\in[L]}\hat{\lambda}_i^{e_k}\zeta_\star^{e_k}(\boldsymbol{y}^n)\right)_{i\in[I]}$ 的新约束;
   求解问题 (4.42) 得到新的解 $(\boldsymbol{y}^\diamond, \boldsymbol{\theta}^\diamond)$;

$$\mathrm{LB}_L\leftarrow \boldsymbol{f}^\top\boldsymbol{y}^\diamond + \sum_{k\in[K]} p_k\theta_k^\diamond; \quad //\text{更新下界}$$

3. 如果 UB $\leqslant$ LB, 那么
   $\boldsymbol{y}^\star\leftarrow\boldsymbol{y}^\diamond$;
   停止算法.
   否则
   $n\leftarrow n+1$; $\boldsymbol{y}^n\leftarrow\boldsymbol{y}^\diamond$; 回到第一步,

输出: 最优解 $\boldsymbol{y}^\star$.

4.4.2　分离问题

接下来, 我们构建算法的分离问题 (separation problem). 给定选址决策 $\boldsymbol{y}$, 根据 RS-FLP 模型的等价形式 (4.28), 我们可知极值概率分布下第二阶段期望成本可以表示为如下形式:

$$\sum_{k\in[K]} p_k \min_{\overline{\boldsymbol{\beta}}_k,\underline{\boldsymbol{\beta}}_k,\boldsymbol{\rho}_k} \Bigg\{ \max_{(\boldsymbol{d},\boldsymbol{w})\in\mathcal{W}_k} \Big(\mathrm{Q}(\boldsymbol{y},\boldsymbol{d}) - (\overline{\boldsymbol{\beta}}_k - \underline{\boldsymbol{\beta}}_k)^\top \boldsymbol{d} - \boldsymbol{w}^\top \boldsymbol{\rho}_k\Big)$$
$$+ \overline{\boldsymbol{\beta}}_k^\top \boldsymbol{d}_k^+ - \underline{\boldsymbol{\beta}}_k^\top \boldsymbol{d}_k^- + \boldsymbol{\rho}_k^\top \boldsymbol{\sigma}_k \Bigg\},$$

其本质上为求解 K 个 “最小-最大-最小” (min-max-min) 问题.

首先, 对于松弛主问题 (4.33) 通过算法 2 计算得到的最优解 $\boldsymbol{y}^\star$, 我们考虑松弛主问题等价形式 (4.34) 且固定其变量 $\boldsymbol{y}=\boldsymbol{y}^\star$. 接下来, 我们求解线性锥规划问题 (4.34), 得到固定 $\boldsymbol{y}=\boldsymbol{y}^\star$ 下问题 (4.34) 的最优解 $(\alpha_k^\star, \overline{\boldsymbol{\beta}}_k^\star, \underline{\boldsymbol{\beta}}_k^\star, \boldsymbol{\rho}_k^\star)_{k\in[K]}$. 最后, 我们构造一个分离问题来检验解 $((\alpha_k^\star, \overline{\boldsymbol{\beta}}_k^\star, \underline{\boldsymbol{\beta}}_k^\star, \boldsymbol{\rho}_k^\star)_{k\in[K]}, \boldsymbol{y}^\star)$ 是否为原问题 (4.28) 的一个可行解, 即我们要验证该解是否满足问题 (4.28) 中的所有约束:

$$\alpha_k^\star \geqslant \mathrm{Q}(\boldsymbol{y}^\star,\boldsymbol{d}) - (\overline{\boldsymbol{\beta}}_k^\star - \underline{\boldsymbol{\beta}}_k^\star)^\top \boldsymbol{d} - \boldsymbol{w}^\top \boldsymbol{\rho}_k^\star, \quad \forall (\boldsymbol{d},\boldsymbol{w})\in\mathcal{W}_k,\ k\in[K]. \tag{4.43}$$

若满足, 则解 $((\alpha_k^\star, \overline{\boldsymbol{\beta}}_k^\star, \underline{\boldsymbol{\beta}}_k^\star, \boldsymbol{\rho}_k^\star)_{k\in[K]}, \boldsymbol{y}^\star)$ 为原问题 (4.28) 的最优解.

对于每个 $k\in[K]$, 上述不等式 (4.43) 等价于 $\alpha_k^\star \geqslant \Omega_k$, $k\in[K]$, 其中

$$\Omega_k := \max_{(\boldsymbol{d},\boldsymbol{w})\in\mathcal{W}_k} \Bigg\{ \min_{\boldsymbol{x}\in\mathcal{X}(\boldsymbol{y}^\star,\boldsymbol{d})} \Bigg\{ \sum_{i\in[I]}\sum_{j\in[J]} (c_{ij}-r_j)x_{ij} \Bigg\}$$
$$+ (\boldsymbol{r} - \overline{\boldsymbol{\beta}}_k^\star + \underline{\boldsymbol{\beta}}_k^\star)^\top \boldsymbol{d} - \boldsymbol{w}^\top \boldsymbol{\rho}_k^\star \Bigg\}. \tag{4.44}$$

接下来, 我们利用线性规划的互补松弛条件 (complementary slackness conditions) 推出上述问题等价于一个混合整数线性锥规划问题. 此外, 我们还给出有效不等式 (valid inequality) 来避免问题中含有大 M(big-M) 项. 我们有如下两个命题.

命题 4.6　对于每个 $k\in[K]$, 给定松弛主问题 (4.34) 在决策变量 $\boldsymbol{y}=\boldsymbol{y}^\star$ 下的解 $(\overline{\boldsymbol{\beta}}_k^\star, \underline{\boldsymbol{\beta}}_k^\star, \boldsymbol{\rho}_k^\star, \boldsymbol{y}^\star)$, 问题 (4.44) 等价于如下混合整数线性锥规划问题:

$$
\begin{aligned}
\Omega_k = \max_{\substack{\boldsymbol{d},\boldsymbol{w},\boldsymbol{x}\\ \boldsymbol{\nu},\boldsymbol{\lambda},\boldsymbol{\pi}}} \quad & \sum_{i\in[I]}\sum_{j\in[J]}(c_{ij}-r_j)x_{ij} + (\boldsymbol{r}-\overline{\boldsymbol{\beta}}_k^\star + \underline{\boldsymbol{\beta}}_k^\star)^\top \boldsymbol{d} - \boldsymbol{w}^\top \boldsymbol{\rho}_k^\star,\\
\text{s. t.} \quad & \sum_{i\in[I]} x_{ij} \leqslant d_j, \forall j\in[J],\\
& \sum_{j\in[J]} x_{ij} \leqslant q_i y_i^\star, \forall i\in[I],\\
& \nu_j + \lambda_i \geqslant r_j - c_{ij}, \forall i\in[I],\ j\in[J],\\
& \nu_j \leqslant \overline{r}\pi_j^1, \forall j\in[J],\\
& \lambda_i \leqslant \overline{r}\pi_i^2, \forall i\in[I],\\
& x_{ij} \leqslant \overline{q}(1-\pi_{ij}^3), \forall i\in[I],\ j\in[J],\\
& d_j - \sum_{i\in[I]} x_{ij} \leqslant \overline{d}(1-\pi_j^1), \forall j\in[J],\\
& \sum_{j\in[J]} x_{ij} \geqslant q_i y_i^\star - \overline{q}(1-\pi_i^2), \forall i\in[I],\\
& \nu_j + \lambda_i \leqslant (2\overline{r}-\underline{r})\pi_{ij}^3 + r_j - c_{ij}, \forall i\in[I],\ j\in[J],\\
& \boldsymbol{A}_k\boldsymbol{d} \leqslant \boldsymbol{b}_k, (\boldsymbol{d},\boldsymbol{w},1)\in\mathcal{K}_k,\\
& \boldsymbol{\nu}\in\mathbb{R}_+^J, \boldsymbol{\lambda}\in\mathbb{R}_+^I, \boldsymbol{x}\in\mathbb{R}_+^{I\times J},\\
& \boldsymbol{\pi}^1\in\{0,1\}^J, \boldsymbol{\pi}^2\in\{0,1\}^I, \boldsymbol{\pi}^3\in\{0,1\}^{I\times J},
\end{aligned}
\tag{4.45}
$$

其中, $\overline{d}, \overline{q}, \overline{r}, \underline{r}$ 分别为如下给定的常数

$$
\overline{q} = \max_{i\in[I]}\{q_i y_i^\star\}, \quad \overline{d} = \max_{j\in[J],k\in[K]}\{\ \overline{d}_{jk}\},
$$

$$
\overline{r} = \max_{i\in[I],j\in[J]}\{0, r_j - c_{ij}\}, \quad \underline{r} = \min_{i\in[I],j\in[J]}\{0, r_j - c_{ij}\},
$$

其中, 对于任意 $k\in[K],\ j\in[J],\ \overline{d}_{jk} := \max\{\boldsymbol{e}_j^\top \boldsymbol{d} \mid \boldsymbol{A}_k\boldsymbol{d}\leqslant \boldsymbol{b}_k,\ \boldsymbol{d}\geqslant \boldsymbol{0}\}$.

命题 4.7 考虑松弛主问题 (4.34) 的最优解 $((\alpha_k^\star, \overline{\boldsymbol{\beta}}_k^\star, \underline{\boldsymbol{\beta}}_k^\star, \boldsymbol{\rho}_k^\star)_{k\in[K]}, \boldsymbol{y}^\star)$ 及其最优值 Π^L, 令 Ω_k 为给定 $((\overline{\boldsymbol{\beta}}_k^\star, \underline{\boldsymbol{\beta}}_k^\star, \boldsymbol{\rho}_k^\star)_{k\in[K]}, \boldsymbol{y}^\star)$ 下问题 (4.45) 的最优值. 则有以下结论:

(1) $\Omega_k \geqslant \alpha_k^\star, \forall k\in[K]$.

(2) 原问题 (4.7) 与松弛主问题 (4.34) 的最优性间隙 (optimality gap) 存在一个上界

$$
\Pi - \Pi^L \leqslant \sum_{k\in K} p_k(\Omega_k - \alpha_k^\star). \tag{4.46}
$$

对于每个 $k \in [K]$, 通过求解问题 (4.45), 我们不仅可以验证解 $(\alpha_k^\star \geqslant \Omega_k, k \in K)$ 的最优性, 也可以得到给定 $(\boldsymbol{y}^\star, \boldsymbol{d}^\star)$ 下问题 (4.19) 的最优解 $(\boldsymbol{\nu}^\star, \boldsymbol{\lambda}^\star)$. 这有利于我们获得集合 $\mathcal{H}$ 的极值点 $(\boldsymbol{\nu}^{\star\star}, \boldsymbol{\lambda}^{\star\star})$, 以更新松弛主问题 (4.34). 我们有如下命题.

命题 4.8 令 $(\boldsymbol{x}^\star, \boldsymbol{\nu}^\star, \boldsymbol{\lambda}^\star, \boldsymbol{d}^\star, \boldsymbol{w}^\star, \boldsymbol{\pi}^\star)$ 为给定 $((\alpha_k^\star, \overline{\boldsymbol{\beta}}_k^\star, \underline{\boldsymbol{\beta}}_k^\star, \boldsymbol{\rho}_k^\star)_{k\in[K]}, \boldsymbol{y}^\star)$ 下问题 (4.45) 的最优解. 则 $(\boldsymbol{\nu}^\star, \boldsymbol{\lambda}^\star)$ 是给定 $(\boldsymbol{y}^\star, \boldsymbol{d}^\star)$ 下对偶子问题 (dual subproblem)(4.19) 的最优解. 进一步, 下述线性规划问题

$$\min_{(\boldsymbol{\nu},\boldsymbol{\lambda})} \left\{ \boldsymbol{\nu}^\top \boldsymbol{d}^\star + \sum_{i\in[I]} q_i y_i^\star \lambda_i \,\middle|\, (\boldsymbol{\nu}, \boldsymbol{\lambda}) \in \mathcal{H}, [\boldsymbol{d}^\star]^\top \boldsymbol{\nu} + \sum_{i\in[I]} q_i y_i^\star \lambda_i = [\boldsymbol{d}^\star]^\top \boldsymbol{\nu}^\star + \sum_{i\in[I]} q_i y_i^\star \lambda_i^\star \right\} \tag{4.47}$$

的基本可行解 (basic feasible solution) 是集合 $\mathcal{H}$ 的极值点.

4.4.3 整体算法流程

现在, 我们总结求解 RS-FLP 模型 (4.30) 的整体算法流程.

(1) 首先, 我们需要集合 $\mathcal{H}$ 极值点集的一个子集 $[L]$. 给定任意可行解 $\boldsymbol{y}$ 和需求 $\boldsymbol{d}$, 我们可以通过求解一个线性规划问题

$$\max_{(\boldsymbol{\nu},\boldsymbol{\lambda})\in\mathcal{H}} \left\{ -\boldsymbol{\nu}^\top \boldsymbol{d} - \sum_{i\in I} q_i y_i \lambda_i \right\}$$

得到集合 $\mathcal{H}$ 的一个极值点 $(\widehat{\boldsymbol{\lambda}}^0, \widehat{\boldsymbol{\nu}}^0)$.

(2) 给定集合 $[L]$, 我们利用 4.4.1 小节中介绍的分解算法 (算法 2) 来求解松弛主问题 (4.34), 并得到解 $(\boldsymbol{y}^\star, \{\alpha_k^\star, \overline{\boldsymbol{\beta}}_k^\star, \underline{\boldsymbol{\beta}}_k^\star, \boldsymbol{\rho}_k^\star\}_{k\in K})$.

(3) 然后, 我们求解 K 个分离子问题 (4.45), 并使用不等式 (4.43)(或使用最优性间隙的上界(4.46)) 来验证解 $\boldsymbol{y}^\star$ 的最优性.

(4) 假设有 K' 个不等式 (4.43) 不成立, 则通过线性规划问题 (4.47) 的基本可行解获得集合 $\mathcal{H}$ 的 K' 个极值点 $(\widehat{\boldsymbol{\lambda}}^k, \widehat{\boldsymbol{\nu}}^k), k \in [K']$. 利用这些极值点, 可以更新松弛主问题.

(5) 当找到满足不等式 (4.43) 的解时, 算法停止.

算法 3 嵌套 Benders 分解算法

输入: 模型参数, 最优性间隙容差 (optimality gap tolerance)ϵ.

初始化: 找到集合 $\mathcal{H}$ 的一个极值点 $e_0 = (\hat{\boldsymbol{\nu}}^0, \hat{\boldsymbol{\lambda}}^0)$, 并令 $\mathcal{L} := \{e_0\}$.

1. 执行算法 2 来求解松弛主问题 (4.33), 并得到一个可行解 $\boldsymbol{y}^\star$.
2. 固定问题 (4.34) 的变量 $\boldsymbol{y} = \boldsymbol{y}^\star$, 并求解 K 个线性锥规划问题获得解 $(\alpha^\star, \overline{\boldsymbol{\beta}}_k^\star, \underline{\boldsymbol{\beta}}_k^\star, \boldsymbol{\rho}_k^\star)_{k\in[K]}$.
3. 对于每个 $k \in [K]$, 求解分离子问题 (4.45) 得到 Ω_k 和极值点 $(\widehat{\boldsymbol{\lambda}}^k, \widehat{\boldsymbol{\nu}}^k)$.

---

4. 如果 $\sum_{k\in[K]} p_k\left(\Omega_k-\alpha_k^{\star}\right)\leqslant\epsilon$, 那么
   $\boldsymbol{y}^{\star\star}\leftarrow\boldsymbol{y}^{\star}$;
   停止算法.
5. $\mathcal{L}\leftarrow\mathcal{L}\cup\{(\widehat{\boldsymbol{\lambda}}_k,\widehat{\boldsymbol{\nu}}_k)\mid\alpha_k^{\star}<\Omega_k,k\in K\}$;
6. 使用 $[L]$ 更新松弛主问题 (4.34), 并返回第一步;

输出: 最优选址决策 $\boldsymbol{y}^{\star\star}$.

最后, 可以证明所提出的嵌套 Benders 分解算法是有限步收敛的.

定理 4.1 算法 3 可以在有限迭代步数内终止.

4.5 数值实验

本节设计数值实验, 从不同方面验证所介绍的 RS-FLP 模型 (4.7) 处理状态依赖需求分布不确定性问题的有效性. 具体来说, 主要验证: ① 状态依赖分布信息的价值; ② 嵌套 Benders 分解精确算法的性能. 此外, 第一个数值实验 ① 考虑 $I=10$ 个备选工厂地址和 $J=20$ 个客户需求点的产能限制型工厂选址问题. 客户需求数据是 20 个市 13 年 (2008 年 ~ 2020 年) 的季度零售数据, 这些数据来自 CEIC 数据库①. 图 4.2 (a) 给出了三个城市从 2014 年 3 月到 2020 年 3 月的季度零售额时间序列数据.

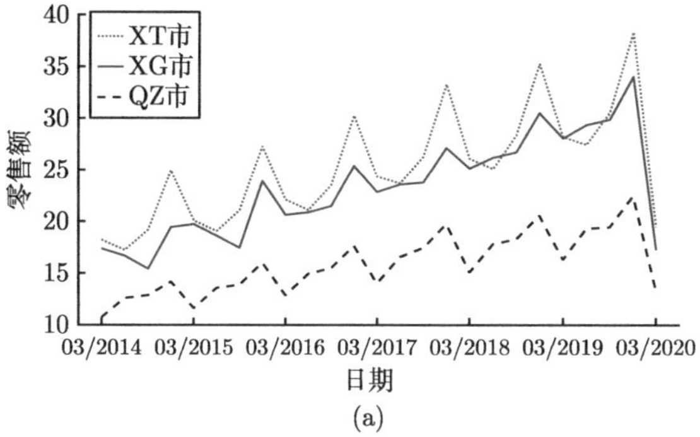

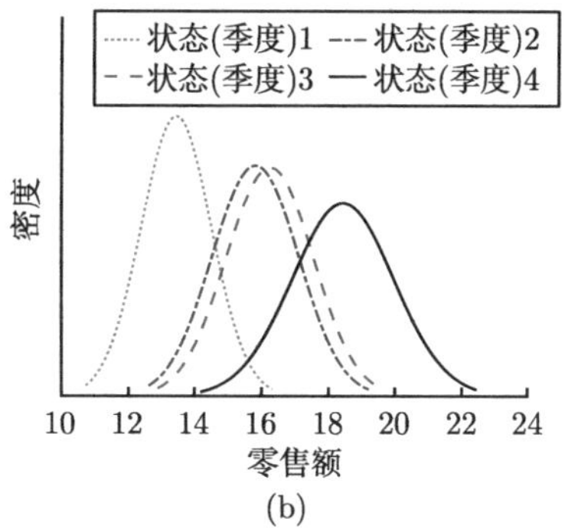

图 4.2 三个城市季度零售额的时间序列数据 (a) 和 QZ 市不同季度 (状态) 零售额分布 (b)

从图 4.2 (b) QZ 市不同季度的零售额分布可以看出, 销售数据呈现出显著的季度特征. 因此, 我们将季度状态信息纳入到状态分布不确定集中来构造 RS-FLP 模型, 其中的四个状态 ($K=4$) 分别代表 4 个季度. 在所有实验中, 状态依赖分

---

① CEIC 数据是一个宏观和微观经济数据库, 提供最广泛和最准确的数据集, 包含 660 万个时间序列数据, 涵盖 200 多个经济体、20 个行业和 18 个宏观经济部门.

布不确定集 $\mathcal{F}$ 由不确定需求的期望值、例 10 给出的期望 Mahalanobis 距离和支撑集信息组成. 特别地, 我们选择 20 个城市作为需求点, 并使用这些城市的销售数据估计了不同季度 $k=1,2,3,4$ 下的商品需求期望值的上下界 $[\boldsymbol{d}_k^-, \boldsymbol{d}_k^+]$ 和期望 Mahalanobis 距离的上界 σ_k①. 最后, 状态概率 p_k 的取值均为 1/4. 读者可以参考 Ebery 等 (2000) 的研究获取更多计算细节.

最后, 为了更简洁地展示实验结果, 我们将基于状态依赖分布不确定集的 RS-FLP 模型记为 "S-DRO" 模型, 一般的分布鲁棒工厂选址模型, 即没有考虑状态依赖信息的分布鲁棒工厂选址模型, 记为 "DRO" 模型, 以及随机工厂选址模型, 记为 "SAA" 模型.

4.5.1 状态依赖分布信息的价值

本小节主要通过比较 S-DRO 模型及 DRO 模型的样本外实验效果来评估状态分布信息的价值. 首先, 我们使用 20 个城市 2009 年 $\sim$ 2018 年零售额的季度增长率及 2018 年季度零售额数据, 构造一组 20 个城市每个季度 $k=1,2,3,4$ 下的需求预测样本. 其次, 估计每个季度下参数 $\bar{\boldsymbol{d}}_k$, $\boldsymbol{d}_k^+$, $\boldsymbol{d}_k^-$ 和 σ_k 的值来构造 S-DRO 模型的状态依赖分布不确定集 $\mathcal{F}$. 另外, 我们忽略状态信息, 将不同季度的需求样本汇总在一起, 来构造 DRO 模型的分布不确定集 $\mathcal{F}_0$:

$$
\mathcal{F}_0 := \left\{ \mathbb{P} \in \mathcal{P}(\mathbb{R}^J) \;\middle|\; \begin{array}{l} \tilde{\boldsymbol{d}} \sim \mathbb{P}, \\ \mathbb{E}_{\mathbb{P}}\left[\tilde{\boldsymbol{d}}\right] \in [\boldsymbol{d}_0^-, \boldsymbol{d}_0^+], \\ \mathbb{E}_{\mathbb{P}}\left[\sqrt{\left(\boldsymbol{d}-\bar{\boldsymbol{d}}_0\right)^\top \boldsymbol{\Sigma}_0^{-1}\left(\boldsymbol{d}-\bar{\boldsymbol{d}}_0\right)}\right] \leqslant \sigma_0, \\ \mathbb{P}\left[\tilde{\boldsymbol{d}} \in \mathcal{D}_0\right] = 1 \end{array} \right\},
$$

其中, 用所有的需求预测样本估计上述分布不确定集参数的 $\boldsymbol{d}_0^+$, $\boldsymbol{d}_0^-$, σ_0, $\bar{\boldsymbol{d}}_0$ 和 $\mathcal{D}_0$. 而且, 对于每个季度 k, 我们使用基于估计均值 $\bar{\boldsymbol{d}}_k$ 和需求预测样本方差-协方差矩阵 $\boldsymbol{\Sigma}_k$ 的多元高斯分布 $\mathcal{N}(\bar{\boldsymbol{d}}_k, \boldsymbol{\Sigma}_k)$ 来生成每个季度下的需求测试样本.

表 4.1 展示 DRO 模型和 S-DRO 模型的求解结果, 其中缺货成本是指未满足客户需求的惩罚成本. 结果表明: 虽然两个模型都选择建设 4 个工厂, 但 DRO 模型有 2 个选址与 S-DRO 模型不同, 前者具有更高的总产能来满足极值概率分布下最大需求, 这导致其具有更高的固定成本和总成本. 上述结果表明 S-DRO 模型提供了比 DRO 模型不太保守的选址决策.

---

① 具体而言, 分布不确定集参数 $\boldsymbol{d}_k^-$ 和 $\boldsymbol{d}_k^+$ 分别取为每个状态 k 下的需求样本均值 $\bar{\boldsymbol{d}}_k$ 减去和加上状态 k 下的样本标准差, 而期望离差参数 σ_k 取为每个状态 k 的经验 Mahalanobis 距离. 最后, 支撑集 $\mathcal{D}_k$ 取为每个状态 k 下的最小和最大需求观测值形成的区间.

表 4.1 DRO 模型和 S-DRO 模型的解

模型	最优选址	总容量	期望 总成本	期望 运输成本	期望 缺货成本	固定成本
DRO	$\{2,5,9,10\}$	1013	1.67×10^6	1.13×10^6	0	0.54×10^6
S-DRO	$\{2,4,5,8\}$	1000	1.52×10^6	1.00×10^6	0	0.52×10^6

样本外测试下的性能评估 为了验证 S-DRO 模型使用的状态分布式信息的价值, 我们通过样本外测试比较了 S-DRO 模型与 DRO 模型处理状态依赖需求不确定性问题的效果. 我们共进行 100 次样本外实验, 每次实验生成 2000 个测试样本, 其中每个样本以 1/4 的概率由 4 个需求生成高斯分布 $\mathcal{N}(\bar{\boldsymbol{d}}_k,\boldsymbol{\Sigma}_k), k\in[4]$ 之一生成. 对于每次样本外测试, 我们分别计算 S-DRO 模型和 DRO 模型的最优选址决策对应的每季度样本外总成本分布. 最后, 计算 100 次样本外测试的平均样本外总成本.

接下来从样本外保守性 (out-of-sample conservativeness) 的角度探讨 S-DRO 模型和 DRO 模型最优选址决策的效果, 它衡量了名义最优总成本 (表 4.1 中的期望总成本) 与样本外总成本之差. 具体而言, 样本外保守值定义如下

$$\text{Conservativeness} := c^*_{\text{NOM}} - c^*_{\text{OUT}},$$

其中, c^*_{NOM} 和 c^*_{OUT} 分别代表名义最优总成本和样本外总成本. 保守值较大意味着选址决策可能过于保守, 而绝对值较大的负保守值则表示选址决策过于乐观.

图 4.3 给出了 S-DRO 模型和 DRO 模型的最优选址决策对应的样本外总成本、样本外保守值和样本外运输成本在每个季度 (状态) 下的分布. 结果表明:

① S-DRO 模型的最优选址决策对应的样本外总成本在所有状态 (季度) 下都低于 DRO 模型最优选址决策所对应的样本外总成本. 这说明状态相关信息提高了选址决策的质量. ② 与 DRO 模型的最优选址决策相比, S-DRO 模型的最优选址决策在状态 1~3 下呈现出了更低的保守值, 但在状态 4 下有更高的保守值, 即样本外总成本大于名义总成本. 这意味着尽管在样本外测试中, DRO 模型在大多数状态下比 S-DRO 模型更保守, 但它可以表现出更高的可靠性. ③ S-DRO 模型最优决策对应的样本外运输成本略低于 DRO 模型最优决策对应的, 这表明 S-DRO 模型在样本外测试上的效果好于 DRO 模型, 同样验证了状态依赖信息的价值.

为了更加全面地评估 S-DRO 模型和 DRO 模型最优选址决策的质量, 下面使用 2019 年需求的真实观测值验证两个模型在实际情况下处理不确定性问题的效果, 如表 4.1 所示. 图 4.4 给出了 2019 年每个季度下 S-DRO 和 DRO 模型对应的总成本和样本外保守值.

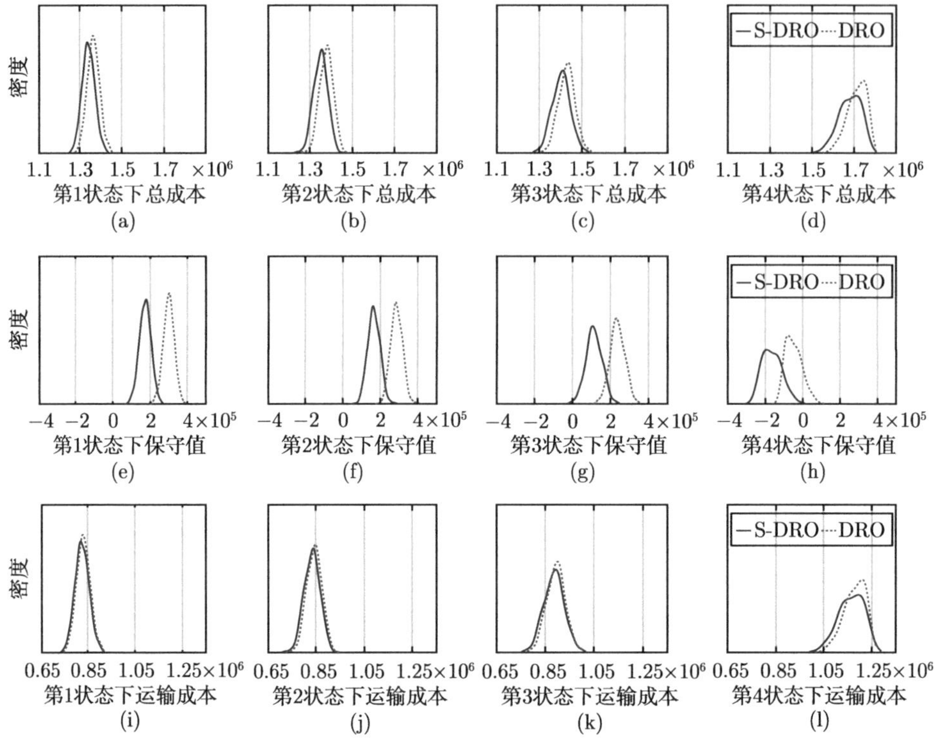

图 4.3　每个状态 (季度) 下, DRO 模型和 S-DRO 模型最优选址决策对应的样本外总成本 ((a)~(d))、保守值 ((e)~(h)) 和运输成本 ((i)~(l)) 的分布

从图 4.4 中可以看出, 2019 年各季度的总成本和保守值结果与之前样本外测试的结果一致. 从图 4.4 (b) 和 (c), 可以看出 DRO 模型的名义总成本与其样本外总成本偏离较大, 即 DRO 模型更保守, 而 S-DRO 模型的名义总成本可以

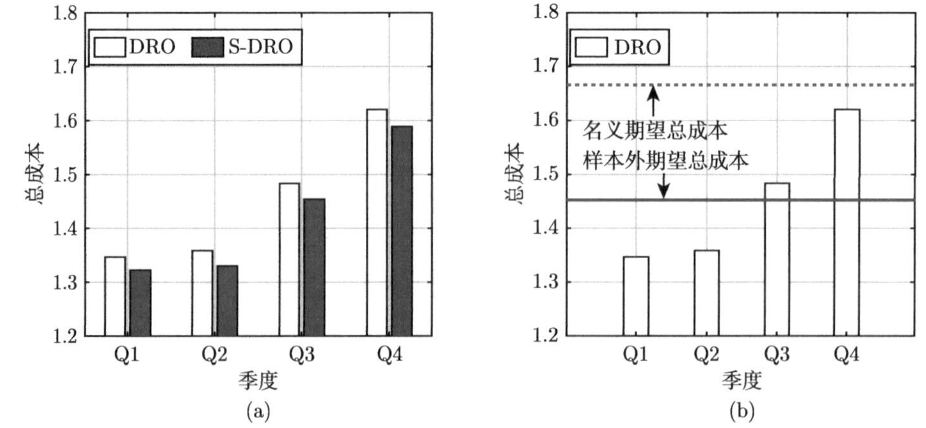

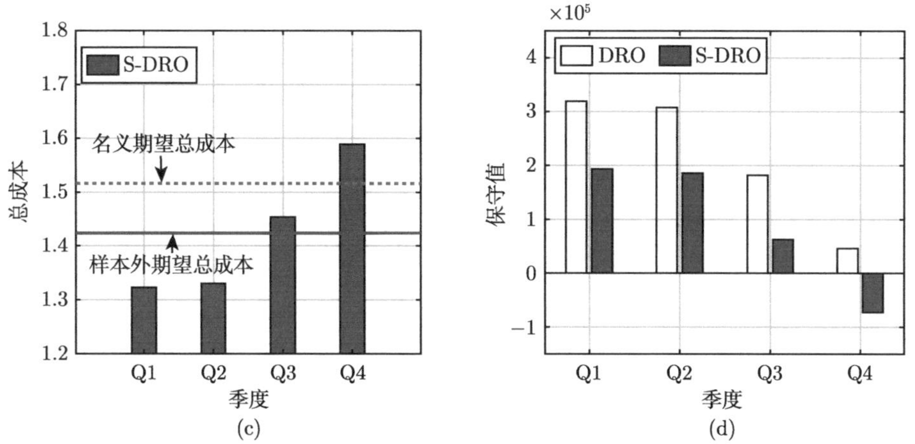

图 4.4 每个季度下使用 2019 年需求观测的 DRO 和 S-DRO 模型选址决策对应的总成本和保守值

看作一个对未来总成本更加准确的鲁棒预测 (robust prediction). 为了更加直观地展示这些结果, 我们计算 S-DRO 和 DRO 模型期望保守值比率, 定义为 "(名义期望总成本 − 样本外期望总成本)/样本外期望总成本", 结果分别为 6.5% 和 14.7%. 此外, 我们还计算了不同状态下 S-DRO 模型最优选址决策相对于 DRO 模型最优选址决策对应的总成本降低的百分比, 分别为 1.8%、2.1%、2.0% 和 1.9%. 这些结果进一步表明使用状态信息的 S-DRO 模型在降低总成本和减轻决策保守性方面的价值.

4.5.2 计算性能

最后, 我们通过实验验证 4.4 节提出的精确算法的计算性能. 特别地, 我们考虑备选工厂地址个数为 $I \in \{10, 20\}$ 和需求点个数为 $J \in \{10, 30, 50\}$ 的产能限制型工厂选址问题. 对于每组 (I, J) 问题组合, 我们考虑多个状态 $K \in \{1, 3, 5\}$. 记嵌套 Benders 分解算法为 "S-NBD" 算法. 在实验中, 我们考虑基于 Mahalanobis 距离的状态依赖分布不确定集. 为了验证 S-NBD 算法的高效性能, 我们将其与一般的 Benders 分解方法 (称为 "BD" 算法) 进行比较. BD 算法可以看作将 Zeng 和 Zhao (2013) 的研究中求解基于多面体不确定集的两阶段鲁棒优化模型的约束和列生成算法, 直接应用于基于锥结构分布不确定集的 RS-FLP 模型的求解.

两个模型的计算结果总结在表 4.2 中. 我们观察到, 在最优性间隙为 $\{2\%, 1\%, 0.5\%, 0\%\}$ 的所有算例中, S-NBD 算法几乎都优于 BD 算法. 特别是随着问题规模变大时, S-NBD 算法相对于 BD 算法的计算性能优势更加明显. 比如, 当 $(I, J, K) = (10, 30, 1)$ 时, S-NBD 算法可以在 29 秒内求出精确解, 而 BD 算法需要 116 秒; 当 $(I, J, K) = (10, 30, 5)$ 时, S-NBD 算法可以在 220 秒内找到精确解,但 BD 算法需要 14935 秒; 最后, 我们观察到当 $I=20$, $J\geqslant 10$ 和 $K\geqslant 3$ 时, 我们的 S-NBD 算法仍然可以在可接受的时间内找到最优解, 而 BD 算法会耗尽内存. 上述结果验证了本章所介绍的嵌套 Benders 分解算法的高效性.

表 4.2 S-NBD 算法和 BD 算法求解 RS-FLP 模型的计算性能比较

(I,J,K)	S-NBD 算法				BD 算法			
	Gap=2%	Gap=1%	Gap=0.5%	Gap=0%	Gap=2%	Gap=1%	Gap=0.5%	Gap=0%
(10, 10, 1)	1s	1s	1s	2s	1s	1s	1s	2s
(10, 10, 3)	2s	2s	3s	4s	1s	1s	3s	10s
(10, 10, 5)	2s	2s	2s	6s	7s	7s	8s	86s
(10, 20, 1)	2s	3s	4s	9s	1s	2s	3s	7s
(10, 20, 3)	5s	7s	9s	27s	8s	15s	24s	152s
(10, 20, 5)	5s	11s	11s	30s	38s	117s	118s	541s
(10, 30, 1)	3s	9s	15s	29s	3s	12s	40s	116s
(10, 30, 3)	10s	35s	52s	120s	77s	596s	1138s	3935s
(10, 30, 5)	13s	22s	82s	220s	262s	705s	5452s	14935s
(10, 50, 1)	88s	121s	147s	272s	126s	164s	189s	302s
(10, 50, 3)	390s	611s	713s	1086s	2443s	5873s	9296s	11389s
(10, 50, 5)	560s	2017s	2696s	5509s	19326s	OOM	OOM	OOM
(20, 10, 1)	7s	10s	13s	23s	12s	30s	47s	119s
(20, 10, 3)	7s	9s	18s	36s	26099s	OOM	OOM	OOM
(20, 10, 5)	12s	17s	35s	67s	OOM	OOM	OOM	OOM
(20, 20, 1)	43s	105s	157s	293s	27s	467s	4335s	OOM
(20, 20, 3)	15s	42s	115s	327s	10561s	OOM	OOM	OOM
(20, 20, 5)	12s	29s	415s	865s	OOM	OOM	OOM	OOM
(20, 30, 1)	104s	196s	303s	805s	94s	1312s	9209s	OOM
(20, 30, 3)	27s	53s	127s	1137s	994s	6096s	OOM	OOM
(20, 30, 5)	30s	106s	176s	2820s	13738s	OOM	OOM	OOM
(20, 50, 1)	366s	510s	966s	1279s	440s	830s	3608s	6273s
(20, 50, 3)	224s	1147s	2703s	4215s	12816s	OOM	OOM	OOM
(20, 50, 5)	172s	900s	3282s	7389s	2464s	17196s	OOM	OOM

注: Gap $=(1-\mathrm{LB/UB})\times 100\%$, OOM 表示 Out-of-Memory.

4.6 本章小结

本章介绍具有状态依赖需求分布不确定性的产能限制型工厂选址问题的建模方法. 为了准确地描述未来需求的分布不确定性, 基于状态依赖分布不确定集构建鲁棒随机工厂选址模型. 该模型可以减轻鲁棒选址决策的保守度, 并提升鲁棒选址决策对抗不确定环境的决策质量, 分析了分布不确定集参数的变化以及选址决策的变化对极值概率分布下期望成本的影响, 并给出了鲁棒敏感性界限. 特别地, 证明了极值概率分布下第二阶段期望成本对选址决策变化的敏感度界限可以由极值概率分布下期望影子产能成本刻画. 最后, 介绍了一种有限步迭代精确求解鲁棒随机工厂选址模型的嵌套 Benders 分解算法.

关于不确定环境下选址问题的研究, 读者可参考更多相关文献. 例如, Lu 等 (2015) 考虑了一个具有不确定中断概率的产能充足型工厂选址问题, 其中不同工厂的中断概率具有相关性. Liu 等 (2019) 考虑了一个具有分布鲁棒机会约束的紧急医疗服务站选址问题. Wang 等 (2020) 研究了需求和成本不确定下的数据驱动分布鲁棒工厂选址问题. 基于 Wasserstein 球定义的需求分布不确定集, Saif 和 Delage (2021) 研究了单阶段和两阶段分布鲁棒容量限制型工厂选址问题, 其中 Wasserstein 距离为 ℓ_1 范数. Shehadeh 和 Sanci (2021) 考虑了一个具有双峰需求分布不确定的工厂选址问题. Basciftci 等 (2021) 研究了一个分布鲁棒工厂选址问题, 其研究考虑了选址决策对客户需求的影响, 并通过一个分段线性函数来刻画选址决策对客户需求的影响.

第 5 章　鲁棒性指标与两阶段选址优化

本章主要介绍不确定环境下资源回收系统 (resource recovery systems) 的容量配置与能源转化技术组合问题. 在模型结构上, 本章的问题与第 3 章和第 4 章介绍的两阶段选址优化模型相同. 该问题的不确定性主要包含两个方面: 原料质量 (feedstock volume) 不确定与原料组成成分 (feedstock composition) 不确定. 首先, 结合决策相关 (decision-dependent) 动态鲁棒优化以及逆优化思想, 介绍一种系统鲁棒性指标 (robustness index), 它是资源回收系统的决策优化指标. 此外, 该指标的计算建立在求解一个资源回收系统运营的两阶段动态鲁棒优化问题基础之上, 而该鲁棒优化模型的不确定集是决策相关的 (decision-dependent uncertainty set). 进一步, 在该指标体系下, 我们介绍一套针对资源回收系统运营环境 (operating environment) 的多目标鲁棒性分析框架, 同时介绍该框架对应的多项式时间计算模型. 最后, 设计数值实验来验证模型性能. 本章主要结构如下: 5.1 节介绍资源回收问题背景. 5.2 节介绍确定性资源回收模型以及原料状态不确定集的构建. 5.3 节介绍用于评估资源回收系统经济可行性 (economic feasibility) 的鲁棒性函数 (robustness function) 的构建. 5.4 节介绍基于鲁棒性函数的复合鲁棒指标 (composite robustness index) 以及鲁棒资源回收系统模型. 最后, 5.5 节介绍实验及结果分析.

5.1　背　　景

随着人口快速增长, 废弃物和生活垃圾的产生量急剧增加, 社会正面临环境污染、垃圾填埋场空间减少等挑战. 因此, 可持续的废弃物管理 (sustainable waste management) 是现代社会面临的最紧迫问题之一. 许多国家利用多种资源回收技术从各种废弃物中回收可用资源, 常见的资源回收技术包括厌氧消化 (anaerobic digestion)、气化 (gasification)、热解 (pyrolysis) 与高效焚烧 (highly enhanced incineration) 等, 这些技术通常比传统的废弃物直接焚烧方法更环保、更节能 (Yılmaz and Selim, 2013; McConnell, 2014). 上述废弃物转化为能源 (waste-to-energy) 的回收技术或系统一般称为资源回收系统 (resource recovery system), 而待回收的废弃物和垃圾统称为资源回收系统的原料 (feedstock). 本章考虑多种能源回收技术相结合的混合资源回收系统 (hybrid resource recovery system), 以从不同种类的原料中回收有价值的资源.

建设资源回收系统通常需要在回收技术、设备和基础设施上投入大量资金. 与其他行业相比, 资源回收系统的利润较低且成本回收周期较长. 因此, 经济可行性 (economic feasibility) 是资源回收系统建设的一个关键因素. 如果一个资源回收系统在经济上无法持续运行下去, 则无法发挥其能源转化再利用的功能. 在实际生活中, 影响资源回收系统经济可持续性的关键因素是原料状态的不确定性 (feedstock condition uncertainty), 即原料质量的不确定性 (volume uncertainty) 和原料组成成分的不确定性 (composition uncertainty). 一方面, 若实际原料质量低于资源回收系统的预期原料质量, 则会导致资源回收系统出售能源转化产品的利润大幅减少 (Guerrero et al., 2013; Fujii et al., 2014). 反之, 若实际原料质量高于预期原料质量, 则会增加资源回收系统的运行压力并产生额外的运营成本. 此外, 系统的能源转化效率与原料组成成分高度相关. 例如, 不可回收原料 (non-recyclables) 或高转化率原料的组成若不适用于系统的处理能力, 则会导致资源回收处理不充分, 且大幅增加原料的处理成本和废料的填埋成本. 另一方面, 可靠准确的原料状态数据在实际问题中很难获取 (Guerrero et al., 2013; Hannan et al., 2015), 这对精确估计原料状态形成挑战, 并可导致资源回收系统经济可行性评估的不准确. 例如, 在 2011 年, 新加坡最大的厨余垃圾回收公司 IUT 环球有限公司在成立三年后倒闭, 主要原因是对原料质量和组成成分预测不准确. 事实上, 一些资源回收企业倒闭的根本原因为正是原料状态分析不充分, 以及对原料质量和组成成分的预测值过度信任 (World Bank Report, 1999).

5.2 资源回收模型与原料状态不确定性建模

本节首先介绍一个混合资源回收系统, 其由一组资源回收单元组成, 每个单元有特定的原料处理技术和容量水平. 原料收集者 (collectors) 将原料运送到资源回收系统, 并向资源回收企业支付回收服务费 (service tipping fee). 根据不同种类的原料状态 (质量和组成成分), 原料会分配给不同的资源回收单元以充分处理不同种类的原料 (图 5.1).

模型基本假设:

(1) 在原料送到资源回收系统前, 可以根据原料的来源 (例如, 收集地区) 进行粗略的分类, 但在原料进入资源回收处理单元前, 不再进一步细分其组成成分.

(2) 资源回收单元从原料中回收可利用的资源, 且总收益为各种原料组成成分转化为能源产品所产生的收益之和.

(3) 我们假设资源回收企业采取谨慎的 (cautious) 运营策略, 而不是投机 (opportunistic) 的运营策略, 并以系统的经济可行性为目标, 以对抗原料状态的不确定性.

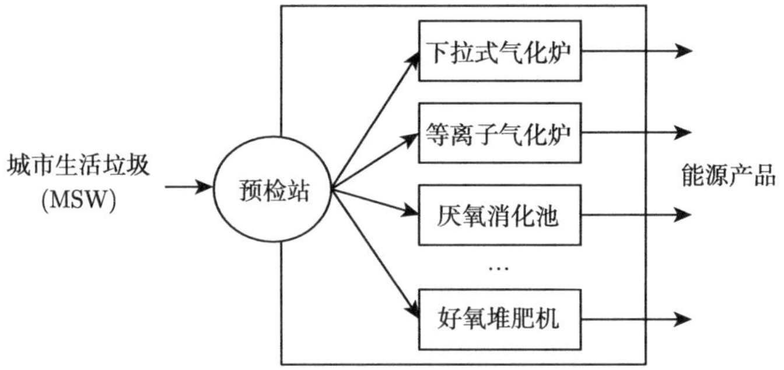

图 5.1 混合资源回收系统示意图 (原料首先被检测然后直接分配给资源回收单元产生能源产品)

5.2.1 确定型资源回收优化模型

考虑有 I 个原料收集点、J 个备选资源回收单元和 M 种原料组成成分. 我们记 v_i 为来自收集点 $i \in [I]$ 的原料质量且向量 $\boldsymbol{v} = (v_1, v_2, \cdots, v_I)^\top$, 并且记 $u_{mi} \in [0,1]$ 为收集点 i 的原料中第 m 种组成成分的质量占收集点 i 原料总质量的比例且向量 $\boldsymbol{u} = (u_{mi})_{M\times I} \in \mathbb{R}_+^{M\times I}$. 此外, 我们记 h_i 为收集点 i 的超量原料或废弃原料的单位处理成本 (disposal cost), c_j 为启用资源回收单元 j 的年化固定成本 (annualized fixed cost), s_j 为资源回收单元 j 的最大原料处理质量或处理容量 (processing capacity). 该问题包含三个决策变量: x_j 为资源回收单元选择决策变量, 它是一个 0-1 二元决策变量, 如果资源回收单元 j 开启, 则 $x_j = 1$, 否则为 0; y_{ij} 为从收集点 i 向资源回收单元 j 配送的原料质量; z_i 为收集点 i 处未被处理的或剩余的原料质量. 根据上述决策变量, 我们可知资源回收模型与第 3 章和第 4 章介绍的两阶段选址模型具有相同结构.

在资源回收系统中, 每个资源回收单元 j 具有特定的处理技术和容量 (capacity), 因此, 每个处理单元都不是完全相同的. 给定决策变量 $(\boldsymbol{y}, \boldsymbol{z})$, 总运营利润为总收益减去原料残留物的处理成本和减排成本 (emission abated cost) 之和:

$$
\begin{aligned}
&\sum_{i\in[I]}\sum_{j\in[J]}\sum_{m\in[M]} u_{mi}q^o_{mj}y_{ij} - D_L\sum_{i\in[I]}\sum_{j\in[J]} y_{ij}\beta_j - \sum_{j\in[J]}\sum_{m\in[M]}\left(\sum_{i\in[I]} y_{ij}u_{mi}\right)\rho_{mj} \\
=& \sum_{i\in[I]}\sum_{j\in[J]}\sum_{m\in[M]}\left(u_{mi}q^o_{mj} - \frac{D_L\beta_j}{M} - u_{mi}\rho_{mj}\right)y_{ij} \\
=& \sum_{i\in[I]}\sum_{j\in[J]}\sum_{m\in[M]} u_{mi}q_{mj}y_{ij},
\end{aligned} \tag{5.1}
$$

其中, β_j 为使用资源回收单元 j 处理原料的残留率 (residual ratio), D_L 为处理原料残留的单位填埋成本 (landfill cost), q^o_{mj} 和 ρ_{mj} 分别为使用资源回收单元 j 处

理原料组成成分 m 的单位净收益和单位减排成本, q_{mj} 为使用资源回收单元 j 处理单位原料成分 m 的总利润系数. 为了简洁地表述, 我们记

$$\kappa_{ij}(\boldsymbol{u}) := \sum_{m\in[M]} u_{mi}q_{mj}$$

为原料从收集点 i 到资源回收单元 j 的单位利润系数, 则总运营利润 (5.1) 可以写成如下形式:

$$\sum_{i\in[I]}\sum_{j\in[J]} \kappa_{ij}(\boldsymbol{u})y_{ij}.$$

最后, 资源回收单元无法处理的过量原料的总成本为 $\sum_{i\in[I]} h_i z_i$. 这里假设处理成本 h_i 依赖于收集点 i. 特别地, z_i 被看作混合原料来处理.

给定资源回收单元选择决策 $\boldsymbol{x}$, 原料的质量 $\boldsymbol{v}$ 与原料的组成成分 $\boldsymbol{u}$, 以最大化年化运营总利润, 我们可以构建如下确定型资源回收优化模型:

$$\begin{aligned}
\phi(\boldsymbol{x};\boldsymbol{v},\boldsymbol{u}) := \max_{\boldsymbol{y},\boldsymbol{z}} \quad & \sum_{i\in[I]}\sum_{j\in[J]} \kappa_{ij}(\boldsymbol{u})\boldsymbol{y}_{ij} - \sum_{i\in[I]} h_i z_i - \boldsymbol{c}^\top \boldsymbol{x} \\
\text{s.t.} \quad & \sum_{j\in[J]} y_{ij} + z_i = v_i, \forall\, i \in [I], \\
& \sum_{i\in[I]} y_{ij} \leqslant s_j x_j, \forall\, j \in [J], \\
& \boldsymbol{y},\boldsymbol{z} \geqslant \boldsymbol{0},
\end{aligned} \tag{5.2}$$

其中, 模型 (5.2) 的目标函数为总运营利润, 是原料接收和资源回收运营所产生的利润, 减去原料残留的成本 $\sum_{i\in[I]} h_i z_i$ 以及固定成本 $\boldsymbol{c}^\top\boldsymbol{x}$. 第一个约束要求所有来自收集点的原料都要进行处理, 即原料的处理质量 $\sum_{j\in[J]} y_{ij}$ 同超出系统回收能力的原料质量 z_i 与收集点 i 的原料总质量 v_i 相等. 第二个约束是系统处理单元的容量约束, 同时确保只有在处理单元 j 开启的情况下才能将原料送到该单元. 最后一个约束是决策变量约束, 即保证原料的配送量 $\boldsymbol{y}$ 和残留量 $\boldsymbol{z}$ 是非负的. 此外, 值得注意的是, 在模型 (5.2) 中, 原料质量 $\boldsymbol{v}$ 仅出现在约束中, 而原料的组成成分 $\boldsymbol{u}$ 仅出现在目标函数中. 可以看出, 模型 (5.2) 与容量限制型选址优化模型具有相同结构, 因此, 可视为后者的应用.

5.2.2 经济可行性与原料状态不确定性

为了评估回收系统对原料状态变化的适应能力 (the ability to adapt to variations in the feedstock conditions), 本小节介绍资源回收系统的经济可行性, 并构建原料状态不确定性下资源回收模型.

1. 经济可行性

假设资源回收企业的利润目标为 τ_0, 它是资源回收系统可以运行的最低利润, 即

$$\phi(\boldsymbol{x};\boldsymbol{v},\boldsymbol{u}) \geqslant \tau_0, \tag{5.3}$$

其中, $\phi(\boldsymbol{x};\boldsymbol{v},\boldsymbol{u})$ 为问题 (5.2) 最优运营利润. 从经济可行性的角度, 在原料状态 $(\boldsymbol{v},\boldsymbol{u})$ 不确定下, 如果在绝大部分原料状态 $(\boldsymbol{v},\boldsymbol{u})$ 实现值下, 资源回收单元决策 $\boldsymbol{x}$ 都可以保证系统运营利润达到 τ_0, 则被称为经济可行决策. 这也表明经济可行决策 $\boldsymbol{x}$ 具有鲁棒性.

2. 可调节原料不确定集合

为了更好建模经济可行性, 我们引入可调节原料不确定集合 (adjustable feedstock uncertainty sets). 记原料质量 $\boldsymbol{v}$ 和原料组成成分 $\boldsymbol{u}$ 的不确定集合分别为 $V(\beta_1)$ 和 $U(\beta_2)$, 且集合的大小取决于参数 $\beta_1,\beta_2 \in [0,1]$. 这里称 $V(\beta_1)$ 和 $U(\beta_2)$ 为可调节不确定集. 因为原料质量和组成成分的不确定性产生原因可能不同 (Beigl et al., 2008; Guerrero et al., 2013), 所以分别建模这两个不确定变量.

首先, 构建原料质量不确定性模型. 我们考虑一般情况, 即来自不同收集点的原料质量可能存在相关性. 给定收集点 i 的原料质量 v_i, 仿射因子模型 (affine factor model) 将 v_i 视为一组因子 (factors) $\boldsymbol{\xi} = [\xi_1,\xi_2,\cdots,\xi_K]^\top$ 的仿射函数:

$$v_i = \boldsymbol{\alpha}_i^\top \boldsymbol{\xi}, \quad \forall\, i \in [I],$$

其中, $\boldsymbol{\alpha}_i = [\alpha_{i1},\alpha_{i2},\cdots,\alpha_{iK}]^\top$ 为待估计的因子系数. 因子 $\boldsymbol{\xi}$ 为解释变量, 如家庭人数或经济状况等 (Beigl et al., 2008; Navarro-Esbrı et al., 2002). 基于仿射因子模型, 我们定义如下可调节原料质量不确定集:

$$V(\beta_1) := \left\{ \boldsymbol{v} \in \mathbb{R}_+^I \;\middle|\; \begin{array}{l} v_i = \boldsymbol{\alpha}_i^\top \boldsymbol{\xi}, \forall i \in [I], \\ \xi_k \in \left[\xi_k^-(\beta_1), \xi_k^+(\beta_1)\right], \forall k \in [K] \end{array} \right\}, \tag{5.4}$$

其中, $\left[\xi_k^-(\beta_1),\xi_k^+(\beta_1)\right]$ 是不确定变量 ξ_k 变化的上下界, 且对于任意 $\beta_1 \leqslant \beta_1'$, 有 $\left[\xi_k^-(\beta_1),\xi_k^+(\beta_1)\right] \subseteq \left[\xi_k^-(\beta_1'),\xi_k^+(\beta_1')\right]$. 特别地, 当因子 $\boldsymbol{\xi}$ 的数据或概率分布已知时, 可以将集合 $\left[\xi_k^-(\beta_1),\xi_k^+(\beta_1)\right]$ 设为关于 ξ_k 的预测区间 (prediction interval), 其置信水平 (confidence level) 为 β_1 或由参数 β_1 控制. 因此, 更大的 β_1 表示不确定集对 $\boldsymbol{v}$ 变化的容忍度 (variation allowance) 更高.

与原料质量 $\boldsymbol{v}$ 相比, 原料组成成分受很多复杂因素影响, 这使得准确估计原料组成成分十分困难 (Guerrero et al., 2013; Fujii et al., 2014). 因此, 我们使用原

料相对回收效率来建模原料组成成分 $\boldsymbol{u}$ 的不确定性. 原料组成成分 m 的相对平均能源回收率 (relative average energy recovery ratio) 定义如下:

$$\delta_m := \frac{\sum_{j\in[J]} q_{mj}}{\max\limits_{1\leqslant m\leqslant M}\left\{\sum_{j\in[J]} q_{mj}\right\}} \tag{5.5}$$

其中, 在所有资源回收单元 (或技术) 下, 资源回收率最高的原料组成成分为 $m^* := \arg\max_{1\leqslant m\leqslant M}\ \{\sum_{j\in[J]} q_{mj}\}$. 因此, δ_m 为第 m 种原料组成成分在所有资源回收单元中的资源回收效率相对于原料组成成分 m^* 资源回收效率的比例. 基于 δ_m, 我们定义如下关于原料组成成分的可调节不确定集:

$$U(\beta_2) := \left\{\boldsymbol{u}\in\mathbb{R}_+^{M\times I}\ \middle|\ \sum_{m\in[M]} u_{mi} = 1,\ \forall\ i\in[I];\ \sum_{i\in[I]}\sum_{m\in[M]}\delta_m u_{mi} \geqslant I(1-\beta_2)\right\}, \tag{5.6}$$

其中, $1-\beta_2\in[0,1]$ 可以表示使用不同资源回收技术处理原料的平均资源回收效率或平均原料纯度 (purity) 水平. $1-\beta_2$ 值越高, 则混合原料的回收效率越高. 因此, β_2 表示混合原料的杂质水平 (impurity level). 特别地, $\beta_2=0$ 时, $\delta_{m^*}=1$, 因此 $u_{m^*i}=1, \forall i\in[I]$, 即对于所有原料收集点 $i\in[I]$, 原料只包含回收效率最高的组成成分 m^*.

5.3 资源回收系统的鲁棒性分析

本节分析在原料状态不确定性下资源回收系统的鲁棒性. 特别地, 我们定义两个鲁棒性函数, 并构建对应的整数线性规划模型. 最后, 介绍鲁棒性函数与实现资源回收利润目标之间的概率关系.

5.3.1 鲁棒性函数

根据 5.2.2 小节的经济可行性约束 (5.3), 在 β_1 和 β_2 取值较高时, 资源回收系统可以实现利润目标, 即 $\phi(\boldsymbol{x};\boldsymbol{v},\boldsymbol{u})\geqslant\tau_0$, 则认为此系统具有较高的鲁棒性来对抗原料状态的不确定性. 为了详细描述系统的鲁棒性, 我们定义如下原料质量鲁棒性函数和原料组成成分鲁棒性函数.

定义 5.1 给定系统回收单元选择决策 $\boldsymbol{x}$ 与目标利润 τ_0, $\phi(\boldsymbol{x};\boldsymbol{v},\boldsymbol{u})$ 为模型 (5.2) 的最优资源回收系统运营利润. 针对给定的 β_2 和 $\boldsymbol{x}$, 资源回收系统的原料质量鲁棒性函数 $\beta_1^*[\boldsymbol{x}|\beta_2]$ 定义如下:

$$\beta_1^*\Big[\boldsymbol{x}\Big|\beta_2\Big] := \max_{\beta_1\in[0,1]}\Big\{\beta_1\Big|\phi(\boldsymbol{x};\boldsymbol{v},\boldsymbol{u})\geqslant\tau_0,\ \forall\ \boldsymbol{v}\in V(\beta_1), \boldsymbol{u}\in U(\beta_2)\Big\}, \tag{5.7}$$

其中, 如果不存在可行的 $\beta_1 \in [0,1]$, 则 $\beta_1^*[\boldsymbol{x}|\beta_2] := 0$. 类似地, 针对给定的 β_1, 原料组成成分鲁棒性函数 $\beta_2^*[\boldsymbol{x}|\beta_1]$ 定义如下:

$$\beta_2^*\left[\boldsymbol{x}\middle|\beta_1\right] := \max_{\beta_2\in[0,1]}\left\{\beta_2\middle|\phi\left(\boldsymbol{x};\boldsymbol{v},\boldsymbol{u}\right) \geqslant \tau_0,\ \forall\ \boldsymbol{v} \in V(\beta_1), \boldsymbol{u} \in U(\beta_2)\right\}. \tag{5.8}$$

因此, 在原料组成成分相对平均回收效率不低于 $1-\beta_2$ 且原料质量变化不超过 β_1 控制范围的情况下, β_1^* 和 β_2^* 是给定系统设计决策 $\boldsymbol{x}$ 以及利润目标 τ_0 下, 所能承受的最高原料状态不确定性水平.

引理 5.1　给定资源回收系统的回收单元选择决策 $\boldsymbol{x}$, 当 $\beta_2 \leqslant \beta_2'$ 时, 则 $\beta_1^*[\boldsymbol{x}|\beta_2] \geqslant \beta_1^*[\boldsymbol{x}|\beta_2']$. 类似地, 当 $\beta_1 \leqslant \beta_1'$ 时, 则 $\beta_2^*[\boldsymbol{x}|\beta_1] \geqslant \beta_2^*[\boldsymbol{x}|\beta_1']$.

上述引理与定义 5.1 中系统针对不同原料状态鲁棒性的直观感受一致: 系统不能同时最大化针对原料组成成分不确定性与原料质量不确定性的鲁棒性. 此外, 根据 β_1^* 和 β_2^* 的定义, 原料质量和组成成分的不确定集 V 与 U 的组合可以记为

$$V \times U \in \left\{V(\beta_1) \times U\left(\beta_2^*[\boldsymbol{x}|\beta_1]\right), \beta_1 \in [0,1]\right\}, \tag{5.9}$$

或者

$$V \times U \in \left\{V\left(\beta_1^*[\boldsymbol{x}|\beta_2]\right) \times U(\beta_2), \beta_2 \in [0,1]\right\}. \tag{5.10}$$

集合 $V \times U$ 表示在资源回收系统满足利润目标 τ_0 的前提下, 原料质量和组成成分的最大变化范围.

5.3.2　鲁棒性函数的计算

本节介绍在分别给定 β_2 和 β_1 下, 鲁棒性函数 $\beta_1^*[\boldsymbol{x}|\beta_2]$ 和 $\beta_2^*[\boldsymbol{x}|\beta_1]$ 的高效计算形式. 因为 β_1^* 和 β_2^* 的结构相同, 所以本节仅介绍鲁棒性函数 β_1^* 的高效计算形式. β_1^* 可以写成如下形式:

$$\beta_1^*\left[\boldsymbol{x}\middle|\beta_2\right] = \max\{\beta_1 \mid \beta_1 \in [0,1], \mathcal{Z}(\boldsymbol{x};\beta_1,\beta_2) \geqslant \tau_0\}, \tag{5.11}$$

这里 $\mathcal{Z}(\boldsymbol{x};\beta_1,\beta_2)$ 是一个的最小最大化 (min-max) 利润函数:

$$\mathcal{Z}(\boldsymbol{x};\beta_1,\beta_2) = \min_{(\boldsymbol{v},\boldsymbol{u})\in V(\beta_1)\times U(\beta_2)}\left[\max_{(\boldsymbol{y},\boldsymbol{z})\in\mathcal{H}(\boldsymbol{x},\boldsymbol{v})}\sum_{i\in[I]}\sum_{j\in[J]}\kappa_{ij}(\boldsymbol{u})y_{ij} - \sum_{i\in[I]}h_i z_i - \boldsymbol{c}^\top\boldsymbol{x}\right], \tag{5.12}$$

其中,

$$\mathcal{H}(\boldsymbol{x},\boldsymbol{v})=\left\{(\boldsymbol{y},\boldsymbol{z})\in\mathbb{R}_+^{IJ}\times\mathbb{R}_+^{I}\ \middle|\ \begin{array}{l}\sum\limits_{j\in[J]}y_{ij}+z_i=v_i,\forall\ i\in[I],\\ \sum\limits_{i\in[I]}y_{ij}\leqslant s_jx_j,\forall\ j\in[J]\end{array}\right\}. \tag{5.13}$$

根据式 (5.11), 可知在给定回收单元选择决策 $\boldsymbol{x}$ 下, $\mathcal{Z}(\boldsymbol{x};\beta_1,\beta_2)$ 是不确定集 $V(\beta_1)\times U(\beta_2)$ 上资源回收系统的可保证利润水平 (guaranteed profit level). 容易验证 $\mathcal{Z}(\boldsymbol{x};\beta_1,\beta_2)$ 关于 $\beta_1\in[0,1]$ 和 $\beta_2\in[0,1]$ 是单调不增的 (non-increasing). 因此, $\beta_1^*[\boldsymbol{x}|\beta_2]$ 可以通过二分法求解. 给定误差容忍度 (error tolerance) ϵ, 最多只需要 $\log(1/\epsilon)$ 步便可求得 $\beta_1^*[\boldsymbol{x}|\beta_2]$, 其中每一步都需要求解最小最大化问题 (5.12).

为了表述简洁, 我们记

$$\bar{\xi}_k(\beta_1):=\frac{\xi_k^-(\beta_1)+\xi_k^+(\beta_1)}{2},\quad \underline{\xi}_k(\beta_1):=\frac{\xi_k^+(\beta_1)-\xi_k^-(\beta_1)}{2},\quad \forall\ k\in[K]. \tag{5.14}$$

在一般情况下, 我们证明问题 (5.12) 可以转化为混合整数线性规划问题.

命题 5.1 给定回收单元设计决策 $\boldsymbol{x}$ 和 (β_1,β_2), 问题 (5.12) 等价于如下混合整数线性规划问题:

$$\begin{aligned}\min_{\boldsymbol{\lambda},\boldsymbol{d},\boldsymbol{\pi},\boldsymbol{\gamma},\boldsymbol{u}}\quad & \sum_{k\in[K]}\bar{\xi}_k(\beta_1)\left(\sum_{i\in[I]}\alpha_{ik}d_i\right)+\sum_{j\in[J]}s_jx_j\pi_j-\sum_{k\in[K]}\gamma_k-\boldsymbol{c}^\top\boldsymbol{x}\\ \text{s.t.}\quad & d_i+\pi_j\geqslant\kappa_{ij}(\boldsymbol{u}),\forall i\in[I],j\in[J],\\ & d_i+h_i\geqslant 0,\forall i\in[I],\\ & \gamma_k\leqslant\underline{\xi}_k(\beta_1)\sum_{i\in[I]}\alpha_{ik}d_i+W\lambda_k,\forall k\in[K],\\ & \gamma_k\leqslant-\underline{\xi}_k(\beta_1)\sum_{i\in[I]}\alpha_{ik}d_i+W(1-\lambda_k),\forall k\in[K],\\ & \lambda_k\in\{0,1\},\gamma_k\geqslant 0,d_i\in\mathbb{R},\pi_j\geqslant 0,\forall i\in[I],j\in[J],k\in[K],\\ & \boldsymbol{u}\in U(\beta_2),\end{aligned} \tag{5.15}$$

其中, $\boldsymbol{\lambda}$ 为二元整数变量, $\boldsymbol{d}$, $\boldsymbol{\pi}$, $\boldsymbol{\gamma}$, $\boldsymbol{u}$ 为实数变量. 此外, W 为一个足够大的正数, $\bar{\xi}_k(\beta_1)$ 和 $\underline{\xi}_k(\beta_1)$ 在式 (5.14) 中给出.

在一些与实际情况相关的假设下, 问题 (5.12) 可以进一步简化为一个线性规划问题. 我们考虑下面三种情形.

1. 情形一: 原料质量固定

考虑原料质量 $\boldsymbol{v}$ 固定的情况, 此时问题 $\mathcal{Z}(\boldsymbol{x};\beta_1,\beta_2)$ 可以简化为 $\mathcal{Z}(\boldsymbol{x};\beta_2)$. 它等价于一个线性规划问题.

命题 5.2 给定回收单元设计决策 $\boldsymbol{x}$, β_2 并固定原料质量 $\boldsymbol{v}$, 问题 $\mathcal{Z}(\boldsymbol{x};\beta_2)$ 等价于如下线性规划问题:

$$\begin{aligned}\mathcal{Z}(\boldsymbol{x};\beta_2)=\max_{\boldsymbol{y},\boldsymbol{z},\gamma}\quad & \gamma-\sum_{i\in[I]}h_iz_i-\boldsymbol{c}^\top\boldsymbol{x}\\ \text{s.t.}\quad & (\boldsymbol{y},\boldsymbol{z},\gamma)\in\mathcal{H}(\boldsymbol{x},\boldsymbol{v})\cap\Gamma(\beta_2),\end{aligned}\tag{5.16}$$

这里, $\Gamma(\beta_2)$ 为多面体集合:

$$\Gamma(\beta_2):=\left\{\boldsymbol{y},\gamma\ \middle|\ \begin{array}{l}\gamma\leqslant\sum_{i\in[I]}d_i+I\beta_2d_0,\\ d_i+d_0\delta_m\leqslant\sum_{j\in[J]}y_{ij}q_{mj},\ d_0\geqslant0,\ d_i\in\mathbb{R},\quad\forall\, i\in[I],m\in[M]\end{array}\right\},\tag{5.17}$$

其中, δ_m 是式 (5.5) 中原料组成成分 m 的相对回收效率.

在给定回收单元设计决策 $\boldsymbol{x}$ 与原料组成成分不确定集 $U(\beta_2)$, 且有一组未来原料质量的预测值时, 决策者可以参考情形一对问题 $\mathcal{Z}(\boldsymbol{x};\beta_2)$ 进行分析.

2. 情形二: 原料质量互不相关, 放松原料处理约束

问题 (5.2) 中第一个约束条件要求所有原料都必须分配给回收单元进行处理, 如果将这一约束放松:

$$\sum_{j\in[J]}y_{ij}\leqslant v_i,\quad\forall\, i\in[I],\tag{5.18}$$

也就是说, 在获得原料的准确状态信息 $\boldsymbol{v}$ 和 $\boldsymbol{u}$ 后, 能源回收企业可以对要回收处理原料进行挑选和分类. 在这种情况下, 能源回收系统没有未被处理或剩余原料质量 $\boldsymbol{z}$. 此外, 假设不同收集点的原料质量 $\boldsymbol{v}$ 互不相关, 我们有如下结果.

命题 5.3 假设不同收集点的原料质量 $\boldsymbol{v}$ 互不相关. 给定 $\boldsymbol{x}$ 和原料组成成分 $\boldsymbol{u}$, 若将问题 $\phi(\boldsymbol{x};\boldsymbol{v},\boldsymbol{u})$ 中第一个约束替换为约束 (5.18), 则问题 $\mathcal{Z}(\boldsymbol{x};\beta_1,\beta_2)$ 可以转化为线性规划问题 (5.16), 其中, $\boldsymbol{z}\equiv\boldsymbol{0}$, $\boldsymbol{v}\equiv\boldsymbol{v}^-(\beta_1):=\left[v_1^-(\beta_1),v_2^-(\beta_1),\cdots,v_I^-(\beta_1)\right]^\top$, 且

$$v_i^-(\beta_1):=\sum_{k\in[K]}\left\{\xi_k^-(\beta_1)\left[\alpha_{ik}\right]^+-\xi_k^+(\beta_1)\left[\alpha_{ik}\right]^-\right\},\tag{5.19}$$

这里 $[\cdot]^+$ 和 $[\cdot]^-$ 分别为正部和负部算子.

不同收集点原料质量互不相关的情况是普遍存在的, 例如, 当原料的收集点相距较远远 (Guerrero et al., 2013; Hannan et al., 2015). 这种情况下, 决策者可以参考情形二对利润进行分析.

3. 情形三: 原料质量互不相关, 正收益原料组成成分

给定原料组成成分 $\boldsymbol{u}$ 和回收单元选择决策 $\boldsymbol{x}$, 我们定义如下集合:

$$\mathcal{J}_{\boldsymbol{x}}(i,\boldsymbol{u}) := \{j \mid x_j = 1, \kappa_{ij}(\boldsymbol{u}) > 0\}, \tag{5.20}$$

上述集合表示给定原料组成成分 $\boldsymbol{u}$, 在所有已选择的处理单元中, 处理来自收集点 i 的原料产生正净收益的单元序号 j. 进一步, 我们记

$$\mathcal{J}_{\boldsymbol{x}}(\boldsymbol{u}) := \bigcup_{i\in[I]} \mathcal{J}_{\boldsymbol{x}}(i,\boldsymbol{u}) = \{\, j \mid x_j = 1,\ \exists\, i, \text{s.t. } \kappa_{ij}(\boldsymbol{u}) > 0\}. \tag{5.21}$$

基于上述集合, 给定 $\boldsymbol{x}$ 和原料质量不确定集 $V(\beta_1)$, 定义如下原料组成成分集合 $\mathcal{O}(\boldsymbol{x},\beta_1)$:

$$\mathcal{O}(\boldsymbol{x},\beta_1) := \left\{ \boldsymbol{u} \;\middle|\; \begin{array}{l} \mathcal{J}_{\boldsymbol{x}}(i,\boldsymbol{u}) \neq \varnothing, \forall\, i \in [I], \\ v_i^+(\beta_1) \leqslant \displaystyle\sum_{j\in\mathcal{J}_{\boldsymbol{x}}(i,\boldsymbol{u})} s_j, \forall\, i \in [I], \\ \displaystyle\sum_{i\in[I]} v_i^+(\beta_1) \leqslant \sum_{j\in\mathcal{J}_{\boldsymbol{x}}(\boldsymbol{u})} s_j \end{array} \right\}, \tag{5.22}$$

其中,

$$v_i^+(\beta_1) := \sum_{k\in[K]} \left\{ \xi_k^+(\beta_1)\,[\alpha_{ik}]^+ - \xi_k^-(\beta_1)\,[\alpha_{ik}]^- \right\}. \tag{5.23}$$

上述集合 (5.22) 中, 第一组约束要求在能源回收系统选择决策 $\boldsymbol{x}$ 中, 至少有一个已选择的处理单元在处理来自收集点 i 的原料时有正净收益. 其余两个约束确保有正净收益的处理单元具有足够的容量水平或处理能力. 因此, 集合 $\mathcal{O}(\boldsymbol{x},\beta_1)$ 是给定原料质量 $\boldsymbol{v} \in V(\beta_1)$ 与 $\boldsymbol{x}$, 能够产生非负净收益的所有原料组成成分的集合.

命题 5.4 给定 $\boldsymbol{x}$, 原料质量不确定集 $V(\beta_1)$ 和原料组成成分不确定集 $U(\beta_2)$, 如果有 $U(\beta_2) \subseteq \mathcal{O}(\boldsymbol{x},\beta_1)$, 则问题 $\mathcal{Z}(\boldsymbol{x};\beta_1,\beta_2)$ 退化为线性规划问题 (5.16), 其中, $\boldsymbol{z} \equiv \boldsymbol{0}$ 且 $\boldsymbol{v} \equiv \boldsymbol{v}^-(\beta_1)$, $\boldsymbol{v}^-(\beta_1)$ 是由式 (5.19) 所定义的 $\boldsymbol{v}$ 下界.

上述命题 5.4 说明: 如果原料组成成分的回收效率足够高且回收系统的容量也足够大, 即 $U(\beta_2) \subseteq \mathcal{O}(\boldsymbol{x},\beta_1)$, 则处理原料总能产生非负净利润, 并且可保证利

润水平 $\mathcal{Z}(\boldsymbol{x};\beta_1,\beta_2)$ 可以通过求解给定 $\boldsymbol{v}\equiv\boldsymbol{v}^-(\beta_1)$ 的线性规划问题来获得. 此外, 如果 β_1 和 β_2 满足 $U(\beta_2)\subseteq\mathcal{O}(\boldsymbol{x},\beta_1)$, 那么总是可以通过处理更多的原料来获得更高的利润.

5.3.3 系统鲁棒性前沿

接下来, 我们利用系统鲁棒性图对资源回收系统的鲁棒性进行分析. 通过计算不同 β_2 值下的 $\beta_1^*[\boldsymbol{x}|\beta_2]$ 得到图 5.2. 此图将资源回收系统受原料质量和组成成分不确定性的影响进行可视化展示. 对于一个给定回收单元选择决策 $\boldsymbol{x}$ 的资源回收系统, 图 5.2 给出所有可以保证资源回收系统经济可行性的参数 (β_1,β_2) 组合. 每组 β_1 和 β_2 的值都代表了一种原料回收方案, 例如, 图 5.2 中的 A 点至 E 点. 此外, 鲁棒性前沿曲线之外的点 (β_1,β_2) 不能保证系统的经济可行性, 例如, 图 5.2 中的 E 点.

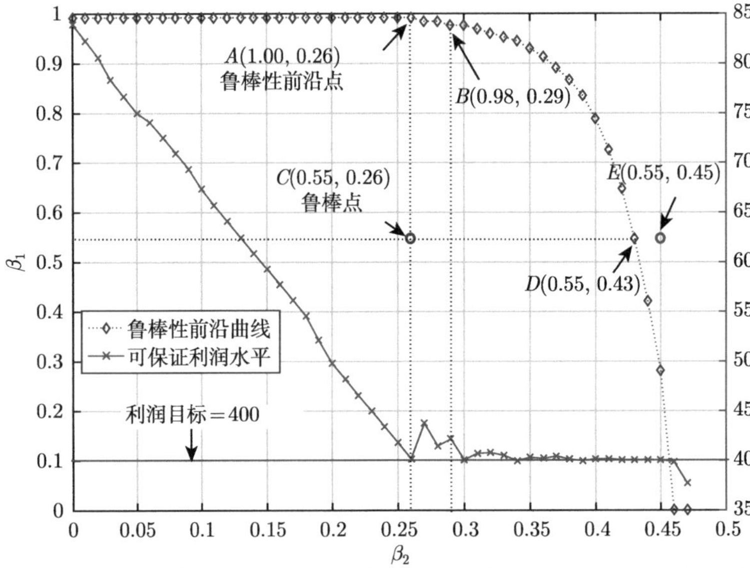

图 5.2 给定回收单元选择决策 $\boldsymbol{x}_I$, 资源回收系统在利润目标 $\tau_0=400$ 下的系统鲁棒性图

点 C $(\beta_1=0.55,\beta_2=0.26)$ 可以进一步优化为点 A $(1.00,0.26)=(\beta_1^*[\boldsymbol{x}_I|0.26],0.26)$ 和点 D $(0.55,0.43)=(0.55,\beta_2^*[\boldsymbol{x}_I|0.55])$, 以及点 A 和点 D 之间鲁棒前沿上所有其他点. 点 E (0.55, 0.45) 落在鲁棒前沿曲线之外, 表示其未达到利润目标. 此外, 图中给出在 β_2 不同取值下对应的利润保证 $\mathcal{Z}(\boldsymbol{x}_I;\beta_1^*[\boldsymbol{x}_I|\beta_2],\beta_2)$

通过分析给定回收单元选择决策 $\boldsymbol{x}_I$ 下系统鲁棒性图 5.2, 我们得到如下结论.

(1) 从图 5.2 中可以看出, 如果 $\beta_2>0.46$, 即原料平均资源回收效率 $1-\beta_2<54\%$, 即使原料质量不存在不确定性 $(\beta_1=0)$, 系统也无法实现利润目标 (当 $\beta_2>0.46$ 时, 利润低于 $\tau_0=400$). 此外, 当原料回收效率足够高, 即 $\beta_2\leqslant 0.26$

$(1-\beta_2 \geqslant 74\%)$ 时，系统可以实现 $\tau_0 = 400$ 的利润目标，且允许原料质量在不确定集 $V(1)$ 内任意变化. 因此，为了保证资源回收系统在运行期内的经济可行性，原料的平均资源回收效率应不低于 54%. 此外，如果原料平均资源回收效率高于 74%，则进一步提高回收效率原料组成成分的纯度也不会提升系统的鲁棒性.

(2) 虽然强制要求原料平均回收效率 $1-\beta_2 \geqslant 74\%$ 可以使系统在原料质量不确定性上具有完全的鲁棒性 (图 5.2 中点 A 的 $\beta_1^*[\boldsymbol{x}_I|\beta_2 = 0.26] = 1.00$)，但是，在实际生产中，对原料平均回收效率的高标准会带来更高的原料组成成分的分离或提纯成本. 如果放宽原料平均回收效率至 65%，虽然系统在原料质量不确定性上的鲁棒性会略微降低，但是其仍保持较高的鲁棒性水平 0.93 (图 5.2 中的点 B 的 $\beta_1^*[\boldsymbol{x}_I|\beta_2 = 0.65] = 0.93$). 因此，这种权衡的选择对资源回收系统节约成本是有利的.

(3) 在不同原料平均回收效率 $1-\beta_2$ 下，资源回收系统的利润保证水平 $\mathcal{Z}(\boldsymbol{x}_I; \beta_1^*[\boldsymbol{x}_I|\beta_2], \beta_2)$ 可以为决策者提供额外的决策信息. 从图 5.2 中可以看出鲁棒前沿曲线上的点对应的可保证利润水平 (guaranteed profit level) 不是关于 $1-\beta_2$ 单调的. 例如，鲁棒性前沿曲线上的点 A (1.00, 0.26) 和点 B (0.98, 0.29)，与点 B 相比，点 A 的原料平均回收效率更高，但是其对应的成本也可能更高. 如果只基于上述信息无法区分两个原料回收方案的优劣，则根据两个方案对应的利润水平，我们可能更倾向于 B 点所对应的方案，因为在 B 点系统的可保证利润更高 (A 点和 B 点对应的系统可保证利润分别为 401.9 和 421.6).

5.3.4 鲁棒性函数与经济可行性实现概率

本小节介绍鲁棒性函数 $\beta_1^*[\boldsymbol{x}|\beta_2]$ 和 $\beta_2^*[\boldsymbol{x}|\beta_1]$ 与经济可行实现概率 $\mathbb{P}\big[\phi(\boldsymbol{x};\boldsymbol{v},\boldsymbol{u}) \geqslant \tau_0\big]$ 之间的关系. 特别地，本小节证明鲁棒性值函数可用来定义经济可行概率的下界.

命题 5.5 给定鲁棒性函数 $\beta_2^*[\boldsymbol{x}|\beta_1]$，假设不同收集点间的原料组成成分 $\boldsymbol{u}_{\cdot i}, i \in [I]$ 分布相互独立 (independently distributed) 且 $\mathbb{E}[u_{mi}] \leqslant \bar{\mu}_{mi}$，则有

$$\min_{\boldsymbol{v}\in V(\beta_1)} \mathbb{P}\Big[\phi(\boldsymbol{x};\boldsymbol{v},\boldsymbol{u}) \geqslant \tau_0\Big] \geqslant \left[1 - \frac{(I-1)\Omega^2 + \Omega}{I(\beta_2^*[\boldsymbol{x}|\beta_1])^2}\right]^+, \quad \forall\, \beta_1 \in (0,1), \tag{5.24}$$

其中，$\Omega := \sum_{m\in[M]}(1-\delta_m)\bar{\mu}_{mi}$ 和 δ_m 在式 (5.5) 中定义，$[\cdot]^+$ 是正部算子.

上述结果表明：如果不同收集点之间的原料组成成分是独立的，并且可以有效估计原料组成成分的期望上界，则鲁棒性函数 $\beta_2^*[\boldsymbol{x}|\beta_1]$ 可以用来定义经济可行概率的下界. 此外，通过将原料质量不确定集 $V(\beta_1)$ 中每个因子 ξ_k 的变化区间 $\big[\xi_k^-(\beta_1), \xi_k^+(\beta_1)\big]$ 构建成与参数 β_1 相关的预测区间 (prediction interval)，也可以得到由鲁棒性函数 $\beta_1^*[\boldsymbol{x}|\beta_2]$ 定义的经济可行概率下界.

命题 5.6 给定鲁棒性函数 $\beta_1^*[\boldsymbol{x}|\beta_2]$, 若集合 $V(\beta_1)$ 中因子 ξ_k 的每个区间 $\left[\xi_k^-(\beta_1), \xi_k^+(\beta_1)\right]$ 满足对于任意 $k \in [K]$ 都有 $\mathbb{P}\left[\xi_k \in \left[\xi_k^-(\beta_1), \xi_k^+(\beta_1)\right]\right] \geqslant 1-(1-\beta_1)/K$ 成立, 那么

$$\min_{\boldsymbol{u}\in U(\beta_2)} \mathbb{P}\Big[\phi(\boldsymbol{x};\boldsymbol{v},\boldsymbol{u}) \geqslant \tau_0\Big] \geqslant \beta_1^*\left[\boldsymbol{x}|\beta_2\right], \quad \forall\, \beta_2 \in (0,1). \tag{5.25}$$

5.4 复合鲁棒性指标及高效计算形式

本节讨论如何选择最优的回收单元决策, 并引入复合鲁棒性指标 (composite robustness index) 作为优化准则, 它同时包括鲁棒性函数 β_1^* 和 β_2^*, 而且是高效计算的 (computationally tractable).

5.4.1 复合鲁棒性指标

在 5.3.3 节中, 每个回收单元选择方案 $\boldsymbol{x}$ 都对应一个系统鲁棒性图. 当 (β_1,β_2) 位于鲁棒前沿曲线的下方时, 决策 $\boldsymbol{x}$ 是经济可行的. 因此, 最优决策 $\boldsymbol{x}^\star$ 对应的鲁棒前言曲线应包含 "最多的" (β_1,β_2) 场景. 但是, 这种最优决策无法通过求解一个优化问题获得[①]. 因此, 我们介绍如下复合鲁棒性指标.

定义 5.2 给定回收单元选择决策 $\boldsymbol{x}$, 复合鲁棒性指标 $\pi(\boldsymbol{x})$ 定义如下:

$$\pi(\boldsymbol{x}) := \max_{\beta_2\in[0,1]} \min\Big\{\theta_1\beta_1^*[\boldsymbol{x}|\beta_2], \theta_2\beta_2\Big\}, \tag{5.26}$$

其中, $\beta_1^*[\boldsymbol{x}|\beta_2]$ 是由式 (5.7) 定义的鲁棒性函数, $\theta_1,\theta_2 \geqslant 1$ 是偏好系数 (preference coefficients) 且满足 $\min\{\theta_1,\theta_2\}=1$.

在上述复合鲁棒性指标的定义中, 算子 "$\min\{\theta_1\beta_1,\theta_2\beta_2\}$" 反映了系统对 $\boldsymbol{v}$ 和 $\boldsymbol{u}$ 的鲁棒性水平, 并返回其最大的下界. 其中, 参数 θ_1 和 θ_2 是对 $\boldsymbol{v}$ 和 $\boldsymbol{u}$ 鲁棒性的偏好系数, 取值范围为 $[1,\infty)$, $\min\{\theta_1,\theta_2\}=1$ 是正则条件 (regularity condition). 例如, 当 $\theta_2>1$ 时, 系统更注重对不确定原料质量 $\boldsymbol{v}$ 的鲁棒性. 复合鲁棒性指标 $\pi(\boldsymbol{x})$ 表示在给定的偏好系数 θ_1 和 θ_2 下, 系统的最大鲁棒性水平, 见图 5.3. 此外, 复合鲁棒性指标属于拟凹的满意测度 (satisficing measure with quasi-concavity), 符合目标决策理论中一些常用决策准则的性质 (Brown and Sim, 2009; Lam et al., 2013).

命题 5.7 复合鲁棒性指标 $\pi(\boldsymbol{x})$ 等价于如下形式:

$$\pi(\boldsymbol{x}) = \max_{\alpha\in[0,1]} \left\{\alpha \,\middle|\, \phi(\boldsymbol{x};\boldsymbol{v},\boldsymbol{u}) \geqslant \tau_0,\ \forall\,(\boldsymbol{v},\boldsymbol{u}) \in V\left(\frac{\alpha}{\theta_1}\right) \times U\left(\frac{\alpha}{\theta_2}\right)\right\}, \tag{5.27}$$

① 寻找最优的回收单元选择决策需要针对 $\boldsymbol{x}\in\{0,1\}^J$ 优化积分函数 $\int_{\beta_2\in[0,1]}\beta_1^*[\boldsymbol{x}|\beta_2]\mathrm{d}\beta_2$. 实际上, 对于给定的 $\boldsymbol{x}$, 计算这种积分函数已经相当复杂, 而优化此函数是更复杂的.

如果不存在可行的 $\alpha \in [0,1]$, 则令 $\pi(\boldsymbol{x}) = 0$.

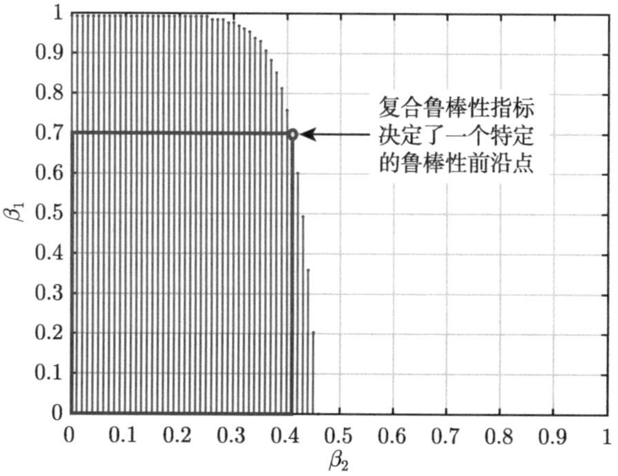

图 5.3 复合鲁棒性指标 π 在系统鲁棒性图上的示意图. $\theta_2 = 1.75$ 定义鲁棒性前沿点 $(\beta_1^*[\boldsymbol{x}|\beta_2^*], \beta_2^*) = (0.7, 0.4)$ 和 $\pi(\boldsymbol{x}) = \min\{\beta_1^*[\boldsymbol{x}|\beta_2^*], \theta_2\beta_2^*\} = 0.7$

上述式 (5.27) 与鲁棒性函数 $\beta_1^*[\boldsymbol{x}|\beta_2]$ 具有相似的结构, 这表明可以使用类似的计算方法来求解问题 $\pi(\boldsymbol{x})$. 特别地, 注意到

$$\begin{aligned}\pi(\boldsymbol{x}) &= \max_{\alpha\in[0,1]}\left\{\alpha \,\middle|\, \phi(\boldsymbol{x};\boldsymbol{v},\boldsymbol{u}) \geqslant \tau_0,\ \forall\,(\boldsymbol{v},\boldsymbol{u}) \in V\left(\frac{\alpha}{\theta_1}\right) \times U\left(\frac{\alpha}{\theta_2}\right)\right\} \\ &= \max_{\alpha\in[0,1]}\left\{\alpha \,\middle|\, \mathcal{Z}\left(\boldsymbol{x};\frac{\alpha}{\theta_1},\frac{\alpha}{\theta_2}\right) \geqslant \tau_0\right\},\end{aligned}$$

其中, $\mathcal{Z}(\boldsymbol{x};\alpha/\theta_1,\alpha/\theta_2)$ 是最小最大化利润函数 (5.12), 它关于 $\alpha \in [0,1]$ 是单调不增的. 因此, 利用式 (5.27) 和二分法, 可以计算复合鲁棒性指标 $\pi(\boldsymbol{x})$.

5.4.2 基于复合鲁棒性指标的资源回收问题

本节考虑如下以复合鲁棒性指标 (5.7) 为目标函数的资源回收问题:

$$\max_{\boldsymbol{x}\in\mathcal{X}} \pi(\boldsymbol{x}) = \max_{\boldsymbol{x}\in\mathcal{X},\alpha\in[0,1]}\left\{\alpha \,\middle|\, \phi(\boldsymbol{x};\boldsymbol{v},\boldsymbol{u}) \geqslant \tau_0, \forall\,(\boldsymbol{v},\boldsymbol{u}) \in V\left(\frac{\alpha}{\theta_1}\right) \times U\left(\frac{\alpha}{\theta_2}\right)\right\}, \tag{5.28}$$

其中, 假设可行集 $\mathcal{X} := \{\boldsymbol{x} \in \{0,1\}^J | \ \boldsymbol{Bx} = b\}$ 且 $\{\boldsymbol{x} \geqslant \boldsymbol{0} | \boldsymbol{Bx} = b\}$ 为多面体集. 问题 (5.28) 与两阶段自适应鲁棒优化问题密切相关 (Ben-Tal et al., 2009; Bertsimas et al., 2011), 而两个问题的不同之处在于问题 (5.28) 中的不确定集参数也是优化变量. 此外, 虽然两阶段鲁棒优化问题通常是难以求解的, 但是在本章

的设定下, 问题 (5.28) 有高效计算形式. 首先, 问题 (5.28) 可以等价地转化为如下问题:

$$\begin{aligned}\max_{\alpha\in[0,1]} \quad & \alpha \\ \text{s.t.} \quad & \varpi(\alpha):=\max_{\boldsymbol{x}\in\mathcal{X}}\left\{\min_{(\boldsymbol{v},\boldsymbol{u})\in V\left(\frac{\alpha}{\theta_1}\right)\times U\left(\frac{\alpha}{\theta_2}\right)}\phi\left(\boldsymbol{x};\boldsymbol{v},\boldsymbol{u}\right)\right\}\geqslant\tau_0.\end{aligned} \tag{5.29}$$

上述问题 (5.29) 有两个特性: 首先, 因为问题 $\varpi(\alpha)$ 关于 α 是非增的, 所以可以利用二分法求得参数 $\alpha\in[0,1]$ 的最优值; 其次, 问题 (5.29) 可以写成如下的最大化最小化最大化 (max-min-max) 问题:

$$\begin{aligned}&\max_{\boldsymbol{x}\in\mathcal{X}}\left\{\min_{(\boldsymbol{v},\boldsymbol{u})\in V(\beta_1)\times U(\beta_2)}\phi\left(\boldsymbol{x};\boldsymbol{v},\boldsymbol{u}\right)\right\}\\ &=\max_{\boldsymbol{x}\in\mathcal{X}}\min_{(\boldsymbol{v},\boldsymbol{u})\in V(\beta_1)\times U(\beta_2)}\left[\max_{\boldsymbol{y},\boldsymbol{z}\in\mathcal{H}(\boldsymbol{x},\boldsymbol{v})}\sum_{i\in[I]}\sum_{j\in[J]}\kappa_{ij}(\boldsymbol{u})y_{ij}-\sum_{i\in[I]}h_iz_i-\boldsymbol{c}^\top\boldsymbol{x}\right],\end{aligned} \tag{5.30}$$

其中, (β_1,β_2) 定义为 $\beta_1=\dfrac{\alpha}{\theta_1}$ 和 $\beta_2=\dfrac{\alpha}{\theta_2}$. 问题 (5.30) 具有两阶段自适应鲁棒优化问题结构, 其中第一阶段决策是二元变量 $\boldsymbol{x}$, 第二阶段决策 $\boldsymbol{y}$ 与不确定参数 $(\boldsymbol{v},\boldsymbol{u})\in V(\beta_1)\times U(\beta_2)$ 有关. 接下来, 介绍如何在给定 (β_1,β_2) 下求解问题 (5.30).

对于 5.3.2 小节中的三种特殊情形, 利用命题 5.2 中的线性规划问题 (5.16) 可以将问题 (5.30) 转化为如下混合整数线性规划问题:

$$\begin{aligned}\max_{\boldsymbol{x},\boldsymbol{y},\boldsymbol{z},\gamma} \quad & \gamma-\boldsymbol{c}^\top\boldsymbol{x}\\ \text{s.t.} \quad & (\boldsymbol{y},\boldsymbol{z},\gamma)\in\mathcal{H}(\boldsymbol{x},\boldsymbol{v})\cap\Gamma(\beta_2),\\ & \boldsymbol{x}\in\mathcal{X},\end{aligned}$$

其中, 针对情形二和三, 原料质量 $\boldsymbol{v}\equiv\boldsymbol{v}^-(\beta_1)$ 且 $\boldsymbol{z}\equiv\boldsymbol{0}$. 此外, 针对情形三, 最优的 $(\boldsymbol{x},\beta_1,\beta_2)$ 还需满足 $U(\beta_2)\subseteq\mathcal{O}(\boldsymbol{x},\beta_1)$.

对于求解一般情况下的问题 (5.30), 我们需要将问题 (5.15) 中的

$$\min_{(\boldsymbol{v},\boldsymbol{u})\in V(\beta_1)\times U(\beta_2)}\phi\left(\boldsymbol{x};\boldsymbol{v},\boldsymbol{u}\right)$$

转化为如下形式.

命题 5.8 给定 $\boldsymbol{x}$ 和 (β_1,β_2), 利润函数 $\min_{(\boldsymbol{v},\boldsymbol{u})\in V(\beta_1)\times U(\beta_2)}\phi\left(\boldsymbol{x};\boldsymbol{v},\boldsymbol{u}\right)$ 等价于如下线性规划问题:

$$
\begin{aligned}
\max\quad & r-\boldsymbol{c}^{\top}\boldsymbol{x} \\
\text{s.t.}\quad & r\leqslant\gamma^{e}-\sum_{i\in[I]}h_{i}z_{i}^{e},\forall\, e\in[E], \\
& \sum_{j\in[J]}y_{ij}^{e}+z_{i}^{e}=\boldsymbol{\alpha}_{i}^{\top}\hat{\boldsymbol{\xi}}^{e},\forall\, i\in[I],e\in[E], \\
& \sum_{i\in[I]}y_{ij}^{e}\leqslant s_{j}x_{j},\forall\, j\in[J],e\in[E], \\
& (\boldsymbol{y}^{e},\gamma^{e})\in\Gamma^{e}(\beta_{2}),\forall\, e\in[E],
\end{aligned} \tag{5.31}
$$

其中 $[E]:=\{1,\cdots,E=2^{K}\}$, 每个 $\hat{\boldsymbol{\xi}}^{e}=[\hat{\xi}_{1}^{e},\hat{\xi}_{2}^{e},\cdots,\hat{\xi}_{K}^{e}]^{\top}$ 且 $\hat{\xi}_{k}^{e}\in\{\xi_{k}^{-}(\beta_{1}),\xi_{k}^{+}(\beta_{1})\}$, 每个 $\Gamma^{e}(\beta_{2})$ 是由式 (5.17) 定义的多面体集.

因此, 求解问题 (5.30) 等价于求解线性规划问题 (5.31). 虽然问题 (5.31) 是一个线性规划问题, 但是它的变量和约束的规模相对问题规模是指数级的 (具体见 (Wang and Ng, 2019)). 因此, 为了高效求解问题 (5.31), 我们可以采用切平面算法来迭代求解此问题.

5.5 数值实验

本节基于新加坡城市废弃物数据, 设计数值验证实验模型处理原料质量和组成成分不确定问题的效果. 考虑有 $I=10$ 个收集点和 $J=40$ 个资源处理单元的资源回收问题, 其中, 对于每个收集点 i, 记 $[v_{i}^{-}(\beta_{1}),v_{i}^{+}(\beta_{1})]=[\mu_{i}-\Theta(1-\beta_{1})\sigma_{i},\mu_{i}+\Theta(1-\beta_{1})\sigma_{i}]$ 为仿射因子估计范围 (affine factor estimation region), 其中 μ_{i},σ_{i} 通过原料历史数据估计得到, 且 $[\mu_{i}-\Theta\sigma_{i},\mu_{i}+\Theta\sigma_{i}]$ 为原料质量 v_{i} 的支撑集. 此外, 考虑四种原料组成组分: 餐厨废弃物、木料废弃物、庭院废弃物和包装废弃物. 同时, 考虑四种处理单元: 两种气化炉 (gasifiers), 分别为等离子体辅助气化 (plasma-assisted gasifications) 和下降式气化 (dropdown gasifications), 分别记为 "G-1" 和 "G-2". 两种厌氧消化器 (anaerobic digesters), 分别为单级和两级厌氧消化 (single-stage and two-stage anaerobic digestions), 分别记为 "D-1" 和 "D-2". 模型的主要参数在表 5.1 中给出.

表 5.1 实验参数

资源回收技术	G-1	G-2	D-1	D-2
最大容量 (处理能力) (t)	3100	3200	3200	3500
年化固定成本 (SGD)	20500	25200	31600	35000
单位资源回收系数 (SGD/kg)				
餐厨废弃物 ($m=1$)	0.11	0.13	0.27	0.32
木料废弃物 ($m=2$)	0.17	0.19	0.08	0.09
庭院废弃物 ($m=3$)	0.12	0.15	0.02	0.02
包装废弃物 ($m=4$)	0.003	0.002	0.004	0.003

5.5.1 基于系统鲁棒性前沿评估资源回收系统

本节对一组资源回收系统进行了鲁棒性分析, 以比较它们在保证经济可行性方面的效果. 表 5.2 给出了四个资源回收系统, 它们的总容量介于 10500 吨到 11000 吨之间.

表 5.2 四种资源回收系统设计

决策 $\boldsymbol{x}_{\mathrm{I}}$		决策 $\boldsymbol{x}_{\mathrm{II}}$		决策 $\boldsymbol{x}_{\mathrm{III}}$		决策 $\boldsymbol{x}_{\mathrm{IV}}$	
类型	容量	类型	容量	类型	容量	类型	容量
G-1	2000	G-1	3000	G-1	2000	G-1	2500
G-2	2500	G-2	2500	G-2	3000	G-2	2000
D-1	3000	D-1	3000	D-1	3500	D-1	3000
D-2	3000	D-2	2000	D-2	2500	D-2	3500

图 5.4(a) 展示了四个系统 ($\boldsymbol{x}_{\mathrm{I}}, \boldsymbol{x}_{\mathrm{II}}, \boldsymbol{x}_{\mathrm{III}}$ 和 $\boldsymbol{x}_{\mathrm{IV}}$) 在目标利润 $\tau_0 = 800$ 的鲁棒性图上的表现. 通过这些图, 可以很容易地对系统效果排序, 即 $\boldsymbol{x}_{\mathrm{III}} \succ \boldsymbol{x}_{\mathrm{II}} \succ \boldsymbol{x}_{\mathrm{I}} \succ \boldsymbol{x}_{\mathrm{IV}}$, 其中, “$\succ$” 表示更好的选择. 图 5.4(b) 是根据复合鲁棒性指标 $\pi(\boldsymbol{x})$ 的排序结果. 这些结果表明: 资源回收系统的回收单元选择决策在确保系统的经济可行性方面起着关键作用. 特别地, 虽然 $\boldsymbol{x}_{\mathrm{III}}$ 和 $\boldsymbol{x}_{\mathrm{IV}}$ 具有相同的总容量, 但是两个系统的处理单元容量配置不同, 导致它们的经济可行性有显著差异.

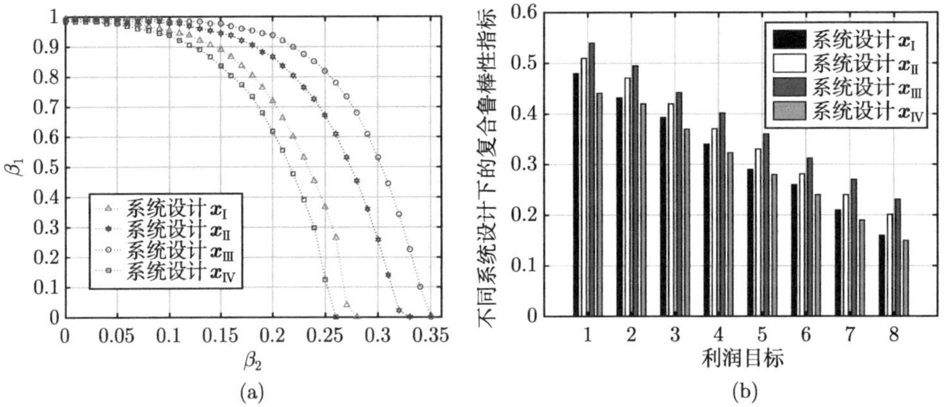

图 5.4 (a) 四种回收系统 ($\boldsymbol{x}_{\mathrm{I}}, \boldsymbol{x}_{\mathrm{II}}, \boldsymbol{x}_{\mathrm{III}}$ 和 $\boldsymbol{x}_{\mathrm{IV}}$) 在利润目标 $\tau_0 = 800$ 下的鲁棒性图. (b) 不同利润目标 ($\tau_0 = 300 \sim 1000$ 美元) 下四种回收系统的复合鲁棒性指标比较

接下来, 对数据样本量分别为 1000, 3000, 5000 和 8000 的四种不同的情况展示概率模型

$$\mathbb{P}\left[\phi\left(\boldsymbol{x};\boldsymbol{v},\boldsymbol{u}\right) \geqslant \tau_0\right]$$

的表现. 具体来说, 本节随机从集合 $V(1)$ 和 $U(1)$ 中生成原料质量 $\boldsymbol{v}$ 和组成成分 $\boldsymbol{u}$ 的样本, 然后对于每个样本 $(\hat{\boldsymbol{v}}, \hat{\boldsymbol{u}})$, 通过求解线性规划问题 (5.2) 来计算利润值 $\phi(\boldsymbol{x}; \hat{\boldsymbol{v}}, \hat{\boldsymbol{u}})$. 针对不同的利润目标 (从 300 美元到 1000 美元) 使用成功概率的样本估计值来进行排序, 进而评估不同系统设计方案的效果.

表 5.3 给出了不同样本量和利润目标下, 概率模型和复合鲁棒性指标模型的排序结果. 结果表明: ① 在不同利润目标下, 如 $\tau_0 = 800$ 和 900, 概率模型给出的排序与鲁棒性模型给出的排序是一致的 (即 $\boldsymbol{x}_{\mathrm{III}} \succ \boldsymbol{x}_{\mathrm{II}} \succ \boldsymbol{x}_{\mathrm{I}} \succ \boldsymbol{x}_{\mathrm{IV}}$). ② 在不同的样本规模下, 不同回收系统的成功概率会有很大的差异. 例如, 当样本容量在 1000, 3000 和 5000 上变化时, 系统设计决策 $\boldsymbol{x}_{\mathrm{II}}$ 在利润目标为 800 下的成功概率分别为 0.950, 0.887 和 0.894. 这表明模型的效果在很大程度上依赖于样本量.

表 5.3 不同模型的回收系统排序结果

模型	回收系统	$\tau_0 = 300$	400	500	600	700	800	900	1000
复合鲁棒性 π	$\boldsymbol{x}_{\mathrm{I}}$	0.480	0.432	0.393	0.340	0.290	0.260	0.210	0.160
	$\boldsymbol{x}_{\mathrm{II}}$	0.510	0.471	0.420	0.371	0.330	0.281	0.240	0.201
	$\boldsymbol{x}_{\mathrm{III}}$	0.540	0.495	0.442	0.402	0.360	0.312	0.270	0.231
	$\boldsymbol{x}_{\mathrm{IV}}$	0.441	0.420	0.370	0.323	0.280	0.240	0.190	0.150
成功概率 (样本量 8000)	$\boldsymbol{x}_{\mathrm{I}}$	1.000	1.000	0.998	0.980	0.881	0.543	0.185	0.035
	$\boldsymbol{x}_{\mathrm{II}}$	1.000	1.000	1.000	1.000	0.989	**0.921**	0.643	0.245
	$\boldsymbol{x}_{\mathrm{III}}$	1.000	1.000	1.000	1.000	0.995	0.952	0.746	0.369
	$\boldsymbol{x}_{\mathrm{IV}}$	1.000	1.000	0.997	0.968	0.815	0.441	0.124	0.020
成功概率 (样本量 5000)	$\boldsymbol{x}_{\mathrm{I}}$	1.000	1.000	1.000	0.982	0.860	0.528	0.170	0.024
	$\boldsymbol{x}_{\mathrm{II}}$	1.000	1.000	1.000	1.000	0.998	**0.894**	0.622	0.232
	$\boldsymbol{x}_{\mathrm{III}}$	1.000	1.000	1.000	1.000	1.000	0.928	0.730	0.374
	$\boldsymbol{x}_{\mathrm{IV}}$	1.000	1.000	1.000	0.968	0.796	0.442	0.110	0.012
成功概率 (样本量 3000)	$\boldsymbol{x}_{\mathrm{I}}$	1.000	1.000	1.000	0.980	0.837	0.543	0.133	0.027
	$\boldsymbol{x}_{\mathrm{II}}$	1.000	1.000	1.000	1.000	0.983	**0.887**	0.610	0.233
	$\boldsymbol{x}_{\mathrm{III}}$	1.000	1.000	1.000	1.000	0.997	0.937	0.710	0.350
	$\boldsymbol{x}_{\mathrm{IV}}$	1.000	1.000	1.000	0.970	0.780	0.450	0.093	0.013
成功概率 (样本量 1000)	$\boldsymbol{x}_{\mathrm{I}}$	1.000	1.000	1.000	0.990	0.930	0.650	0.220	0.020
	$\boldsymbol{x}_{\mathrm{II}}$	1.000	1.000	1.000	1.000	0.990	**0.950**	0.750	0.260
	$\boldsymbol{x}_{\mathrm{III}}$	1.000	1.000	1.000	1.000	0.990	0.980	0.830	0.410
	$\boldsymbol{x}_{\mathrm{IV}}$	1.000	1.000	1.000	0.980	0.920	0.530	0.110	0.020

注: 带有下划线的单元格表示无法区分的结果.

5.5.2 基于复合鲁棒性指标的资源回收系统的价值

接下来, 利用给定 $\theta_1 = \theta_2 = 1$ 下复合鲁棒性指标 (5.26) 建模资源回收系统, 并得到其最优决策 $\boldsymbol{x}^{\star}$. 在复合鲁棒性指标下, 最优资源回收系统设计决策 $\boldsymbol{x}^{\star}$ 是在原料不确定性集合 $V(\pi(\boldsymbol{x}^{\star})) \times U(\pi(\boldsymbol{x}^{\star}))$ 上保证利润目标 τ_0 的资源回收单元

设计决策. 进一步, 比较复合鲁棒性指标模型与最大化利润目标实现概率的概率模型:

$$\max_{\boldsymbol{x}\in\{0,1\}^J} \mathbb{P}\left[\phi\left(\boldsymbol{x};\boldsymbol{v},\boldsymbol{u}\right)\geqslant\tau_0\right] \tag{5.32}$$

之间的效果.

表 5.4 给出了复合鲁棒性指标模型和概率模型的比较结果. 结果表明: 对于较高的利润目标 τ_0, 概率模型和复合鲁棒性指标模型的鲁棒性水平较为接近. 但是, 当利润目标 τ_0 减小时, 概率模型的表现变差. 例如, 当 $\tau_0=300$ 时, 概率模型的解的鲁棒性水平仅为复合鲁棒性指标模型的三分之一. 本节还进行了样本外测试, 以评估两个模型在实现利润目标概率水平上的表现, 如表 5.4 所示. 在许多情况下, 特别是 300 到 600 之间的利润目标, 复合鲁棒性指标模型的回收系统在样本外的成功概率都高于概率模型的成功概率, 这进一步验证了复合鲁棒性处理原料不确定问题的有效性.

表 5.4 复合鲁棒性指标模型和概率模型的比较

模型与评价标准	$\tau_0=300$	400	500	600	700	800	900	1000
复合鲁棒性指标模型								
鲁棒性水平	0.623	0.573	0.522	0.445	0.422	0.374	0.327	0.280
平均成功概率 *	0.946	0.940	0.930	0.924	0.905	0.912	0.711	0.512
成功概率的标准差 **	0.51%	0.03%	0.26%	0.58%	1.00%	1.58%	10.58%	20.62%
概率模型								
鲁棒性水平	0.198	0.225	0.385	0.310	0.385	0.360	0.295	0.265
平均成功概率	0.863	0.811	0.922	0.879	0.954	0.932	0.853	0.656
成功概率的标准差	13.63%	11.96%	9.77%	8.16%	1.48%	2.46%	7.33%	17.06%

* “平均成功概率” 的每个值是一组实验的成功概率的平均值, 其中每个成功概率使用 1000 个具有相应回收系统的样本计算.

** “成功概率的标准差” 为给定回收系统计算的所有样本外成功概率的标准差.

5.6 本 章 小 结

本章利用两阶段选址优化模型结构介绍了考虑原料总量和组成成分不确定性下的资源回收问题, 构建了鲁棒性函数, 并给出了系统鲁棒性图来帮助决策者分析资源回收系统的经济可行性, 进一步给出了鲁棒性函数的高效求解形式. 在上述基础上, 介绍了基于复合鲁棒性指标的资源回收系统优化准则, 并构建了复合鲁棒性指标资源回收模型且给出模型的高效求解形式.

关于不确定环境下资源回收系统问题的研究, 读者可以参考更多相关文献. 例如, Zhu 和 Huang (2011) 将废弃物产生率建模为已知的离散分布, 并将城市生活废弃物回收系统设计问题建模为考虑需求和容量概率约束的随机线性规划模型, 并给出了迭代求解方法. Sun 等 (2012) 以及 Srivastava 和 Nema (2012) 利用均匀分布估计不确定变量, 并分别研究了单目标和多目标下的资源回收系统的容量规划问题. Šomplák 等 (2013) 在概率分布函数已知的假设下, 考虑了资源回收设施的容量选择问题. Xiong 等 (2016) 开发了一个概率模型, 用于优化废弃物产生率不确定性下分散式资源回收系统的经济可行性. Ghiani 等 (2014) 的综述研究探讨了不确定性环境下资源回收问题的建模方法.

第 6 章 Stackelberg 博弈与双层选址优化

第 5 章基于两阶段选址优化模型研究了复合鲁棒性指标的资源回收系统优化问题, 本章从协同的角度, 讨论在 PPP 框架下, 考虑带有地方政府 (local authority, LA) 与资源回收系统私人运营商 (private operator, PO) 的资源回收系统联合运营选址优化问题. 该问题考虑原料状态 (质量和组成成分) 的分布不确定性, 以及政府源头分类 (sorting-at-source scheme) 或初处理方案对原料状态不确定性的影响. 本章通过构造一个决策-依赖型分布不确定集 (decision-dependent ambiguity set) 来建模政府干预对不确定原料分布的影响, 并在分布鲁棒优化的框架下将上述问题建模为带有不确定分布的斯塔克尔伯格 (Stackelberg) 博弈模型. 通过此模型, 对不确定性环境下地方政府决策-能源运营商决策进行协同决策分析. 最后, 设计基于真实数据的半仿真实验来说明政府干预对保证能源回收企业可持续性运营的重要作用. 本章的主要结构如下: 6.1 节介绍 PPP 框架下资源回收系统优化问题的背景. 6.2 节介绍分布鲁棒双层 (bilevel) 资源回收系统选址模型并分析模型结构. 6.3 节介绍决策依赖型原料分布不确定集的构建. 6.4 节介绍分布鲁棒双层资源回收系统选址模型求解. 6.5 节介绍实验及结果分析.

6.1 背　　景

在第 5 章, 我们介绍了资源回收系统优化问题中原料质量和组成成分不确定性产生的原因及其决策影响. 在此基础上, 本章重点关注地方政府干预对原料状态不确定性及决策的影响. 在资源回收管理问题中, 地方政府和资源回收系统私人运营商之间通常建立公私合作伙伴关系 (public-private partnership). 这种伙伴关系通过不同的成本分摊机制, 决定不同的资源回收模式 (Song et al., 2013). 公私合作伙伴关系下资源回收系统优化问题主要包括两个关键阶段: 原料源头分类 (sorting-at-source) 和原料回收处理, 其中, 资源回收系统的建设成本由地方政府和私人运营商分摊. 地方政府通过制定预算, 确定分摊系统建设成本的比例, 并决定所有原料收集点的分类方案. 因此, 地方政府的问题是根据私人运营商的经济 (或财务) 状况, 决定原料的分类方案和建设成本预算水平, 以最大限度地减少地方政府支出. 基于地方政府的预算和分类方案, 私人运营商负责原料的回收处理和能源产品出售. 私人运营商通过制定资源回收设施的选址和原料分配决策, 最大化系统的净利润. 从优化建模的角度, 资源回收系统优化问题具有双层结构

(bilevel structure), 其中, 地方政府的分类方案选择和预算制定问题为第一层问题. 在给定第一层问题的决策后, 私人运营商面临两阶段回收系统设施选址和原料分配问题 (two-stage recovery system facility location and feedstock flow allocation problem), 这是第二层问题.

在资源回收系统问题中, 原料状态 (质量和组成成分) 的不确定性会影响资源回收系统的能源产品转化效率和盈利水平 (已在第 5 章详细分析). 此外, 在公私合作伙伴关系下, 原料状态直接受到地方政府分类方案的影响. 因此, 原料状态对分类方案的依赖性是公私合作伙伴关系下资源回收系统优化问题中一个关键特征及挑战. 例如, 如果原料本身能源转化效率高 (例如, 食物垃圾中的碳水化合物含量高), 则可以轻松实现资源回收系统的经济可行性 (economic feasibility). 因此, 地方政府无需对分类方案投入过多资金, 并制定较低的分摊建设成本预算. 此外, 如果原料能源转化效率低或地方政府预算有限, 私人运营商可能无法保证资源回收系统的经济可行性. 在这种情况下, 地方政府需要决定是制定更高的资源回收系统建设预算, 还是更多地投资于原料源头分类方案来提升原料的能源转化效率. 当原料状态高度不确定时, 上述问题的处理将更具挑战性. 因此, 原料状态对分类方案的依赖性是本章建模处理的重点.

6.2 分布鲁棒双层资源回收规划模型

本章考虑一个包括地方政府和私人运营商两方的资源回收系统优化运营问题. 假设双方计划对一项大规模资源回收计划进行公私合作. 该合作主要包括两个关键阶段: 原料源头分类和原料回收转化为能源产品. 图 6.1 给出一个生活垃圾回收系统示意图. 这项合作的一个核心因素是原料回收转化为能源产品阶段的生产效率受到原料源头分类质量的影响. 因此, 资源回收项目的成功实施取决于上述两个阶段的高效规划与管理.

在公私合作伙伴关系中, 地方政府决定原料源头分类方案, 且承担所有分类方案产生的费用. 此外, 地方政府还执行成本分摊政策, 以配比基金 (matching fund) 的形式, 由地方政府承担部分私人运营商建设资源回收设施的成本. 成本分摊在 PPP 实践中被认为是一种财务激励机制, 能够吸引私人运营商投资资源回收项目 (Cointreau-Levine, 1994; Sadka, 2007; Kim et al., 2015; Sharafi et al., 2018). 地方政府对成本分摊的预算, 被称为配比基金预算 (matching fund budget), 其一般由成本分摊比率 (cost-share ratio) 来衡量, 它影响资源回收系统的数量和类型, 进而影响私人运营商的资源回收能力 (resource recycling capability). 此外, 原料源头分类方案会影响原料的能源转化效率, 从而影响私人运营商的收益. 根据地方政府的成本分摊比率和分类方案, 私人运营商决定资源回收系统设施的位置和

处理能力 (或容量), 并管理回收系统的运营, 以最大化利润. 为了维护公私合作关系的稳定性, 双方的决策必须保证双方的经济可行性 (economic feasibility). 因此, 地方政府的问题是确定最优的原料源头分类方案和成本分摊比率, 保证私人运营商的资源回收系统是经济可行的 (私人运营商的利润应不小于某个目标水平[①]), 同时最小化自己的成本. 特别地, 地方政府在分类方案上投资越大, 则其成本分摊比率就越小, 反之亦然.

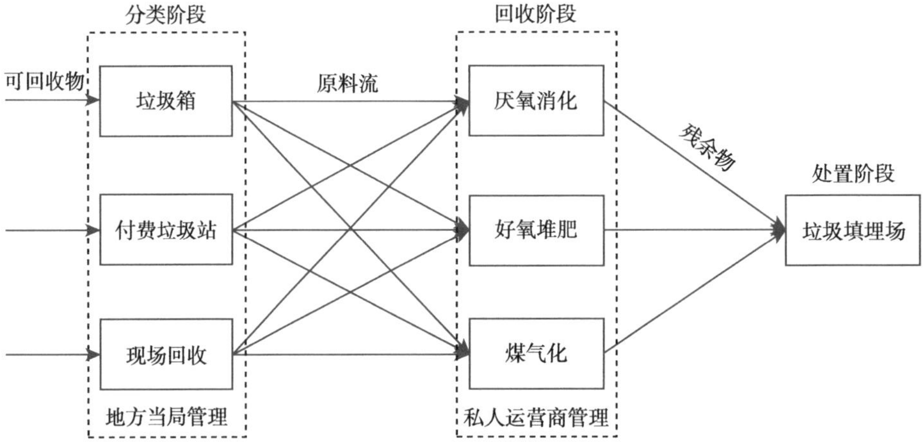

图 6.1　PPP 合作模式下 3×3 生活垃圾资源回收系统示意图

本章考虑地方政府只与一个私人运营商合作的公私合作伙伴关系, 其中该私人运营商是竞标 PPP 合同的多个合格投标企业中的一个. 基于上述公私合作伙伴关系, 考虑有 I 个原料收集点, H 种原料源头分类方案, J 个资源回收系统备选地址以及 N 种原料组成成分的资源回收系统优化问题. 我们记收集点 i 的原料质量为 $v_i, i\in[I]$, 且向量 $\boldsymbol{v}:=(v_i)_I$. 记收集点 i 处第 n 种原料组成成分的质量占总质量 v_i 的比例为 $u_{in}, i\in[I], n\in[N]$, 其中, $\boldsymbol{u}:=(u_{in})_{I\times N}$ 满足 $\sum_{n\in[N]}u_{in}=1,\forall i\in[I]$ 且 $u_{in}\geqslant 0,\forall i\in[I],n\in[N]$. 地方政府主要有两个决策变量: 原料分类方案的选择决策与成本分摊预算水平 (budget level) 决策. 我们记原料分类方案的选择决策为 $s_{ih}, i\in[I], h\in[H]$, 其为一个 0-1 二元决策变量, 若在原料收集点 i 使用第 h 种分类方案, 则 $s_{ih}=1$, 否则为 0. 记向量 $\boldsymbol{s}:=(s_{ih})_{I\times H}$. 我们记成本分摊预算水平为 ξ, 它是一个连续变量. 此外, 私人运营商的决策变量为资源回收设施选址决策和原料分配决策. 我们记资源回收设施选址决策为 $x_j, j\in[J]$, 其为一个 0-1 二元决策变量, 如果在第 j 个备选地址处建设资源回收系统设施, 则 $x_j=1$. 记向量 $\boldsymbol{x}:=(x_j)_J$. 最后, 记原料分配决策为

① 为了使 PPP 合同在市场上接受, 地方政府通常需要承诺为私人运营商提供财政支持, 以保证其投资回报. 私人运营商的目标水平可以从历史合同数据中估计 (Sharafi et al., 2018).

$y_{ij}, i \in [I], j \in [J]$, 其表示从收集点 i 运输到资源回收设施 j 处的原料质量, 并令向量 $\boldsymbol{y} := (y_{ij})_{I\times J}$. 接下来, 基于上述定义我们分别构建私人运营商问题与地方政府问题.

6.2.1 私人运营商: 资源回收设施选址与运营管理

给定资源回收设施选址决策 $\boldsymbol{x}$, 在收集区被分类处理过的原料将被分配到不同的资源回收设施进行回收处理. 在实际问题中, 私人运营商收益主要有两个来源: 一个是向原料收集者收取的服务费或倾倒费 (gate fee), 另一个是出售原料转化的能源产品所产生的收益. 特别地, 对于每个资源回收设施 $j \in [J]$, 其收益为

$$\sum_{i\in[I]} y_{ij}\varsigma_j + \sum_{i\in[I]}\sum_{n\in[N]} \mu_{jn}u_{in}y_{ij}, \tag{6.1}$$

其中, $\varsigma_j, j \in [J]$ 是在资源回收设施 j 处理每单位质量原料的倾倒费或服务费, $\mu_{jn}, j \in [J], n \in [N]$ 是资源回收设施 j 处理一个单位原料组成成分 n 获得的单位收益, 即能源产品的销售收益. 此外, 对于每个回收设施 $j \in [J]$, 其运营成本为

$$\epsilon\left(\sum_{i\in[I]} \vartheta_{ij}y_{ij}\right) + \sum_{i\in[I]} y_{ij}\imath_j + c_D\left(\sum_{i\in[I]}\sum_{n\in[N]} \sigma_{jn}u_{in}y_{ij}\right) + \sum_{i\in[I]}\sum_{n\in[N]} \eta_{jn}u_{in}y_{ij}, \tag{6.2}$$

它由运输成本、原料处理成本、残留物处置成本以及排放成本组成, 其中, 式 (6.2) 的第一项是总运输成本, ϵ 是每单位质量原料运输每单位距离的成本, ϑ_{ij} 是收集点 i 到资源回收设施 j 的距离. 第二项是总原料处理成本, 其中 $\imath_j$ 是资源回收设施 j 处理原料的单位成本. 第三项是残留物处置成本, 其中 c_D 为残留物的单位处置成本[①], σ_{jn} 是在资源回收设施 j 处理每单位质量原料组成成分 n 所产生的残留物比例. 最后一项是总排放成本, 其中 η_{jn} 是在资源回收设施 j 处理单位质量原料组成成分 n 产生的排放成本.

基于式 (6.1) 和式 (6.2), 私人运营商的总利润为

$$\sum_{j\in[J]}\left[\sum_{i\in[I]}(\varsigma_j - \epsilon\vartheta_{ij} - \imath_j)y_{ij} + \sum_{i\in[I]}\sum_{n\in[N]}(\mu_{jn} - c_D\sigma_{jn} - \eta_{jn})u_{in}y_{ij}\right]. \tag{6.3}$$

为了表述简洁, 我们定义 $r_{ijn} := \varsigma_j + \mu_{jn} - \epsilon\vartheta_{ij} - \imath_j - c_D\sigma_{jn} - \eta_{jn}$, 其表示资源回收设施 j 处理从收集点 i 分配的原料组成成分 n 的单位利润. 因为

① 在实际问题中, 未处理的原料和处理原料产生的废物等系统残留物一般通过特定的残留物处理企业进行焚烧或填埋处理, 这导致私人运营商要付出残留物处置成本.

$\sum_{n\in[N]} u_{in} = 1, \forall i \in [I]$, 所以我们可以将式 (6.3) 等价地写成如下形式:

$$\sum_{i\in[I]}\sum_{j\in[J]}\sum_{n\in[N]} r_{ijn}u_{in}y_{ij}. \tag{6.4}$$

给定的设施选址 $\boldsymbol{x}$ 和原料状态 $(\boldsymbol{v},\boldsymbol{u})$, 私人运营商的利润最大化问题如下:

$$\begin{aligned}
\phi(\boldsymbol{x},\boldsymbol{v},\boldsymbol{u}) := \max_{\boldsymbol{y}} \quad & \sum_{i\in[I]}\sum_{j\in[J]}\sum_{n\in[N]} r_{ijn}u_{in}y_{ij} - c_D\sum_{i\in[I]}\Big(v_i - \sum_{j\in[J]} y_{ij}\Big) \\
\text{s.t.} \quad & \sum_{j\in[J]} y_{ij} \leqslant v_i, \forall i\in[I], \\
& \sum_{i\in[I]} y_{ij} \leqslant x_j\varpi_j, \forall j\in[J], \\
& y_{ij} \geqslant 0, \forall i\in[I], j\in[J],
\end{aligned} \tag{6.5}$$

其中, 目标函数的第一项为回收处理原料的利润, 第二项为未处理原料的处置成本 $\sum_{i\in[I]}(v_i - \sum_{j\in[J]} y_{ij})$. 此外, 第一组约束确保由收集点 $i\in[I]$ 运输到不同资源回收设施的原料总质量不能超过收集点 $i\in[I]$ 的原料总质量 $v_i, i\in[I]$. 第二组约束确保由资源回收设施 $j\in[J]$ 处理的总原料质量不超过其容量 (capacity) 或处理能力 ϖ_j.

接下来, 假设地方政府成本分摊预算 ξ 和原料分类方案 $\boldsymbol{s}$ 已确定, 其中 ξ 是地方政府能够给予私人运营商建设资源回收设施的最大成本分摊金额. 此外, 我们假设原料状态 $(\tilde{\boldsymbol{v}},\tilde{\boldsymbol{u}})$ 的概率分布是不确定的, 且受分类方案 $\boldsymbol{s}$ 的影响. 特别地, 原料状态概率分布 $\mathbb{P}$ 属于一个分类方案依赖分布不确定集 (sorting-dependent ambiguity set) $\mathbb{F}(\boldsymbol{s})$, 即 $(\tilde{\boldsymbol{v}},\tilde{\boldsymbol{u}}) \sim \mathbb{P} \in \mathbb{F}(\boldsymbol{s})$, 将在后续介绍. 给定 $(\xi,\boldsymbol{s})$, 构建如下不确定原料状态下私人运营商资源回收设施选址问题:

$$\mathcal{Q}^{\lambda}_{\mathbb{F}(\boldsymbol{s})}(\xi) := \max_{\boldsymbol{x}}\left\{\inf_{\mathbb{P}\in\mathbb{F}(\boldsymbol{s})} \mathbb{E}_{\mathbb{P}}\Big[\phi(\boldsymbol{x},\tilde{\boldsymbol{v}},\tilde{\boldsymbol{u}})\Big] - \lambda\boldsymbol{c}^{\top}\boldsymbol{x} \;\middle|\; (1-\lambda)\boldsymbol{c}^{\top}\boldsymbol{x} \leqslant \xi, \boldsymbol{x}\in\{0,1\}^J\right\}, \tag{6.6}$$

其中, $\boldsymbol{c} := (c_j)_J$ 是建设资源回收设施的固定成本, $\phi(\boldsymbol{x},\tilde{\boldsymbol{v}},\tilde{\boldsymbol{u}})$ 是给定 $\boldsymbol{x}$ 和 $(\tilde{\boldsymbol{v}},\tilde{\boldsymbol{u}})$ 下问题 (6.5) 的最优值. 此外, $\lambda\in[0,1]$ 为成本分摊比例, 其为私人运营商承担的资源回收设施建设成本比例, 而 $(1-\lambda)$ 是地方政府需要承担的比例. 对于私人运营商在建设资源回收设施上花费的每一美元, 地方政府将等比支付 $\dfrac{1-\lambda}{\lambda}$ 美元. 因此, 如果私人运营商承担的建设费用为 $\lambda\boldsymbol{c}^{\top}\boldsymbol{x}$, 则地方政府的分摊费用为 $\left(\dfrac{1-\lambda}{\lambda}\right)\lambda\boldsymbol{c}^{\top}\boldsymbol{x} = (1-\lambda)\boldsymbol{c}^{\top}\boldsymbol{x}$. 因为地方政府的成本分摊上限是 ξ, 所以 $(1-\lambda)\boldsymbol{c}^{\top}\cdot\boldsymbol{x}\leqslant\xi$.

在问题 (6.6) 中, 私人运营商的目标是最大化分布不确定集 $\mathbb{F}(\boldsymbol{s})$ 中极值概率分布下的期望总利润. 该问题是一个两阶段分布鲁棒混合整数规划问题 (Bertsimas et al., 2019), 并且基于决策依赖分布不确定集, 其中设施选址决策 $\boldsymbol{x}$ 是在观测到原料状态 $(\tilde{\boldsymbol{v}}, \tilde{\boldsymbol{u}})$ 之前做出的第一阶段决策, 而原料分配决策 $\boldsymbol{y}$ 是观测到原料状态 $(\tilde{\boldsymbol{v}}, \tilde{\boldsymbol{u}})$ 后的第二阶段决策. 问题 (6.6) 与第 3 章和第 4 章的两阶段选址问题有类似的两阶段结构.

6.2.2 地方政府: 分类方案制定与成本分摊

地方政府主要负责原料在收集点分类方案的选择与成本分摊预算水平 ξ 的制定, 其中, 我们定义 $s_{ih}, i \in [I], h \in [H]$ 为分类方案选择决策变量, 如果在收集点 i 选择方案 h, 则 $s_{ih} = 1$, 否则 $s_{ih} = 0$. 此外, 本章假设每个收集点只选择一种原料分类方案[①], 则原料分类方案选择决策变量 $\boldsymbol{s}$ 的可行域为

$$\boldsymbol{s} \in \mathcal{S} := \left\{ \boldsymbol{s} \in \{0,1\}^{I \times H} \,\middle|\, \sum_{h \in [H]} s_{ih} = 1, \forall i \in [I] \right\}.$$

给定原料质量 $\boldsymbol{v}$ 和分类方案 $\boldsymbol{s}$, 则分类成本 $\ell(\boldsymbol{s}, \boldsymbol{v})$ 为

$$\ell(\boldsymbol{s}, \boldsymbol{v}) := \sum_{i \in [I]} \sum_{h \in [H]} (\varrho_{ih} v_i + \iota_{ih}) s_{ih},$$

其中, ϱ_{ih} 是每单位原料的分类成本, ι_{ih} 是在收集点 i 实施原料分类方案 $h \in [H]$ 的成本.

除了选择原料分类方案, 地方政府还要决定与私人运营商分摊成本的预算水平 ξ. 地方政府决策问题是一个鲁棒双层资源回收规划 (robust bilevel resource recovery planning, BRRP) 问题:

$$\min_{\xi, \boldsymbol{s}} \left\{ \xi + \sup_{\mathbb{P} \in \mathbb{F}(\boldsymbol{s})} \mathbb{E}_{\mathbb{P}} \left[\ell(\boldsymbol{s}, \tilde{\boldsymbol{v}}) \right] \,\middle|\, \mathcal{Q}^{\lambda}_{\mathbb{F}(\boldsymbol{s})}(\xi) \geqslant \tau, \boldsymbol{s} \in \mathcal{S}, \xi \in \mathbb{R}_+ \right\}, \tag{6.7}$$

其中, $\mathcal{Q}^{\lambda}_{\mathbb{F}(\boldsymbol{s})}(\xi)$ 为给定地方政府决策 $(\xi, \boldsymbol{s})$ 和成本分摊比率 λ 下的问题 (6.6). 地方政府的目标是最小化分摊成本预算与极值概率分布下期望分类成本之和, 并保证私人运营商的经济可行性:

$$\mathcal{Q}^{\lambda}_{\mathbb{F}(\boldsymbol{s})}(\xi) \geqslant \tau. \tag{6.8}$$

一方面, 如果地方政府将分摊成本预算 ξ 设定太低, 则私人运营商无法建设足够的资源回收设施以保证经济可行性, 即满足约束 (6.8). 另一方面, 如果地方

① 一般原料分类方案可以由多种相同或不同的分类技术组合而成.

政府选择了太差的 (或廉价的) 原料分类方案 $\boldsymbol{s}$, 这会导致原料回收效率过低并增加私人运营商的期望运营成本, 使其无法实现经济可行性. 因此, 问题 (6.7) 的关键是地方政府在两种决策上的权衡.

此外, 我们假设地方政府对分布不确定性持厌恶的态度, 即分布不确定性厌恶 (ambiguity-averse) 决策者, 这可以提升不确定性环境下决策的鲁棒性, 有利于保证资源回收系统的经济可行性. 最后, 问题 (6.7) 可以被看作一个 Stackelberg 博弈模型或领导者-跟随者 (leader-follower) 博弈模型, 其中, 地方政府是决定最优预算 $\xi^\star$ 和原料分类方案 $\boldsymbol{s}^\star$ 的领导者. 私人运营商是跟随者, 在给定 $\xi^\star$ 的情况下, 确定最优的资源回收设施选址决策 $\boldsymbol{x}^\star$, 并利用分类方案 $\boldsymbol{s}^\star$ 处理后的原料进行能源产品转化以获得收益.

6.2.3　模型参数分析

1. 针对成本分摊比率 λ 的参数分析

本节首先介绍模型 (6.6)-(6.7) 关于成本分摊比率 λ 的分析. 不同的 $\lambda \in [0,1]$ 取值在实际问题中可以被看作不同类型的 PPP 合约.

当 $\lambda = 1$ 时, 私人运营商承担所有资源回收设施建设成本. 这对应于 PPP 合同中的建设-经营-转让 (build-operate-transfer, BOT) 合同. 特别地, 当 $\lambda = 1$ 时, 问题 (6.7) 中的变量 ξ 可以设为 0, 因为约束 $(1-\lambda)\boldsymbol{c}^\top \boldsymbol{x} \leqslant \xi$ 在问题 (6.6) 中是冗余的.

当 $\lambda = 0$ 时, 地方政府将承担全部资源回收设施的建设成本. 这对应集中管理和控制 (central management and control, CMC) 的情况. 此时, 地方政府将拥有整个资源回收系统的所有权. 同时, 地方政府与私人运营商签订管理合同, 即私人运营商在 PPP 合作中只负责资源回收系统的运营与维护. 在这种情况下, 问题 (6.6) 和 (6.7) 都变成了地方政府问题.

当 $\lambda \in (0,1)$ 时, 地方政府提供分摊成本预算 ξ 来承担部分私人运营商建设资源回收系统的成本, 因此, 地方政府承担 $(1-\lambda)$ 比例的建设成本. 这与 PPP 合作中的特许权协议 (concession agreement, CA) 合同相对应, 此合同允许私人运营商根据地方政府的独家许可 (exclusive license), 自行承担部分设施建设成本和资源回收系统运营风险.

命题 6.1　假设 $0 \leqslant \lambda_1 < \lambda_2 < 1$, 令 $(\boldsymbol{s}_1^\star, \xi_1^\star)$ 与 $(\boldsymbol{s}_2^\star, \xi_2^\star)$ 是问题 (6.7) 分别对应 λ_1 与 λ_2 的最优解, 则有

$$\sup_{\mathbb{P} \in \mathbb{F}(\boldsymbol{s}_2^\star)} \mathbb{E}_{\mathbb{P}}\left[\ell(\boldsymbol{s}_2^\star, \tilde{\boldsymbol{v}})\right] \geqslant \sup_{\mathbb{P} \in \mathbb{F}(\boldsymbol{s}_1^\star)} \mathbb{E}_{\mathbb{P}}\left[\ell(\boldsymbol{s}_1^\star, \tilde{\boldsymbol{v}})\right] + \left[\xi_1^\star - \xi_2^\star\left(\frac{1-\lambda_1}{1-\lambda_2}\right)\right]. \tag{6.9}$$

上述结果 (6.9) 表明当私人运营商承担更高比率的资源回收设施建设成本时

(从 λ_1 到 λ_2), 如果资源运营商的最佳选址决策不会产生比之前更多的建设成本, 那么地方政府就需要付出更多的原料源头分类成本, 即

$$\boldsymbol{c}^\top \boldsymbol{x}_1^\star \geqslant \boldsymbol{c}^\top \boldsymbol{x}_2^\star, \quad \text{或} \quad \xi_1^\star - \xi_2^\star \left(\frac{1-\lambda_1}{1-\lambda_2}\right) \geqslant 0.$$

在上述情况下, 虽然地方政府可以减少建设成本的投入, 但是他必须在分类方案上投入更多的资金以保证系统的经济可行性. 这表明, 当地方政府承担较少比例的建设成本时, 他必须在原料源头分类方案上投入更多的资金以提高原料的能源转化率, 来帮助私人运营商提升资源回收系统的性能, 以确保满足系统的经济可行性约束.

2. 中央控制对等问题

通过分析 BRRP 问题 (6.7) 的结构, 政府问题的最优解可以通过求解如下优化问题得到:

$$\begin{aligned} \min_{\boldsymbol{s},\boldsymbol{x}} \quad & (1-\lambda)\boldsymbol{c}^\top\boldsymbol{x} + \sup_{\mathbb{P}\in\mathbb{F}(\boldsymbol{s})} \mathbb{E}_{\mathbb{P}}\left[\ell(\boldsymbol{s},\tilde{\boldsymbol{v}})\right] \\ \text{s.t.} \quad & \mathbb{E}_{\mathbb{P}}\left[\phi\left(\boldsymbol{x},\tilde{\boldsymbol{v}},\tilde{\boldsymbol{u}}\right)\right] - \lambda\boldsymbol{c}^\top\boldsymbol{x} \geqslant \tau, \ \forall\mathbb{P}\in\mathbb{F}(\boldsymbol{s}), \\ & \boldsymbol{s}\in\mathcal{S}, \boldsymbol{x}\in\{0,1\}^J, \end{aligned} \tag{6.10}$$

这里, 我们称上述问题 (6.10) 是问题 (6.6)-(6.7) 的一个中央控制对等问题 (central control counterpart), 即假设有一个中央控制者 (central controller) 来处理设施选址和分类方案选择决策. 问题 (6.6)-(6.7) 最优解的值和对等问题 (6.10) 最优解的值相等是经济可行性约束 (6.8) 所导致的, 它保证私人运营商最优期望利润不少于 τ. 当将经济可行性约束 (6.8) 加入到地方政府问题 (6.7) 中时, 可以保证私人运营商的设施选址决策 $\boldsymbol{x}$ 都是经济可行的. 此外, 地方政府只需要根据设施选址决策 $\boldsymbol{x}$ 设定分摊成本预算 $\xi = (1-\lambda)\boldsymbol{c}^\top\boldsymbol{x}$, 以使其总成本最小化. 这便是问题 (6.10) 目标函数中的第一项.

定理 6.1 给定 λ, 令 $(\boldsymbol{s}^\star, \boldsymbol{x}^\star)$ 是问题 (6.10) 的最优解, 则 $(\boldsymbol{s}^\star, \xi^\star = (1-\lambda)\boldsymbol{c}^\top\boldsymbol{x}^\star)$ 是地方政府问题 (6.7) 的最优解. 对于私人运营商问题 (6.6), 我们有

(1) 如果 $\boldsymbol{c}^\top\boldsymbol{x} \neq \boldsymbol{c}^\top\boldsymbol{x}', \forall\ \boldsymbol{x},\boldsymbol{x}' \in \{0,1\}^J, \boldsymbol{x}\neq\boldsymbol{x}'$, 则 $\boldsymbol{x}^{\star\star} = \boldsymbol{x}^\star$ 是私人运营商在问题 (6.6)-(6.7) 的最优解.

(2) 一般来说, 私人运营商问题的最优解 $\boldsymbol{x}^{\star\star}$ 可以通过求解如下两阶段分布鲁棒优化问题来得到

$$\boldsymbol{x}^{\star\star} \in \arg\max_{\boldsymbol{x}} \left\{ -\lambda\boldsymbol{c}^\top\boldsymbol{x} + \inf_{\mathbb{P}\in\mathbb{F}(\boldsymbol{s}^\star)} \mathbb{E}_{\mathbb{P}}\Big[\phi\left(\boldsymbol{x},\tilde{\boldsymbol{v}},\tilde{\boldsymbol{u}}\right)\Big] \;\Big|\; \boldsymbol{c}^\top\boldsymbol{x} \leqslant \boldsymbol{c}^\top\boldsymbol{x}^\star,\ \boldsymbol{x}\in\{0,1\}^J \right\}. \tag{6.11}$$

3. 最优成本分摊比率 λ

最后, 考虑地方政府可以优化成本分摊比率 λ 的情况, 即

$$\min_{\lambda\in[0,1]}\min_{\xi,\boldsymbol{s}}\left\{\xi+\sup_{\mathbb{P}\in\mathbb{F}(\boldsymbol{s})}\mathbb{E}_{\mathbb{P}}\left[\ell(\boldsymbol{s},\tilde{\boldsymbol{v}})\right]\;\middle|\;\mathcal{Q}^{\lambda}_{\mathbb{F}(\boldsymbol{s})}(\xi)\geqslant\tau,\boldsymbol{s}\in\mathcal{S},\xi\in\mathbb{R}_+\right\},\tag{6.12}$$

注意到, 通过成本分摊约束 $(1-\lambda)\boldsymbol{c}^\top\boldsymbol{x}\leqslant\xi$, 变量 λ 被加入到问题 $\mathcal{Q}^{\lambda}_{\mathbb{F}(\boldsymbol{s})}(\xi)$ 中. 接下来, 我们定义地方政府问题的一个等价新问题:

$$\min_{\xi_{\rm LA},\boldsymbol{s}}\left\{\xi_{\rm LA}+\sup_{\mathbb{P}\in\mathbb{F}(\boldsymbol{s})}\mathbb{E}_{\mathbb{P}}\left[\ell(\boldsymbol{s},\tilde{\boldsymbol{v}})\right]\;\middle|\;\mathcal{Q}_{\mathbb{F}(\boldsymbol{s})}(\xi_{\rm LA})\geqslant\tau,\boldsymbol{s}\in\mathcal{S},\xi_{\rm LA}\in\mathbb{R}_+\right\},\tag{6.13}$$

以及给定 $\xi_{\rm LA}$ 下私人运营商问题的一个等价新问题 $\mathcal{Q}_{\mathbb{F}(\boldsymbol{s})}(\xi_{\rm LA})$:

$$\begin{aligned}\mathcal{Q}_{\mathbb{F}(\boldsymbol{s})}(\xi_{\rm LA}):=\max_{\boldsymbol{x},\xi_{\rm PO}}\Bigg\{&\inf_{\mathbb{P}\in\mathbb{F}(\boldsymbol{s})}\mathbb{E}_{\mathbb{P}}\left[\phi\left(\boldsymbol{x},\tilde{\boldsymbol{v}},\tilde{\boldsymbol{u}}\right)\right]-\xi_{\rm PO}\;\Bigg|\;\boldsymbol{c}^\top\boldsymbol{x}-\xi_{\rm PO}\\&\leqslant\xi_{\rm LA},\boldsymbol{x}\in\{0,1\}^J,\xi_{\rm PO}\in\mathbb{R}_+\Bigg\},\end{aligned}\tag{6.14}$$

其中, 变量 $\xi_{\rm LA}$ 和 $\xi_{\rm PO}$ 分别是地方政府和私人运营商对应的设施建设分摊成本. 我们可以利用问题 (6.13)-(6.14) 的最优解来获得问题 (6.12) 的最优解 $\lambda^\star$. 同样地, 我们可以构建问题 (6.12) 或其等价问题 (6.13)-(6.14) 的中央控制对等问题:

$$\begin{aligned}\min_{\boldsymbol{s},\boldsymbol{x},\xi_{\rm LA},\xi_{\rm PO}}\quad&\xi_{\rm LA}+\sup_{\mathbb{P}\in\mathbb{F}(\boldsymbol{s})}\mathbb{E}_{\mathbb{P}}\left[\ell(\boldsymbol{s},\tilde{\boldsymbol{v}})\right]\\\text{s.t.}\quad&\mathbb{E}_{\mathbb{P}}\left[\phi\left(\boldsymbol{x},\tilde{\boldsymbol{v}},\tilde{\boldsymbol{u}}\right)\right]-\xi_{\rm PO}\geqslant\tau,\ \forall\mathbb{P}\in\mathbb{F}(\boldsymbol{s}),\\&\boldsymbol{c}^\top\boldsymbol{x}\leqslant\xi_{\rm LA}+\xi_{\rm PO},\\&\boldsymbol{s}\in\mathcal{S},\boldsymbol{x}\in\{0,1\}^J,\ \xi_{\rm LA},\xi_{\rm PO}\in\mathbb{R}_+.\end{aligned}\tag{6.15}$$

命题 6.2 令 $(\boldsymbol{s}^\star,\boldsymbol{x}^\star,\xi^\star_{\rm LA},\xi^\star_{\rm PO})$ 是问题 (6.15) 的最优解, 则我们有

$$\left(\boldsymbol{s}^\star,\xi^\star=\xi^\star_{\rm LA},\lambda^\star=\frac{\xi^\star_{\rm PO}}{\xi^\star_{\rm LA}+\xi^\star_{\rm PO}}\right)$$

是地方政府问题 (6.12) 的最优解. 给定 $\lambda^\star$ 后, 私人运营商的最优选址决策 $\boldsymbol{x}^{\star\star}$ 可以根据定理 6.1 来获得.

6.3 分类方案依赖型原料状态分布不确定集

本节主要介绍分类方案依赖型原料状态分布不确定集的构建. 在实际问题中, 原料源头分类方案的选择会影响原料状态 $(\tilde{\boldsymbol{v}},\tilde{\boldsymbol{u}})$ 的概率分布特征. 首先, 我们考

虑原料质量 $\tilde{\boldsymbol{v}}$ 对分类方案的依赖性. 特别地, 分类方案对原料质量的影响与分类方案实施下的原料分类参与率 (participation rate) 有关. 例如, 对广告和原料分类激励政策的投资可以提高当地居民参与原料源头分类活动的意识和兴趣 (Omran et al., 2009; Tai et al., 2011). 与原料组成成分相比, 原料质量对分类方案的敏感性相对较低. 因此, 我们主要考虑平均原料质量对分类方案的依赖性. 此外, 假设在所有分类方案下原料质量支撑集相同. 给定分类方案 $\boldsymbol{s}$, 期望原料质量为

$$\mathbb{E}_{\mathbb{P}}\left[\tilde{v}_i\right] \in \left[\boldsymbol{\alpha}_i^{\top}\boldsymbol{s}_i, \boldsymbol{\beta}_i^{\top}\boldsymbol{s}_i\right], \quad \forall i \in [I], \tag{6.16}$$

其中, $\boldsymbol{\alpha}_i = (\alpha_{i1}, \cdots, \alpha_{iH})^{\top}$ 以及 $\boldsymbol{\beta}_i = (\beta_{i1}, \cdots, \beta_{iH})^{\top}$, 每对 α_{ih} 和 β_{ih} 是在收集点 i 实施原料分类方案 h 时原料质量上下限的估计值, 即

$$\alpha_{ih} \leqslant \mathbb{E}_{\mathbb{P}}\left[\tilde{v}_i\right] \leqslant \beta_{ih}, \quad \forall i \in [I], h \in [H].$$

此外, 利用历史原料质量数据, 可以构建 (已被源头分类的) 原料质量 $\tilde{\boldsymbol{v}}$ 的支撑集 $\mathcal{V}$:

$$\mathcal{V} := \left\{\boldsymbol{v} \in \mathbb{R}^I : \boldsymbol{v} = \boldsymbol{v}^0 + \boldsymbol{\Delta}, \forall \boldsymbol{\Delta} \in \operatorname{conv}\left\{\boldsymbol{\Delta}^k = (\Delta_1^k, \cdots, \Delta_I^k), k \in [K]\right\}\right\}. \tag{6.17}$$

特别地, $\boldsymbol{v}^0$ 是所有收集点处最新的原料质量观测值, Δ_i^k 是收集点 i 相邻时间段历史原料质量观测值之间的差值①, K 是观测值的样本量, 而 $\boldsymbol{\Delta}$ 是不确定变量且属于点集 $\left\{\boldsymbol{\Delta}^k, k \in [K]\right\}$ 的凸包.

原料组成成分 $\boldsymbol{u}$ 对分类方案 $\boldsymbol{s}$ 的依赖性可以通过分类过程中实现的分离纯度水平 (separation purity level) 来体现. 例如, 为了实现原料组成成分分类的良好效果, 需要投资高效的分离和处置技术以及员工培训 (Dahlén et al., 2007; Gundupalli et al., 2017; Malik et al., 2015). 因此, 将不确定原料组成成分建模为如下形式:

$$\mathbb{E}_{\mathbb{P}}\left[\boldsymbol{G}\tilde{\boldsymbol{u}}_i\right] \geqslant \boldsymbol{F}\boldsymbol{s}_i, \quad \forall i \in [I], \quad \mathbb{P}\left[\tilde{\boldsymbol{u}} \in \mathcal{U}(\boldsymbol{s})\right] = 1, \tag{6.18}$$

这里, 支撑集 $\mathcal{U}(\boldsymbol{s})$ 与原料分类方案 $\boldsymbol{s}$ 相关, 其定义如下:

$$\mathcal{U}(\boldsymbol{s}) := \left\{\boldsymbol{u} \in \mathbb{R}_+^{I \times N} \middle| \boldsymbol{A}\boldsymbol{u}_i \geqslant \boldsymbol{B}\boldsymbol{s}_i, \forall i \in [I]; \sum_{n \in [N]} u_{in} = 1, \forall i \in [I]\right\}. \tag{6.19}$$

其中, $\boldsymbol{G} \in \mathbb{R}_+^{L_1 \times N}$, $\boldsymbol{F} \in \mathbb{R}_+^{L_1 \times H}$, $\boldsymbol{A} \in \mathbb{R}_+^{L_2 \times N}$ 和 $\boldsymbol{B} \in \mathbb{R}_+^{L_2 \times H}$ 为系数矩阵.

---

① 例如, 取 $\Delta_i^k = v_i^t - v_i^{t-1}$, 其中 $t = k$ 是不同历史时间段.

基于上述关于原料状态 $(\tilde{\boldsymbol{v}}, \tilde{\boldsymbol{u}})$ 概率分布特征模式 (6.16)-(6.19), 构建如下分类方案依赖型原料状态分布不确定集:

$$\mathbb{F}(\boldsymbol{s}) := \left\{ \mathbb{P} \in \mathcal{P}\left(\mathbb{R}_+^I \times \mathbb{R}_+^{I\times N}\right) \left| \begin{array}{l} (\tilde{\boldsymbol{v}}, \tilde{\boldsymbol{u}}) \sim \mathbb{P}, \\ \mathbb{E}_{\mathbb{P}}\left[\tilde{v}_i\right] \in \left[\boldsymbol{\alpha}_i^\top \boldsymbol{s}_i, \boldsymbol{\beta}_i^\top \boldsymbol{s}_i\right], \ \forall i \in [I], \\ \mathbb{E}_{\mathbb{P}}\left[\boldsymbol{G}\tilde{\boldsymbol{u}}_i\right] \geqslant \boldsymbol{F}\boldsymbol{s}_i, \ \forall i \in [I], \\ \mathbb{P}\left[\tilde{\boldsymbol{v}} \in \mathcal{V}, \tilde{\boldsymbol{u}} \in \mathcal{U}(\boldsymbol{s})\right] = 1 \end{array} \right. \right\}. \tag{6.20}$$

6.4　分布鲁棒双层资源回收规划模型的高效计算形式

本节介绍利用问题 (6.10) 构建基于分布不确定集 (6.20) 的分布鲁棒双层资源回收规划模型 (6.6)-(6.7) 的高效计算形式. 首先, 介绍在分布不确定集 $\mathbb{F}(\boldsymbol{s})$ 中极值概率分布下, 地方政府的期望原料分类成本与私人运营商的期望资源回收利润对应的等价高效计算问题. 其次, 构建问题 (6.6)-(6.7) 对应的混合整数线性规划问题. 对于问题 (6.10), 有以下几个计算难点.

首先, 问题 (6.10) 目标函数的估计. 它包含地方政府极值概率分布下的期望分类成本, 即

$$\sup_{\mathbb{P} \in \mathbb{F}(\boldsymbol{s})} \mathbb{E}_{\mathbb{P}}\left[\ell(\boldsymbol{s}, \tilde{\boldsymbol{v}})\right]. \tag{6.21}$$

其次, 问题 (6.10) 约束条件的估计. 它包含私人运营商极值概率分布下的期望回收收益:

$$\begin{aligned} &\inf_{\mathbb{P} \in \mathbb{F}(\boldsymbol{s})} \mathbb{E}_{\mathbb{P}}\left[\phi\left(\boldsymbol{x}, \tilde{\boldsymbol{v}}, \tilde{\boldsymbol{u}}\right)\right] \\ =& \inf_{\mathbb{P} \in \mathbb{F}(\boldsymbol{s})} \mathbb{E}_{\mathbb{P}}\left[\max_{\boldsymbol{y} \in \mathcal{Y}(\boldsymbol{x}, \boldsymbol{v})} \sum_{i \in [I]} \sum_{j \in [J]} \sum_{n \in [N]} r_{ijn} \tilde{u}_{in} y_{ij} - c_D \sum_{i \in [I]} \left(\tilde{v}_i - \sum_{j \in [J]} y_{ij}\right)\right], \end{aligned} \tag{6.22}$$

其中,

$$\mathcal{Y}(\boldsymbol{x}, \boldsymbol{v}) = \left\{ \boldsymbol{y} \in \mathbb{R}_+^{I\times J} \left| \begin{array}{l} \sum\limits_{j \in [J]} y_{ij} \leqslant v_i, \ \forall i \in [I], \\ \sum\limits_{i \in [I]} y_{ij} \leqslant x_j \varpi_j, \ \forall j \in [J], \\ y_{ij} \geqslant 0, \ \forall i \in [I], j \in [J] \end{array} \right. \right\}. \tag{6.23}$$

式 (6.22) 是在给定 $(\boldsymbol{s}, \boldsymbol{x})$ 下私人运营商极值概率分布下的期望回收收益. 此外, 根据定理 6.1, 给定问题 (6.10) 的最优解 $(\boldsymbol{s}^\star, \boldsymbol{x}^\star)$, 还需要通过求解如下自适应分布鲁棒优化问题来得到私人运营商的最优选址决策 $\boldsymbol{x}^{\star\star}$, 即

$$\max_{\boldsymbol{x}\in\{0,1\}^J} \left\{ \inf_{\mathbb{P}\in\mathbb{F}(\boldsymbol{s})} \mathbb{E}_{\mathbb{P}}\left[\phi\left(\boldsymbol{x}, \tilde{\boldsymbol{v}}, \tilde{\boldsymbol{u}}\right)\right] - \lambda \boldsymbol{c}^\top \boldsymbol{x} \;\middle|\; \boldsymbol{c}^\top \boldsymbol{x} \leqslant \boldsymbol{c}^\top \boldsymbol{x}^\star \right\}. \tag{6.24}$$

最后, 整体问题 (6.10) 的求解. 因为分布不确定集 $\mathbb{F}(\boldsymbol{s})$ 受原料分类决策 $\boldsymbol{s}$ 的影响, 所以整个问题 (6.10) 本质上是一个决策依赖的自适应分布鲁棒优化问题. 求解此问题需要权衡极值概率分布下的期望成本或收入与决策 (分类方案 $\boldsymbol{s}$) 对不确定原料状态概率分布的影响.

6.4.1 极值概率分布下地方政府的期望分类成本

因为极值概率分布下期望分类成本只与原料质量 $\tilde{\boldsymbol{v}}$ 不确定性相关, 所以将分布不确定集 $\mathbb{F}(\boldsymbol{s})$ 从 $(\tilde{\boldsymbol{v}}, \tilde{\boldsymbol{u}})$ 投影到 $\tilde{\boldsymbol{v}}$ 上, 并构建投影集合 $\mathbb{F}_{\tilde{\boldsymbol{v}}}(\boldsymbol{s})$:

$$\mathbb{F}_{\tilde{\boldsymbol{v}}}(\boldsymbol{s}) := \left\{ \mathbb{P} \in \mathcal{P}\left(\mathbb{R}_+^I\right) \;\middle|\; \begin{array}{l} \tilde{\boldsymbol{v}} \sim \mathbb{P}, \\ \mathbb{E}_{\mathbb{P}}\left[\tilde{v}_i\right] \in \left[\boldsymbol{\alpha}_i^\top \boldsymbol{s}_i, \boldsymbol{\beta}_i^\top \boldsymbol{s}_i\right], \forall i \in [I]; \mathbb{P}\left[\tilde{\boldsymbol{v}} \in \mathcal{V}\right] = 1 \end{array} \right\}.$$

基于上述集合 $\mathbb{F}_{\tilde{\boldsymbol{v}}}(\boldsymbol{s})$, 问题 (6.21) 可等价地表示为 $\sup_{\mathbb{P}\in\mathbb{F}(\boldsymbol{s})} \mathbb{E}_{\mathbb{P}}\left[\ell(\boldsymbol{s}, \tilde{\boldsymbol{v}})\right] = \sup_{\mathbb{P}\in\mathbb{F}_{\tilde{\boldsymbol{v}}}(\boldsymbol{s})} \mathbb{E}_{\mathbb{P}}\left[\ell(\boldsymbol{s}, \tilde{\boldsymbol{v}})\right]$.

命题 6.3 给定原料分类方案 $\boldsymbol{s}$, 极值概率分布下地方政府的期望分类问题 (6.21) 可以等价地转为如下线性规划问题:

$$\max_{\boldsymbol{v}} \left\{ \sum_{i\in[I]} \sum_{h\in[H]} (\varrho_{ih} v_i + \iota_{ih}) s_{ih} \;\middle|\; \begin{array}{l} v_i \in \left[\boldsymbol{\alpha}_i^\top \boldsymbol{s}_i, \boldsymbol{\beta}_i^\top \boldsymbol{s}_i\right], \forall i \in [I], \\ \boldsymbol{v} = \boldsymbol{v}^0 + \sum_{k\in[K]} \gamma^k \boldsymbol{\Delta}^k, \boldsymbol{e}^\top \boldsymbol{\gamma} = 1, \boldsymbol{\gamma} \in \mathbb{R}_+^K \end{array} \right\} \tag{6.25}$$

其中, $\boldsymbol{e} \in \mathbb{R}^K$ 是元素全为 1 的向量.

6.4.2 极值概率分布下私人运营商的期望运营收益

接下来, 考虑极值概率分布下私人运营商的期望运营收益问题 (6.22). 此问题是一个两阶段分布鲁棒优化问题, 其不确定参数既在目标函数中, 又在第二阶段资源回收收益最大化问题 $\phi\left(\boldsymbol{x}, \tilde{\boldsymbol{v}}, \tilde{\boldsymbol{u}}\right)$ 约束条件的右端项中.

$$\phi(\boldsymbol{x}, \tilde{\boldsymbol{v}}, \tilde{\boldsymbol{u}}) = \max_{\boldsymbol{y}} \quad \sum_{i\in[I]} \sum_{j\in[J]} \sum_{n\in[N]} r_{ijn} \tilde{u}_{in} y_{ij} - c_D \sum_{i\in[I]} \left(\tilde{v}_i - \sum_{j\in[J]} y_{ij} \right)$$

$$\text{s.t.}\quad \sum_{j\in[J]} y_{ij} \leqslant \tilde{v}_i, \sum_{i\in[I]} y_{ij} \leqslant x_j\varpi_j, y_{ij} \geqslant 0,\quad \forall i\in[I], j\in[J].$$

上述两阶段分布鲁棒优化问题已被证明是难以求解的 (Bertsimas et al., 2019). 利用原料质量支撑集 $\mathcal{V}$ 是有限样本组成的凸包 (6.17), 我们可将问题 (6.22) 等价地转化为一个线性规划问题.

命题 6.4 给定分类方案和设施选址决策 $(\boldsymbol{s},\boldsymbol{x})$, 极值概率分布下私人运营商的期望运营收益问题 (6.22) 可以被转化为如下线性规划问题:

$$\max\ \pi + \sum_{i\in[I]} \boldsymbol{q}_i^\top \boldsymbol{F}\boldsymbol{s}_i + \sum_{i\in[I]} \overline{\delta}_i(\boldsymbol{\alpha}_i^\top \boldsymbol{s}_i) + \sum_{i\in[I]} \underline{\delta}_i(\boldsymbol{\beta}_i^\top \boldsymbol{s}_i) \tag{6.26}$$

$$\begin{aligned}
\text{s.t.}\ & \sum_{i\in[I]} c_D \boldsymbol{e}_J^\top \boldsymbol{\nu}_i^k - \sum_{i\in[I]} (\boldsymbol{B}\boldsymbol{s}_i)^\top \boldsymbol{p}_i^k - \boldsymbol{e}_I^\top \boldsymbol{\chi}^k - c_D \boldsymbol{e}_I^\top \boldsymbol{v}^k \\
& -\left(\underline{\boldsymbol{\delta}} + \overline{\boldsymbol{\delta}}\right)^\top \boldsymbol{v}^k \geqslant \pi,\ \forall k\in[K], &(6.27)\\
& \boldsymbol{R}_i \boldsymbol{\nu}_i^k + \boldsymbol{A}^\top \boldsymbol{p}_i^k + \chi_i^k \boldsymbol{e}_N \geqslant \boldsymbol{G}^\top \boldsymbol{q}_i,\quad \forall i\in[I], k\in[K], &(6.28)\\
& \boldsymbol{\nu}^k \boldsymbol{e}_J \leqslant \boldsymbol{v}^k,\ \forall k\in[K], &(6.29)\\
& \left(\boldsymbol{\nu}^k\right)^\top \boldsymbol{e}_I \leqslant \mathrm{Diag}(\boldsymbol{\varpi})\boldsymbol{x},\ \forall k\in[K], &(6.30)\\
& \boldsymbol{\nu}\in\mathbb{R}_+^{I\times J\times K}, \boldsymbol{p}\in\mathbb{R}_-^{I\times L_2\times K}, \boldsymbol{\chi}\in\mathbb{R}^{I\times K}, \boldsymbol{q}\in\mathbb{R}_+^{L_1\times I}, \underline{\boldsymbol{\delta}}\in\mathbb{R}_-^I,\\
& \overline{\boldsymbol{\delta}}\in\mathbb{R}_+^I, \pi\in\mathbb{R}, &(6.31)
\end{aligned}$$

其中, $\boldsymbol{e}_I\in\mathbb{R}^I$, $\boldsymbol{e}_J\in\mathbb{R}^J$ 和 $\boldsymbol{e}_N\in\mathbb{R}^N$ 表示元素全为 1 的向量, $\mathrm{Diag}(\boldsymbol{\varpi})\in\mathbb{R}^{J\times J}$ 是以向量 $\boldsymbol{\varpi}$ 作为对角线元素的对角矩阵, $\boldsymbol{v}^k := \boldsymbol{v}^0 + \boldsymbol{\Delta}^k, \forall k\in[K]$, 其中 $\boldsymbol{\Delta}^k$ 由式 (6.17) 给出.

基于命题 6.4, 可以推出资源回收设施选址问题 (6.24) 的等价混合整数线性规划问题.

推论 6.1 令 $(\boldsymbol{s}^\star, \boldsymbol{x}^\star)$ 为问题 (6.10) 的最优解, 则问题 (6.24) 最优解 $(\boldsymbol{s}^\star, \boldsymbol{x}^{\star\star})$ 中 $\boldsymbol{x}^{\star\star}$ 可以通过求解如下混合整数线性规划问题得到

$$\begin{aligned}
\max\ & \pi + \sum_{i\in[I]} \boldsymbol{q}_i^\top \boldsymbol{F}\boldsymbol{s}_i^\star + \sum_{i\in[I]} \overline{\delta}_i(\boldsymbol{\alpha}_i^\top \boldsymbol{s}_i^\star) + \sum_{i\in[I]} \underline{\delta}_i(\boldsymbol{\beta}_i^\top \boldsymbol{s}_i^\star) - \lambda \boldsymbol{c}^\top \boldsymbol{x}\\
\text{s.t.}\ & (6.27)\text{-}(6.31),\\
& \boldsymbol{c}^\top \boldsymbol{x} \leqslant \boldsymbol{c}^\top \boldsymbol{x}^\star,\\
& \boldsymbol{x}\in\{0,1\}^J,
\end{aligned}$$

6.4.3 分布鲁棒双层资源回收规划问题的高效计算形式

现在, 基于命题 6.3 和命题 6.4, 可以将分布鲁棒双层资源回收规划问题 (6.10) 转化为一个混合整数规划问题.

定理 6.2 分布鲁棒双层资源回收规划问题 (6.10) 可以转化如下混合整数线性规划问题:

$$
\begin{aligned}
\min\quad & (1-\lambda)\boldsymbol{c}^\top\boldsymbol{x}+\sum_{i\in[I]}\left(v_i^0\boldsymbol{\varrho}_i^\top+\boldsymbol{\iota}_i^\top\right)\boldsymbol{s}_i+\sum_{i\in[I]}t_i+\zeta-\left(\underline{\boldsymbol{\rho}}+\overline{\boldsymbol{\rho}}\right)^\top\boldsymbol{v}_0\\
\text{s.t.}\quad & (6.28)\text{-}(6.30),\\
& (\underline{\boldsymbol{\rho}}_i+\overline{\boldsymbol{\rho}}_i)^\top\boldsymbol{\Delta}^k+\zeta\geqslant\sum_{i\in[I]}\boldsymbol{s}_i^\top\boldsymbol{\varrho}_i\Delta_i^k,\ \forall k\in[K],\\
& \boldsymbol{b}_h^\top\boldsymbol{p}_i^k+M(s_{ih}-1)\leqslant\psi_i^k,\ \forall i\in[I],h\in[H],k\in[K],\\
& \sum_{i\in[I]}c_D\boldsymbol{e}_J^\top\boldsymbol{\nu}_i^k-\boldsymbol{e}_I^\top\boldsymbol{\chi}^k-c_D\boldsymbol{e}_I^\top\boldsymbol{v}^k-(\overline{\boldsymbol{\delta}}+\underline{\boldsymbol{\delta}})^\top\boldsymbol{v}^k-\pi\geqslant\boldsymbol{e}_I^\top\boldsymbol{\psi}^k,\quad\forall k\in[K],\\
& \boldsymbol{f}_h^\top\boldsymbol{q}_i+\overline{\delta}_i\alpha_{ih}+\underline{\delta}_i\beta_{ih}+M(1-s_{ih})\geqslant\varphi_i,\quad\forall i\in[I],h\in[H],\\
& \boldsymbol{e}_I^\top\boldsymbol{\varphi}\geqslant\tau+\lambda\boldsymbol{c}^\top\boldsymbol{x}-\pi,\\
& \alpha_{ih}\underline{\rho}_i+\beta_{ih}\overline{\rho}_i-M(1-s_{ih})\leqslant t_i,\ \forall i\in[I],h\in[H],\\
& t_i\leqslant\alpha_{ih}\underline{\rho}_i+\beta_{ih}\overline{\rho}_i+M(1-s_{ih}),\ \forall i\in[I],h\in[H],\\
& \boldsymbol{s}\in\mathcal{S},\boldsymbol{x}\in\{0,1\}^J,\\
& \boldsymbol{\nu}\in\mathbb{R}_+^{I\times J\times K},\boldsymbol{p}\in\mathbb{R}_-^{I\times L_2\times K},\boldsymbol{\chi}\in\mathbb{R}^{I\times K},\boldsymbol{q}\in\mathbb{R}_+^{L_1\times I},\underline{\boldsymbol{\delta}}\in\mathbb{R}_-^I,\overline{\boldsymbol{\delta}}\in\mathbb{R}_+^I,\pi\in\mathbb{R},\\
& \underline{\boldsymbol{\rho}}\in\mathbb{R}_-^I,\overline{\boldsymbol{\rho}}\in\mathbb{R}_+^I,\zeta\in\mathbb{R},\boldsymbol{\phi}\in\mathbb{R}^{I\times K},\boldsymbol{\varphi}\in\mathbb{R}^I,\boldsymbol{t}\in\mathbb{R}^I,
\end{aligned}
$$

其中, $\boldsymbol{e}_I\in\mathbb{R}^I$ 和 $\boldsymbol{e}_J\in\mathbb{R}^J$ 是元素全为 1 的向量, $\boldsymbol{\Delta}^k,k\in[K]$ 由式 (6.17) 给出, $\boldsymbol{f}_h,\forall h\in[H]$ 是矩阵 $\boldsymbol{F}$ 的列, 由式 (6.20) 给出, M 是一个足够大的正数.

6.5 数值实验

本节设计实验验证在 PPP 合作关系下分布鲁棒 BRRP 模型的决策性能. 本实验使用的数据来自新加坡国家环境局 (National Environment Agency). 考虑有 $I=10$ 个原料收集点, $J=10$ 个资源回收设施备选地址以及 $H=3$ 种原料分类方案的资源回收设施运营选址问题. 此外, 假设私人运营商只使用厌氧消化技术从食物原料中回收资源转化为能源产品. 表 6.1 列出了厌氧消化技术设施的参数, 表 6.2 给出了厌氧消化技术设施对收集的食物原料不同组成成分的能源转化效率 (Banks, 2009), 而表 6.3 给出了不同食品原料分类方案的参数.

表 6.1 厌氧消化技术设施的参数

容量	能源产品价格	原料处理成本	残留物	设施固定成本
50000 吨/年	0.2365 新元/(千瓦 · 时)	42.46 新元/吨	77 新元/吨	1558589 新元/年

表 6.2 厌氧消化技术设施回收原料中不同组成成分的能源转化效率

原料种类	食物垃圾 (有机)	园艺垃圾 (有机)	其余杂质 (无机)
能量回收率	347.2 (千瓦 · 时)/吨	198.0 (千瓦 · 时)/吨	0 (千瓦 · 时)/吨
残留率	0.1 吨/吨	0.1 吨/吨	1 吨/吨

表 6.3 不同分类方案的参数

h	分类方案 h	设施运营成本/(新元/吨)	固定投资成本/(新元/年)	期望原料质量范围/(吨/年)	有机物含量
1	回收箱	35.93	164150	[12957, 18140]	50%
2	现场回收	44.91	196980	[16844, 21427]	70%
3	激励回收	53.06	264691	[14252, 22002]	90%

6.5.1 不同 (τ,λ) 下分布鲁棒双层资源回收规划模型的最优解

本节分析在不同参数 (τ,λ) 下分布鲁棒 BRRP 模型的最优解 $(\xi^\star,\boldsymbol{s}^\star,\boldsymbol{x}^\star)$, 其中, τ 是资源回收的收益目标水平, λ 是私人运营商的成本分摊比例, 见表 6.4.

表 6.4 不同收益目标水平 τ 下分布鲁棒 BRRP 模型的最优解 $(\lambda=0.5)$

收益目标 (PO)	极值概率分布下总费用 (LA)	极值概率分布下分类费用 (LA)	极值概率分布下总固定费用 $\boldsymbol{c}^\top\boldsymbol{x}^\star$	配比资金预算与分类方案 † (LA) $\xi^\star$	$s_{i1}^\star=1$	$s_{i2}^\star=1$	$s_{i3}^\star=1$	选址 †† (PO) $x_j^\star=1$
0	12.94	10.60	4.68	2.34	i=3, 4, 7	i=5, 8	i=1, 2, 6, 9, 10	j=4-6†††
0.5	13.16	10.82	4.68	2.34	i=2, 5, 6	i=10	i=1, 3, 4, 7-9	j=2, 4, 5
1.0	13.48	10.36	6.23	3.12	i=7	i=2-4, 6, 8-10	i=1, 5	j=2-5
1.5	13.69	10.57	6.23	3.12	i=2	i=3, 4, 6-9	i=1, 5, 10	j=2, 4-6
2.0	14.13	11.01	6.23	3.12	i=8	i=1, 3-5	i=2, 6, 7, 9, 10	j=4, 5, 7, 10
2.5	14.34	11.22	6.23	3.12	i=9	i=1-3	i=4-8, 10	j=3-5, 7
3.0	14.56	11.44	6.23	3.12	i=6	i=3, 5	i=1, 2, 4, 7-10	j=2, 4-6
3.5	16.66	13.54	6.23	3.12	–	i=1, 7	i=2-6, 8-10	j=2-5
4.0	16.90	13.78	6.23	3.12	–	i=1	i=2-10	j=2, 4-6
⩾ 4.5	不可行							

注: 货币单位为 10^6 新元/年.

† $s_{i1}^\star=1, s_{i2}^\star=1$ 和 $s_{i3}^\star=1$ 分别表示收集点 i 采用回收箱、现场回收和激励回收的分类方案 (表 6.3).

†† $x_j^\star=1$ 表示资源回收设施选择建设在地址 j.

††† 4-6 等表示区间 4 到 6.

首先, 给定 $\lambda=0.5$, 并让 τ 在 $[0,4.0\times10^6]$ 范围内变化. 表 6.4 给出不同 τ

取值下对应的分布鲁棒 BRRP 模型最优解. 结果表明: 地方政府的总成本随着私人运营商的收益目标水平 τ 的增加而增加. 此外, 仅当收益目标由 0.5×10^6 变为 1×10^6 时, 建设回收设施的数量会产生变化, 由 3 个变为 4 个, 这使得固定成本增加. 特别地, 随着 τ 的增加, 地方政府的原料源头分类方案的选择会发生变化: ① 当 τ 相对较小时 ($\tau<0.5$), 地方政府将主要选择廉价且分类效率较低的回收箱方案, 并建设较少的资源回收设施. ② 当 τ 增加到中间水平时 ($1.0\leqslant\tau\leqslant2.5$), 地方政府将提高设施建设的预算, 并将分类方案从回收箱转移到更有效但更昂贵的现场回收方案. ③ 当 τ 增加到一个相对较高的水平时 ($3.0\leqslant\tau\leqslant4.0$), 地方政府将重点投资原料源头分类方案的改进, 即将回收箱和现场回收方案换成更有效的激励回收方案, 这会导致其分类成本增加. 此外, 地方政府将不再提高资源回收设施建设预算以建设更多的设施. 这表明当设施建设投入资金较大时, 进一步改善地方政府的分类方案则产生更有效的收益, 以保证私人运营商的经济可行性.

其次, 给定收益水平 τ, 分析不同成本分摊比例 λ 下分布鲁棒 BRRP 问题的最优解. 根据实际问题情况, 将收益目标 τ 设为 4.0×10^6 新元/年. 结果在表 6.5 给出. 结果表明: ① 在给定目标水平下, 分摊比例 λ 的上限是 50%, 如果私人运营商的成本分摊比例大于这个水平, 则无法实现目标 τ. ② 此外, 当私人运营商具有相对较高的利润目标 (4.0×10^6 新元/年) 时, 为了保证资源回收系统的运营, 随着成本分摊比例 λ 增加, 回收箱和现场回收分类方案逐渐被更有效的激励回收方案所取代, 同时地方政府在分类方案上的投资也在增加. 这表明, 如果地方政府希望私人运营商承担更多的设施建设成本, 则他需要在分类方案上投入更多资金, 以确保原料回收的品质, 从而维持私人运营商的经济可行性.

表 6.5　不同成本分摊比例 λ 下分布鲁棒 BRRP 模型的最优解 ($\tau=4.0\times10^6$ 新元/年)

分摊比例 (PO)	极值概率分布下总费用 (LA)	极值概率分布下分类费用 (LA)	极值概率分布下总固定费用 $\boldsymbol{c}^\top\boldsymbol{x}^\star$	配比资金预算和分类方案 (LA) $\xi^\star$	$s_{i1}^\star=1$	$s_{i2}^\star=1$	$s_{i3}^\star=1$	选址 (PO) $x_j^\star=1$
0 %	16.59	10.36	6.23	6.23	i=5	i=1,3,6-10	i=2,4	j=2-5
10 %	16.39	10.78	6.23	5.61	i=7	i=1,3,4,8,10	i=2,5,6,9	j=2,4-6
20 %	16.00	11.01	6.23	4.99	i=10	i=1,5,8,9	i=2-4,6,7	j=2,4-6
30 %	15.79	11.43	6.23	4.36	i=4	i=3,7	i=1,2,5,6,8-10	j=2,4-6
40 %	15.39	11.65	6.23	3.74	i=3	i=10	i=1,2,4-9	j=4-6,10
50 %	16.90	13.78	6.23	3.12	—	i=1	i=2-10	j=2,4-6
⩾ 60 %	不可行							

注: 货币单位为 10^6 新元/年.

6.5.2 分类依赖型原料状态分布不确定集的价值

本小节设计实验验证分类依赖型原料状态分布不确定集的价值. 特别地, 本小节考虑三个分类依赖分布不确定集 $\mathbb{F}(\boldsymbol{s}_1),\mathbb{F}(\boldsymbol{s}_2)$ 和 $\mathbb{F}(\boldsymbol{s}_3)$, 分别为给定 $\boldsymbol{s}^\star=\boldsymbol{s}_1$(只使用回收箱方案)、$\boldsymbol{s}^\star=\boldsymbol{s}_2$(只使用现场回收方案) 和 $\boldsymbol{s}_3$(只使用激励回收方案). 然后, 我们基于上述分布不确定集合给定 $\boldsymbol{s}^\star$ 的分布鲁棒优化模型, 并将 R_1,R_2 和 R_3 记为三个模型对应的最优解. 表 6.6 和表 6.7 分别给出固定 λ 和 τ 下模型的最优解.

表 6.6 不同分布不确定集和 $\boldsymbol{\tau}$ 下模型的最优解 (给定分类方案 s 与 $\lambda=0.5$)

模型	利润目标 (PO)	极值概率分布下总成本 (LA)	极值概率分布下分类费用 (LA)	极值概率分布下总固定费用 $\boldsymbol{c}^\top\boldsymbol{x}^\star$	预算与分类方案 (LA) $\xi^\star$	$s_{i1}^\star=1$	$s_{i2}^\star=1$	$s_{i3}^\star=1$	选址 (PO) $x_j^\star=1$
R_1	$\geqslant 0$	不可行							
R_2	0-0.5	14.71	11.60	6.23	3.12	–	i=1-10	–	j=2,4-6
	$\geqslant 1.0$	不可行							
R_3	0-2.5	17.44	14.32	6.23	3.12	–	–	i=1-10	j=2,4,5,10
	3.0-3.5	17.44	14.32	6.23	3.12	–	–	i=1-10	j=2,4-6
	$\geqslant 4.0$	不可行							

注: 货币单位为 10^6 新元/年.

表 6.7 不同分布不确定集和 $\boldsymbol{\lambda}$ 下模型的最优解 (给定分类方案与 $\tau=4.0\times10^6$ 新元/年)

模型	分摊比例 (PO)	极值概率分布下总成本 (LA)	极值概率分布下分类费用 (LA)	极值概率分布下总固定费用 $\boldsymbol{c}^\top\boldsymbol{x}^\star$	预算与分类方案 (LA) $\xi^\star$	$s_{i1}^\star=1$	$s_{i2}^\star=1$	$s_{i3}^\star=1$	选址 (PO) $x_j^\star=1$
R_1	$\geqslant 0\%$	不可行							
R_2	0 %	19.39	11.60	7.79	7.79	–	i=1-10	–	j=2,4,5-7
	$\geqslant 10\%$	不可行							
R_3	0 %	20.56	14.32	6.23	6.23	–	–	i=1-10	j=2,4-6
	10 %	19.93	14.32	6.23	5.61	–	–	i=1-10	j=2,4-6
	20 %	19.31	14.32	6.23	4.99	–	–	i=1-10	j=2,4-6
	30 %	18.68	14.32	6.23	4.36	–	–	i=1-10	j=2,4-6
	40 %	18.06	14.32	6.23	3.74	–	–	i=1-10	j=2,4-6
	$\geqslant 50\%$	不可行							

注: 货币单位为 10^6 新元/年.

在相同的参数取值下, 比较表 6.6 和表 6.7 中无分类依赖性的模型最优解

R_1 R_2 和 R_3 与表 6.4 和表 6.5 中存在分类依赖性的模型最优解 (记为 R-BRRP), 结果表明: ① 当 $\tau = 0$, $\lambda = 0.5$ 或 $\tau = 4.0 \times 10^6$ 时, R_1 不存在可行解. ② R_2 仅在某些参数设定下存在可行解. 此外, 可行的 R_2 建设设施的个数变多, 产生更多的固定成本预算, 以及更多的分类成本. 例如, 当 $\tau = 0.5 \times 10^6$ 且 $\lambda = 0.5$ 时, R_2 建设 4 个设施, 总分类成本为 11.60×10^6, 而在表 6.4 中 R-BRRP 仅建设 3 个设施, 总分类成本为 10.82×10^6, 这表明分类方案依赖型模型能灵活选择有成本效益的分类方案. ③ 同样地, R_3 解会产生更高的分类成本或建设更多的设施. 此外, R_3 解对于 $\tau = 4.0 \times 10^6$ 且 $\lambda = 0.5$ 的情况也不可行. 这与实际直觉相违背, 因为 R_3 实施最昂贵的分类方案, 其成本完全由地方政府而不是私人运营商承担. 这种结果是由原料质量不确定性相对较高 (见表 6.3) 且 R_3 采用单一激励回收分类方案所导致的. 当原料质量较大时, 只能通过建设更多的设施来处理过多的原料, 这导致建设设施的固定成本增加, 进而使得私人运营商的利润目标无法实现. 相比之下, R-BRRP 可以更灵活地使用现场回收分类方案, 以防止建设更多设施确保经济可行性 (表 6.4). 这表明了在实践中简单地通过经验选择单一的分类方案存在潜在的风险.

6.6 本章小结

本章介绍了公私合作关系下资源回收系统设施选址双层规划问题. 为了建模原料组成成分和质量的分布不确定性以及原料的分类方案依赖性, 本章在数据驱动分布鲁棒优化框架下构建了基于分类方案依赖型分布不确定集的资源回收系统设施选址双层规划模型. 此外, 通过分析模型在成本分摊比率方面的最优性, 本章阐明了地方政府投资分类方案对可持续资源回收系统的价值. 最后, 本章证明了模型等价于一个规模适中混合整数线性规划问题.

关于不确定环境下资源回收设施选址以及分布鲁棒博弈优化等方面的研究, 读者可参考更多相关文献. 例如, Xiong 等 (2016) 研究了具有多个废物处理运营商的联合废物管理规划问题, 并开发了一个两阶段概率优化模型. Cardin 等 (2017) 和 Gambella 等 (2019) 考虑了不确定废物原料产生率下废物原料处理能力和处理计划的多周期固体废物管理规划问题. Wang 等 (2016) 考虑了原料生成率不确定下资源回收系统多期选址规划问题. 该研究提出了一个基于时间序列预测模型和不同时期废物原料的不确定性预算的废物原料不确定集, 并基于所提出的不确定集构建了自适应鲁棒资源回收系统多期选址模型. Wang 和 Ng (2019) 同样研究了一个不确定环境下的资源回收规划问题, 该问题侧重于分析原料质量和成分不确定性对回收运营商经济可行性的影响. 作者提出了一套基于自适应鲁棒优化的条件和复合鲁棒性指标模型, 用于原料不确定性下的回收性能评估和系统设计优化, 并且探讨模型在求解方面的良好特性. 更多关于不确定环境下资源回收系统规划的相关研究, 读者还可以参考 Li 等 (2008), Xu 等 (2010) 以及 Li 和 Huang (2012) 的研究. 在分布鲁棒博弈优化的研究中, Liu 等 (2018) 提出了分布鲁棒 Stackelberg 博弈模型, 并讨论了模型的数值求解方法.

第 7 章 预测不确定集与多阶段选址优化

第 5 章和第 6 章介绍了两阶段资源回收设施选址问题. 本章介绍不确定性环境下多阶段 (或多期) 资源回收设施选址问题. 本章考虑一个基于预测不确定集的资源回收设施选址问题. 首先, 基于原料增长量预测模型 (feedstock growth forecasting model) 和决策者对预测误差的容忍水平 (user-specified levels of forecasting errors), 构建原料产量预测不确定集 (predictive uncertainty set) 来建模多阶段原料产量不确定性, 并给出基于此不确定集的多阶段鲁棒资源回收设施选址模型. 在给定设施选址决策的情况下, 该模型等价于一个多阶段最小最大化动态规划问题 (multiple stage min-max dynamic optimization problem), 此问题可以被转化为一个混合整数线性规划问题. 其次, 在一般情况下, 我们介绍一种可以精确求解模型最优选址决策的割平面方法 (cutting plane approach). 最后, 设计数值实验来验证模型和算法性能. 本章主要结构如下: 7.1 节介绍多阶段资源回收设施选址问题背景. 7.2 节介绍多阶段确定性资源回收设施选址模型. 7.3 节介绍预测不确定集与多阶段鲁棒资源回收设施选址模型的构建. 7.4 节介绍模型的精确求解算法. 最后, 7.5 节介绍实验及结果分析.

7.1 背　　景

本章的背景是日益增长的小型模块化资源回收设施的商业应用需求. 资源回收设施的模块化意味着运营商可以分阶段决策灵活调整资源回收系统的容量或处理能力, 以达到节约固定成本和运营成本的目的. 具体而言, 运营商可以在原料 (feedstock) 阶段性减少时, 快速关闭多余的资源回收设施, 从而规避长期负担高昂的固定成本所带来的经济风险. 与两阶段资源回收系统选址问题相同, 多阶段资源回收系统的经济可行性主要受原料产量高度不确定性的影响. 例如, 低于预期的原料输入量会影响系统的利润, 而过多的原料输入量会造成系统容量或处理能力紧张, 从而增加原料处理成本. 在实际问题中, 由于无法对原料产量进行准确预测, 运营商可能会产生严重的财务赤字, 甚至破产.

不确定性环境下资源回收系统的经济可行性依赖于原料产量预测模型的准确性和历史原料数据的可靠性. 然而, 在现实生活中, 从多个地区获取可靠的原料数据是非常困难的. 不同的国家和地区、政府和行业之间的数据可用度和数据采集的准确性与规范性存在巨大差异 (Butler and Hooper, 2000; Guerrero et al.,

2013). 此外, 很多地区近年来才开始跟踪记录原料数据, 这导致数据量不足以支撑高准确度预测模型的构建. 因此, 准确地预测未来原料产量, 特别是每一个决策阶段的原料产量是十分困难的. 针对上述问题, 本章介绍一种基于有限预测信息的原料产量不确定集构建方法, 并基于此预测不确定集构建多阶段鲁棒资源回收设施选址模型.

7.2 基于点预测的确定性模型

首先介绍确定性多阶段资源回收设施选址模型. 假设每阶段原料产量的点预测值准确. 考虑一个有 I 个居民区 (residential zones) 或原料收集区, J 个备选资源回收设施地址以及 T 个决策阶段的多阶段资源回收设施选址问题. 我们记 ξ_{it} 为第 t 阶段从居民区 i 收集的原料产量. 该问题主要包括三个决策变量: $x_{j,t-1}$ 为 0-1 二元设施选址决策变量, 当且仅当第 t 阶段开设设施 j 时, $x_{j,t-1}=1$; y_{ijt} 原料运输决策变量, 其表示第 t 阶段从居民区 i 运送到资源回收设施 j 的原料产量; z_{it} 是第 t 阶段居民区 i 处未被处理原料产量.

给定各阶段各居民区的原料产量预测值 $\boldsymbol{\xi}=(\xi_{it})_{I\times T}$, 选址决策 $\boldsymbol{x}=(\boldsymbol{x}_0,\boldsymbol{x}_1,\cdots,\boldsymbol{x}_T)$, 原料运输决策 $\boldsymbol{y}=(y_{ijt})_{I\times J\times T}$ 以及 $\boldsymbol{z}=(z_{it})_{I\times T}$, 第 t 阶段资源回收系统出售能源产品的总收益为

$$\left[\sum_{j\in[J]}\sum_{i\in[I]}y_{ijt}\theta_i\mu\right]q_E,$$

其中, θ_i 为在居民区 i 所收集的可回收原料纯度比例 (purity ratio), μ 为可回收原料的单位能源转化率, q_E 为单位能源价格. 此外, 第 t 阶段资源回收系统的运输成本和运营成本分别为

$$\sum_{j\in[J]}\sum_{i\in[I]}y_{ijt}d_{ij}h,\quad \sum_{j\in[J]}\sum_{i\in[I]}y_{ijt}v,$$

其中, d_{ij} 为居民区 i 与资源回收设施 j 之间的距离, h 为单位产量原料单位距离运输成本, 以及 v 为资源回收设施处理单位原料的运营成本, 最后, 第 t 阶段未处理原料和残留原料的处置成本为

$$\sum_{i\in[I]}c_D\left[z_{ti}+\sum_{j\in[J]}y_{ijt}(1-\theta_i)+y_{ijt}\rho_j\right],$$

其中, ρ_j 为资源回收设施 j 处理原料的残留率, c_D 为单位原料的处置成本. 为了表述简洁, 我们定义

$$r_{ij} := \theta_i \mu q_E - d_{ij} h - v - c_D(1 - \theta_i + \rho_j) \tag{7.1}$$

为从居民区 i 所收集的原料在资源回收设施 j 处回收处理的收益系数. 此外, 记开设资源回收设施 j 的固定成本为 c_j, 则开设资源回收设施的总固定成本为 $\boldsymbol{c}^\top \boldsymbol{x}_{t-1} = \sum_{j\in[J]} c_j x_{j,t-1}$. 给定决策 $(\boldsymbol{x}_{t-1}, \boldsymbol{y}_t, \boldsymbol{z}_t)$, 我们定义第 t 阶段净现值 (net present value) 函数如下:

$$\phi_t(\boldsymbol{x}_{t-1}, \boldsymbol{y}_t, \boldsymbol{z}_t) := \beta^t \sum_{i\in[I]} \left[\sum_{j\in[J]} r_{ij} y_{ijt} - c_D z_{it}\right] - \beta^{t-1} \boldsymbol{c}^\top \boldsymbol{x}_{t-1}, \tag{7.2}$$

其中, $\beta \in (0,1)$ 是固定成本折现率.

给定原料产量的准确点预测值 $\boldsymbol{\xi}$, 基于净现值函数 (7.2), 我们构建如下多阶段确定性资源回收设施选址模型:

$$\begin{aligned}
\max_{\boldsymbol{x},\boldsymbol{y},\boldsymbol{z}} \quad & \phi(\boldsymbol{x},\boldsymbol{y},\boldsymbol{z}) := \sum_{t\in[T]} \phi_t(\boldsymbol{x}_{t-1}, \boldsymbol{y}_t, \boldsymbol{z}_t) \\
\text{s.t.} \quad & \sum_{j\in[J]} y_{ijt} + z_{it} = \widehat{\xi}_{it}, \ \forall i \in [I], t \in [T], \\
& \sum_{i\in[I]} y_{ijt} \leqslant \sum_{\tau\in[t]} x_{j\tau} s_j, \ \forall j \in [J], t \in [T], \\
& x_{j,t} \in \{0,1\}, y_{ijt}, z_{it} \in \mathbb{R}_+, \ \forall i \in [I], j \in [J], t \in [T], \\
& \boldsymbol{x} \in \mathcal{X}_o,
\end{aligned} \tag{7.3}$$

其中, s_j 为资源回收设施 j 的容量或处理能力. 上述问题的目标是最大化资源回收系统总净现值 $\phi(\boldsymbol{x},\boldsymbol{y},\boldsymbol{z})$. 第一组约束保证各阶段各居民区所产生的原料都要被处理, 即资源回收设施处理第 i 个居民区的原料产量 $\sum_{j\in[J]} y_{ijt}$ 和其超出处理能力的残留量 z_{it} 与总原料产量 $\widehat{\xi}_{it}$ 相等; 第二组约束保证各阶段的原料仅通过该阶段内开设的资源回收设施处理, 并且每个资源回收设施处理的原料产量上限为其容量 s_j. 第三组约束为选址决策 $\boldsymbol{x}$ 的二元约束, 并保证决策 $(\boldsymbol{y},\boldsymbol{z})$ 是非负的. 最后一个约束通过设置可行集 $\mathcal{X}_o$ 的不同形式来限制选址决策 $\boldsymbol{x}$. 例如, 如果决策者只允许资源回收设施 j 在所有阶段内只开设一次, 则集合 $\mathcal{X}_o$ 为 $\mathcal{X}_o = \{\boldsymbol{x} \mid \sum_{t\in[T]} x_{j,t} \leqslant 1, \ \forall j \in [J]\}$.

7.3 基于预测不确定集的资源回收系统净现值保证水平

为了处理多阶段原料产量的不确定性, 本节介绍基于原料增长量预测模型与预测误差容忍水平的不确定原料产量预测集的构建, 以及在给定设施选址决策下, 鲁棒资源回收系统总净现值函数的高效计算形式.

7.3.1 原料产量预测不确定集

首先, 考虑如下原料增量模型:

$$\xi_{it} = \xi_{it-1} + \eta_{it}, \quad \forall i \in [I], t \in [T], \tag{7.4}$$

其中, η_{it} 是居民区 i 在第 t 阶段的原料增量, 其值可能为负. 特别地, 记 $\xi_{i0} = \widehat{\xi}_{i0}$ 为居民区 i 在决策初始阶段的原料产量. 给定一组历史原料产量数据: $\widehat{\xi}_{i,-k}, i \in [I], k \in [0:K]$, 其中, $\widehat{\xi}_{i,-k}$ 为居民区 i 在决策初始阶段 $t=0$ 前 k 期的原料产量. 我们可以将它们转化为增量数据: $\widehat{\eta}_{i,-k} = \widehat{\xi}_{i,-k} - \widehat{\xi}_{i,-(k+1)}, i \in [I], k \in [0:K-1]$. 对于每个居民区 $i \in [I]$ 和阶段 $t \in [T]$, 利用数据 $\widehat{\boldsymbol{\eta}}_i := (\widehat{\eta}_{i,0}, \widehat{\eta}_{i,-1}, \cdots, \widehat{\eta}_{i,-(K-1)})^\top$, 我们可以预测第 i 居民区第 t 阶段原料增量, 记为 $\mathcal{F}_{it}$. 决策者可以使用任何预测模型来预测原料增量值 $\mathcal{F}_{it}$.

例 13 (指数加权移动平均模型)　*指数加权移动平均 (exponentially weighted moving average) 模型定义如下:*

$$\mathcal{F}_{it} := \sum_{k\in[0:K]} \left[\frac{\alpha^k}{\sum_{k\in[0:K]} \alpha^k} \right] \widehat{\eta}_{i,-k}, \quad \forall t \in [T], \tag{7.5}$$

其中, $\widehat{\eta}_{i,-k}$ 是在居民区 i, 决策初始阶段 $t=0$ 前 k 期的原料增量, 且 $-k=0,\cdots,-K_0$ $(K_0 < K)$. 此外, $\alpha \in [0,1]$ 是需要估计的贴现因子 (discount factor).

例 14 ((p,q) 阶自回归移动平均模型)　*(p,q) 阶自回归移动平均模型 (autoregressive moving average of order (p,q) model) 定义如下:*

$$\begin{aligned}\mathcal{F}_{it} := \mathbb{I}_{\{t\geqslant 2\}} \left[\sum_{\ell\in[t-1]} \alpha_\ell \mathcal{F}_{i\ell} \right] \sum_{r\in[t:p]} \alpha_r \widehat{\eta}_{i,-(r-t)} + \sum_{\varsigma\in[t:q]} \beta_\varsigma \\ \times \left[\widehat{\eta}_{i,-(\varsigma-t)} - \mathcal{F}^{-\varsigma}_{i,-(\varsigma-t)} \right], \quad \forall t \in [T],\end{aligned} \tag{7.6}$$

其中, α_ℓ, $\ell \in [p]$ 和 β_ς, $\varsigma \in [q]$ 是需要估计的常数, $\mathcal{F}^{-\varsigma}_{i,-(\varsigma-t)}$ 是原料增量 $\widehat{\eta}_{i,-(\varsigma-t)}$ 的预测值, $\mathbb{I}_{\{t\geqslant 2\}}$ 是示性函数, 当 $t \geqslant 2$ 时, 其取值为 1, 其余情况下为 0.

假设每个阶段 t 的历史预测误差 $\eta_{it} - \mathcal{F}_{it}$ 可得. 基于历史预测误差值, 我们可以计算多种统计量 $\mathcal{S}_{it}$, 如平均绝对误差 (mean absolute error) 和样本扰动

(sample variations). 考虑到预测误差的存在, 我们用预测区间 $[\mathcal{F}_{it}-\mathcal{S}_{it},\mathcal{F}_{it}+\mathcal{S}_{it}]$ 代替原料增量的点预测值 $\mathcal{F}_{it}$. 下面介绍两个构建预测区间的例子.

例 15 (平均绝对误差) 利用数据 $\widehat{\boldsymbol{\eta}}_i$ 以及预测模型 $\mathcal{F}_{it}$, 可以计算每个居民区 i 原料增量预测值的 t 步前移平均绝对误差 (t-step ahead mean absolute error) MAE_{it}:

$$\mathrm{MAE}_{it}:=\frac{1}{K_0+1}\sum_{k\in[0:K_0]}\left|\mathcal{F}_{i,-k}^{(-k-t)}-\widehat{\eta}_{i,-k}\right|,\quad \forall i\in[I],t\in[T], \tag{7.7}$$

其中, $\mathcal{F}_{i,-k}^{(-k-t)}$ 是原料增量的从 $-k-t$ 阶段的 t 步向前预测值 (t-step ahead forecast value). 令 $\mathcal{S}_{it}:=\mathrm{MAE}_{it}$, 我们可以将预测区间定义为 $[\mathcal{F}_{it}-\mathrm{MAE}_{it},\mathcal{F}_{it}+\mathrm{MAE}_{it}]$.

例 16 (t 分布) 假设预测误差服从一个均值为 0 的正态分布. 根据时间序列预测理论 (Bowerman et al., 2005; Geisser, 2017), 我们有

$$\frac{\eta_{it}-\mathcal{F}_{it}}{\widehat{v}_{it}^{K}}\sim \mathrm{t}_{K-1},$$

其中, $\widehat{v}_{it}^{K}$ 是使用数量为 K 的样本计算出的前 t 期预测误差 $\eta_{it}-\mathcal{F}_{it}$ 的样本方差, t_{K-1} 是自由度为 $K-1$ 的标准 t 分布 (t-distribution). 通过定义 $\mathcal{S}_{it}:=\widehat{v}_{it}^{K}\mathrm{t}_{K-1,\frac{\alpha}{2}}$, 我们可构建置信度为 $100\times(1-\alpha)\%$ 的预测区间 $[\mathcal{F}_{it}-\widehat{v}_{it}^{K}\mathrm{t}_{K-1,\frac{\alpha}{2}},\mathcal{F}_{it}+\widehat{v}_{it}^{K}\mathrm{t}_{K-1,\frac{\alpha}{2}}]$, 其中, $\mathrm{t}_{K-1,\frac{\alpha}{2}}$ 是自由度为 $K-1$ 的 t 分布的 $100\times\left(1-\dfrac{\alpha}{2}\right)$ 分位数.

接下来, 我们定义如下关于原料增量 η_{it} 的预测区间:

$$\eta_{it}\in[\mathcal{F}_{it}-\mathcal{S}_{it},\mathcal{F}_{it}+\mathcal{S}_{it}]\Leftrightarrow \eta_{it}=\mathcal{F}_{it}+\varrho_{it}\mathcal{S}_{it},\quad \varrho_{it}\in[-1,1],$$

其中, 变量 $\varrho_{it}\in[-1,1]$ 用来表示预测误差的符号 (或增量的变化方向) 并控制误差范围. 基于上述预测区间以及不确定性预算 (uncertainty budget) 的思想 (Bertsimas and Sim, 2004), 我们构建如下原料产量 ξ 预测不确定集 (predictive uncertainty set):

$$\mathcal{U}^{\Gamma}:=\left\{\boldsymbol{\xi}\in\mathbb{R}^{I\times T}\;\middle|\;\begin{array}{l}\xi_{it}=\widehat{\xi}_{i0}+\sum\limits_{\tau\in[t]}\mathcal{F}_{i\tau}+\left[\sum\limits_{\tau\in[t]}\varrho_{i\tau}\mathcal{S}_{i\tau}\right],\ \forall i\in[I],t\in[T],\\ \sum\limits_{i\in[I]}|\varrho_{it}|\leqslant\Gamma_t^{\mathrm{Z}},\ \forall t\in[T],\\ \sum\limits_{t\in[T]}|\varrho_{it}|\leqslant\Gamma_i^{\mathrm{T}},\ \forall i\in[I],\\ \varrho_{it}\in[-1,1],\ \forall i\in[I],t\in[T]\end{array}\right\}, \tag{7.8}$$

其中, 对于任意 $i \in [I], t \in [T]$, 集合 (7.8) 的第一项

$$\xi_{it} = \widehat{\xi}_{i0} + \sum_{\tau \in [t]} \mathcal{F}_{i\tau} + \left[\sum_{\tau \in [t]} \varrho_{i\tau} \mathcal{S}_{i\tau} \right]$$

是基于原料增量模型 (7.4) 和预测集 $[\mathcal{F}_{it} - \mathcal{S}_{it}, \mathcal{F}_{it} + \mathcal{S}_{it}]$ 给出的 ξ_{it} 的预测值. 特别地, 从上式可以看出 ξ_{it} 比 ξ_{it-1} 多一个单位预测误差 $\mathcal{S}_{it}$, 这反映了未来原料产量随时间变化的不确定性扩散效应 (uncertainty propagation effect). 其次, 预算集 (7.8) 中的约束

$$\sum_{t \in [T]} |\varrho_{it}| \leqslant \Gamma_i^{\mathrm{T}}, \quad \forall i \in [I], \quad \sum_{i \in [I]} |\varrho_{it}| \leqslant \Gamma_t^{\mathrm{Z}}, \ \forall t \in [T]$$

限制了预测误差 $\mathcal{S}_{i\tau}$ 的数量, 其中, $\Gamma := \left(\Gamma_t^{\mathrm{Z}}, \Gamma_i^{\mathrm{T}}\right)_{T \times I}$ 为预测误差预算 (forecast error budgets). $\Gamma_i^{\mathrm{T}} \in [0, T], i \in [I]$ 控制居民区 i 在 T 个决策阶段原料增量的总预测误差. 此外, $\Gamma_t^{\mathrm{Z}} \in [0, I], t \in [T]$ 控制所有居民区在第 t 阶段原料增量的总预测误差. 特别地, 我们假设预算参数 Γ 是整数.

7.3.2 资源回收系统的净现值保证水平

给定选址决策 $\boldsymbol{x}$, 本节分析资源回收系统的净现值保证水平 (guaranteed net present value level). 首先, 定义 t 阶段选址决策为 $\boldsymbol{x}_{[t]} = (\boldsymbol{x}_0, \boldsymbol{x}_1, \cdots, \boldsymbol{x}_t)$. 在每个决策阶段 $t \in [T]$, 给定原料产量 $\boldsymbol{\xi}_t$, 我们定义

$$\mathcal{Y}_t\left(\boldsymbol{x}_{[t]}, \boldsymbol{\xi}_t\right) := \left\{ (\boldsymbol{y}_t, \boldsymbol{z}_t) \in \mathbb{R}_+^{I \times J} \times \mathbb{R}_+^I \;\middle|\; \begin{array}{l} \displaystyle\sum_{j \in [J]} y_{ijt} + z_{it} = \xi_{it}, \ \forall i \in [I], \\ \displaystyle\sum_{i \in [I]} y_{ijt} \leqslant \sum_{\tau \in [t]} x_{j\tau} s_j, \ \forall j \in [J] \end{array} \right\}$$

为给定 $\boldsymbol{x}_{[t]}$ 和 $\boldsymbol{\xi}_t$ 下第 t 阶段原料分配决策 $(\boldsymbol{y}_t, \boldsymbol{z}_t)$ 的可行集. 根据问题 (7.3) 的结构, 可知最优的原料分配决策 $(\boldsymbol{y}_t, \boldsymbol{z}_t)$ 将最大化第 t 阶段资源回收系统的净现值, 且受选址决策 $\boldsymbol{x}_{[t]}$ 和原料产量 $\boldsymbol{\xi}_t$ 的影响. 因此, 我们构建如下第 t 阶段净现值最大化问题:

$$\max_{(\boldsymbol{y}_t, \boldsymbol{z}_t) \in \mathcal{Y}_t\left(\boldsymbol{x}_{[t]}, \boldsymbol{\xi}_t\right)} \phi_t\left(\boldsymbol{x}_t, \boldsymbol{y}_t, \boldsymbol{z}_t\right),$$

其中, $\phi_t\left(\boldsymbol{x}_t, \boldsymbol{y}_t, \boldsymbol{z}_t\right)$ 为第 t 阶段净现值函数, 其定义由 (7.2) 给出. 此外, 记

$$\Upsilon_t(\boldsymbol{x}_{[t]}, \boldsymbol{\xi}_t) := \max_{(\boldsymbol{y}_t, \boldsymbol{z}_t) \in \mathcal{Y}_t\left(\boldsymbol{x}_{[t]}, \boldsymbol{\xi}_t\right)} \phi_t\left(\boldsymbol{x}_t, \boldsymbol{y}_t, \boldsymbol{z}_t\right) \tag{7.9}$$

为给定 $\boldsymbol{x}_{[t]}$ 和 $\boldsymbol{\xi}_t$ 下第 t 阶段资源回收设施的最优净现值. 在多阶段优化问题中, 决策者在每阶段初始将确定此阶段的最优决策, 而此阶段的原料产量不确定性取决于前一阶段的原料产量不确定性, 因此, 我们需要将预测不确定集按阶段 t 进行分解. 记 $\boldsymbol{\xi}_{[t-1]} = (\widehat{\boldsymbol{\xi}}_0, \boldsymbol{\xi}_1, \boldsymbol{\xi}_2, \cdots, \boldsymbol{\xi}_{t-1})$ 为 $t-1$ 个阶段的原料产量样本序列, 且 $\boldsymbol{\xi}_{[0]} := (\widehat{\boldsymbol{\xi}}_0)$. 给定样本序列 $\boldsymbol{\xi}_{[t-1]}$, 根据集合 $\mathcal{U}^{\Gamma}$, 构建基于 $\boldsymbol{\xi}_{[t-1]}$ 的原料产量 $\boldsymbol{\xi}_t$ 的预测不确定集:

$$\mathcal{U}_t^{\Gamma}(\boldsymbol{\xi}_{[t-1]}) := \left\{ \boldsymbol{\xi}_t \in \mathbb{R}^I \;\middle|\; \begin{array}{l} \xi_{it} = \xi_{it-1} + \mathcal{F}_{it} + \varrho_{it}\mathcal{S}_{it}, \forall i \in [I], \\ \displaystyle\sum_{i\in[I]} |\varrho_{it}| \leqslant \Gamma_t^{\mathrm{Z}}, \\ |\varrho_{it}| \leqslant \Gamma_i^{\mathrm{T}} - \mathbb{I}_{\{t\geqslant 2\}} \left[\displaystyle\sum_{\tau\in[t-1]} |\varrho_{i\tau}| \right], \forall i \in [I], \\ \varrho_{it} \in [-1, 1], \forall i \in [I] \end{array} \right\}, \tag{7.10}$$

其中, $\mathbb{I}_{\{t\geqslant 2\}}$ 为示性函数. 特别地, 集合 (7.10) 的约束

$$|\varrho_{it}| \leqslant \Gamma_i^{\mathrm{T}} - \mathbb{I}_{\{t\geqslant 2\}} \left[\sum_{\tau\in[t-1]} |\varrho_{i\tau}| \right], \quad \varrho_{it} \in [-1, 1], \quad \forall i \in [I]$$

保证对于每个 $i \in [I]$, 如果 $\sum_{\tau\in[t-1]} |\varrho_{i\tau}|$ 的值已经达到预算限制 Γ_i^{T}, 则 $|\varrho_{it}| = 0 \Leftrightarrow \varrho_{it} = 0$; 否则 $|\varrho_{it}| \leqslant 1$. 基于上述集合 (7.10), 预测集 $\mathcal{U}^{\Gamma}$ 可以被分解为一系列嵌套预测子集 $\mathcal{U}_t^{\Gamma}(\boldsymbol{\xi}_{[t-1]}), t \in [T]$.

假设资源回收设施的投资者或决策者都是谨慎型, 且对原料不确定性的影响持规避态度. 基于上述假设, 我们可以构建如下最后决策阶段 $(t = T)$ 内资源回收设施的净现值保证水平 (guaranteed net present value level) 为

$$\begin{aligned} &\min_{\boldsymbol{\xi}_t \in \mathcal{U}_T^{\Gamma}(\boldsymbol{\xi}_{[T-1]})} \Upsilon_T(\boldsymbol{x}_{[T-1]}, \boldsymbol{\xi}_{T-1}) \\ =& \min_{\boldsymbol{\xi}_T \in \mathcal{U}_T^{\Gamma}(\boldsymbol{\xi}_{[T-1]})} \max_{(\boldsymbol{y}_T, \boldsymbol{z}_T) \in \mathcal{Y}_T\left(\boldsymbol{x}_{[T-1]}, \boldsymbol{\xi}_T\right)} \phi_T\left(\boldsymbol{x}_{[T-1]}, \boldsymbol{y}_T, \boldsymbol{z}_T\right). \end{aligned} \tag{7.11}$$

利用式 (7.11) 向后递归 (recursively backwards) 构建每个阶段 $t \in [T]$ 的净现值保证水平, 我们可得如下资源回收系统总净现值保证水平 $\mathcal{Z}_{\mathrm{NPV}}(\boldsymbol{x})$:

$$\mathcal{Z}_{\mathrm{NPV}}(\boldsymbol{x}) := \min_{\boldsymbol{\xi}_1 \in \mathcal{U}_1^{\Gamma}(\boldsymbol{\xi}_{[0]})} \left[\Upsilon_1(\boldsymbol{x}_{[0]}, \boldsymbol{\xi}_1) + \min_{\boldsymbol{\xi}_2 \in \mathcal{U}_2^{\Gamma}(\boldsymbol{\xi}_{[1]})} \left[\Upsilon_2(\boldsymbol{x}_{[1]}, \boldsymbol{\xi}_2) \right.\right.$$

$$+\cdots+\min_{\boldsymbol{\xi}_T\in\mathcal{U}_T^{\Gamma}(\boldsymbol{\xi}_{[T-1]})}\left[\Upsilon_T(\boldsymbol{x}_{[T-1]},\boldsymbol{\xi}_T)\right]\cdots\Big]\Big], \tag{7.12}$$

其中, $\mathcal{U}_t^{\Gamma}(\boldsymbol{\xi}_{t-1}), t\in[T]$ 为式 (7.10) 所给出的依赖于样本序列的预测集.

问题 (7.12) 是不确定性环境下动态优化问题, 此问题通常难以求解. 利用资源回收系统选址问题 (7.3) 的结构, 可以将问题 (7.12) 转化为一个简单的两阶段优化问题. 为了表述简洁, 给定选址决策 $\boldsymbol{x}$ 和原料产量 $\boldsymbol{\xi}$, 定义如下关于决策 $(\boldsymbol{y},\boldsymbol{z})$ 的可行集:

$$\mathcal{Y}(\boldsymbol{x},\boldsymbol{\xi}):=\left\{(\boldsymbol{y},\boldsymbol{z})\in\mathbb{R}_+^{T\times I\times J}\times\mathbb{R}_+^{T\times I}\;\middle|\;\begin{array}{l}\displaystyle\sum_{j\in[J]}y_{ijt}+z_{it}=\xi_{it},\forall i\in[I],t\in[T],\\ \displaystyle\sum_{i\in[I]}y_{ijt}\leqslant\sum_{\tau\in[t]}x_{j\tau}s_j,\forall j\in[J],t\in[T]\end{array}\right\}. \tag{7.13}$$

命题 7.1　给定选址决策 $\boldsymbol{x}$, 我们有

$$\mathcal{Z}_{\mathrm{NPV}}(\boldsymbol{x})=\min_{\boldsymbol{\xi}\in\mathcal{U}^{\Gamma}}\max_{(\boldsymbol{y},\boldsymbol{z})\in\mathcal{Y}(\boldsymbol{x},\boldsymbol{\xi})}\phi(\boldsymbol{x},\boldsymbol{y},\boldsymbol{z}), \tag{7.14}$$

其中, $\phi(\boldsymbol{x},\boldsymbol{y},\boldsymbol{z})$ 是式 (7.3) 定义的总净现值.

命题 7.2 表明: 给定任意 (可行的) 选址决策 $\boldsymbol{x}$, 存在一个极值情况下的原料产量场景 $\boldsymbol{\xi}^{\dagger}=(\xi_{it}^{\dagger})_{I\times T}\in\mathcal{U}^{\Gamma}$, 当此极值场景真实发生时, 则资源回收系统净现值达到净现值保证水平 $\mathcal{Z}_{\mathrm{NPV}}(\boldsymbol{x})$. 此外, 式 (7.14) 还表明资源回收系统选址问题中的最优总净现值函数可以根据阶段 t 分解, 即

$$\max_{(\boldsymbol{y},\boldsymbol{z})\in\mathcal{Y}(\boldsymbol{x},\boldsymbol{\xi})}\phi(\boldsymbol{x},\boldsymbol{y},\boldsymbol{z})=\sum_{t\in[T]}\Upsilon_t(\boldsymbol{x}_{[t]},\boldsymbol{\xi}_t).$$

上述等价关系成立的主要原因是: 由问题 (7.3) 的约束, 可知每个阶段的原料都会被资源回收系统处理且残留物也会被处置, 不存在原料遗留到下一阶段的情况. 因此, 问题 (7.3) 的可行集关于 $t\in[T]$ 可分. 虽然我们将动态优化问题 (7.12) 转化为一个最小最大化问题 (7.14), 但是此问题通常是难以求解的 (Ben-Tal et al., 2004). 利用不确定预测集 $\mathcal{U}^{\Gamma}$ 的结构, 我们可以将问题 (7.14) 转化为一个混合整数线性规划问题.

命题 7.2　给定选址决策 $\boldsymbol{x}$ 和预算参数 Γ, 问题 (7.14) 等价于如下混合整数

线性规划问题:

$$
\begin{aligned}
\mathcal{Z}_{\text{NPV}}(\boldsymbol{x}) = \min_{\boldsymbol{q},\boldsymbol{g},\boldsymbol{\pi},\boldsymbol{h},\boldsymbol{\gamma}} & \sum_{j\in[J]}\left[\sum_{t\in[T]}\sum_{\tau\in[0:t-1]} x_{j\tau}s_j g_{jt} - \beta^{t-1}\boldsymbol{c}^{\top}\boldsymbol{x}_{t-1}\right] \\
& + \sum_{i\in[I]}\sum_{t\in[T]} q_{it}\left\{\left[\widehat{\xi}_{i0} + \sum_{\tau\in[t]}\mathcal{F}_{i\tau}\right]\right. \\
& \left. + \left[\sum_{\tau\in[t]}\left(h^{+}_{it\tau} - h^{-}_{it\tau}\right)\mathcal{S}_{i\tau}\right]\right\} - \sum_{i\in[I]}\sum_{t\in[T]}\left\{\widehat{\xi}_{i0} + \sum_{\tau\in[t]}\mathcal{F}_{i\tau}\right. \\
& \left. + \left[\sum_{\tau\in[t]}\left(\pi^{+}_{i\tau} - \pi^{-}_{i\tau}\right)\mathcal{S}_{i\tau}\right]\right\}\beta^t c_D \\
\text{s.t.}\quad & q_{it} + g_{jt} \geqslant \beta^t r_{ij} + \beta^t c_D, \forall\, i\in[I], j\in[J], t\in[T], \\
& \sum_{i\in[I]}(\pi^{+}_{it} + \pi^{-}_{it}) \leqslant \Gamma^{\mathrm{Z}}_t, \forall\, t\in[T], \\
& \sum_{t\in[T]}(\pi^{+}_{it} + \pi^{-}_{it}) \leqslant \Gamma^{\mathrm{T}}_i, \forall\, i\in[I], \\
& h^{+}_{it\tau} \leqslant q_{it}, \forall\, i\in[I], t\in[T], \tau\in[1:t], \\
& h^{+}_{it\tau} \geqslant q_{it} + (\pi^{+}_{i\tau} - 1)M, \quad \forall\, i\in[I], t\in[T], \tau\in[1:t], \\
& h^{+}_{it\tau} \leqslant \pi^{+}_{i\tau}M, \forall\, i\in[I], t\in[T], \tau\in[1:t], \\
& h^{-}_{it\tau} \leqslant q_{it}, \forall\, i\in[I], t\in[T], \tau\in[1:t], \\
& h^{-}_{it\tau} \geqslant q_{it} + (\pi^{-}_{i\tau} - 1)M, \forall\, i\in[I], t\in[T], \tau\in[1:t], \\
& h^{-}_{it\tau} \leqslant \pi^{-}_{i\tau}M, \forall\, i\in[I], t\in[T], \tau\in[1:t], \\
& \boldsymbol{q}\in\mathbb{R}^{I\times T}_{+}, \boldsymbol{g}\in\mathbb{R}^{J\times T}_{+}, \boldsymbol{\pi}^{+}, \boldsymbol{\pi}^{-}\in\{0,1\}^{I\times T}, \boldsymbol{h}^{+}, \boldsymbol{h}^{-}\in\mathbb{R}^{I\times T^2}_{+},
\end{aligned}
\tag{7.15}
$$

其中, M 为一个足够大的正数.

通过求解上述混合整数线性规划问题 (7.15), 可以得到极值情况下的原料产量 $\boldsymbol{\xi}^{\dagger}$.

命题 7.3 给定选址决策 $\boldsymbol{x}$ 和预算参数 Γ, 居民区 $i\in[I]$ 在阶段 $t\in[T]$ 的极值原料产量为

$$
\xi^{\dagger}_{it} = \widehat{\xi}_{i0} + \sum_{\tau\in[t]}\mathcal{F}_{i\tau} + \left[\sum_{\tau\in[t]}\left(\bar{\pi}^{+}_{i\tau} - \bar{\pi}^{-}_{i\tau}\right)\mathcal{S}_{i\tau}\right], \tag{7.16}
$$

其中, $\bar{\pi}_{i\tau}^{+}, \bar{\pi}_{i\tau}^{-}, i \in [I], \tau \in [1:t]$ 是预算参数 Γ 下求解问题 (7.15) 得到的二元决策变量.

7.4 多阶段鲁棒资源回收设施选址模型

基于预测不确定集 (7.10) 和资源回收设施净现值保证水平 (7.14), 本节介绍多阶段资源回收设施选址模型的构建. 为了表述简洁, 定义如下选址决策可行集:

$$\mathcal{X} := \left\{ \boldsymbol{x} \in \mathbb{R}^{T \times J} \mid \boldsymbol{x} \in \mathcal{X}_o, x_{t-1,j} \in \{0,1\},\ \forall\, t \in [T],\ j \in [J] \right\}. \tag{7.17}$$

利用命题 7.1, 我们构建如下多阶段鲁棒资源回收设施选址模型:

$$\max_{\boldsymbol{x} \in \mathcal{X}} \mathcal{Z}_{\mathrm{NPV}}(\boldsymbol{x}). \tag{7.18}$$

该模型的目标是最大化资源回收系统的总净现值保证水平, 且上述模型可以被转化为一个两阶段最大最小最大化问题 (two-stage max-min-max optimization problem):

$$\max_{\boldsymbol{x} \in \mathcal{X}} \mathcal{Z}_{\mathrm{NPV}}(\boldsymbol{x}) = \max_{\boldsymbol{x}} \left\{ \min_{\boldsymbol{\xi} \in \mathcal{U}^{\Gamma}} \max_{(\boldsymbol{y},\boldsymbol{z}) \in \mathcal{Y}(\boldsymbol{x},\boldsymbol{\xi})} \phi(\boldsymbol{x},\boldsymbol{y},\boldsymbol{z}) \,\middle|\, \boldsymbol{x} \in \mathcal{X} \right\}, \tag{7.19}$$

其中, $\phi(\boldsymbol{x},\boldsymbol{y},\boldsymbol{z})$ 和可行集 $\mathcal{Y}(\boldsymbol{x},\boldsymbol{\xi})$ 由式 (7.3) 和式 (7.13) 给出. 上述问题 (7.13) 是基于预算不确定集的两阶段自适应鲁棒优化问题, 此问题已被证明是 NP 难的 (Atamtürk and Zhang, 2007). 接下来, 本节介绍精确求解此问题的一个割平面算法 (cutting plane algorithm).

7.4.1 割平面精确求解算法

首先, 利用命题 7.2, 可以给出割平面的显式表达形式 (closed-form). 定义如下等价的原料产量预测集合 $\mathcal{V}(\Gamma)$:

$$\mathcal{V}(\Gamma) := \left\{ \boldsymbol{\xi} \in \mathbb{R}^{I \times T} \,\middle|\, \begin{array}{l} \xi_{it} = \widehat{\xi}_{i0} + \sum\limits_{\tau \in [t]} \mathcal{F}_{i\tau} + \left[\sum\limits_{\tau \in [t]} \left(\pi_{i\tau}^{+} - \pi_{i\tau}^{-} \right) \mathcal{S}_{i\tau} \right], \forall\, i \in [I], t \in [T], \\ \sum\limits_{i \in [I]} (\pi_{it}^{+} + \pi_{it}^{-}) \leqslant \Gamma_t^{\mathrm{Z}}, \forall t \in [T], \\ \sum\limits_{t \in [T]} (\pi_{it}^{+} + \pi_{it}^{-}) \leqslant \Gamma_i^{\mathrm{T}}, \forall i \in [I], \\ \pi_{it}^{+}, \pi_{it}^{-} \in \{0,1\}, \forall i \in [I], t \in [T] \end{array} \right\}. \tag{7.20}$$

通过在问题 (7.1) 的对偶形式中将集合 $\mathcal{U}^{\Gamma}$ 替换为集合 $\mathcal{V}(\Gamma)$, 可以得到命题 7.2 中给定 $\boldsymbol{x}$ 下问题 $\mathcal{Z}_{\mathrm{NPV}}(\boldsymbol{x})$ 的等价混合整数线性规划问题 (7.15). 此外, 为了表述简洁, 我们定义给定 $\boldsymbol{x} \in \mathcal{X}$ 与 $\boldsymbol{\xi} \in \mathcal{U}^{\Gamma}$ 下的集合 $\mathcal{K}(\boldsymbol{x}, \boldsymbol{\xi})$:

$$\mathcal{K}(\boldsymbol{x}, \boldsymbol{\xi}) := \left\{ \gamma \;\middle|\; \begin{array}{l} \gamma \leqslant \sum\limits_{t \in [T]} \left[\beta^t \sum\limits_{i \in \mathcal{I}} \sum\limits_{j \in [J]} r_{ij} y_{ijt} - \sum\limits_{i \in [I]} c_D z_{it} \right], \\ \sum\limits_{j \in [J]} y_{ijt} + z_{it} = \xi_{it}, \forall i \in [I], t \in [T], \\ \sum\limits_{i \in [I]} y_{ijt} \leqslant \sum\limits_{\tau \in [t]} x_{j\tau} s_j, \forall j \in [J], t \in [T], \\ y_{ijt}, z_{it} \in \mathbb{R}_+, \forall i \in [I], j \in [J], t \in [T] \end{array} \right\}. \tag{7.21}$$

基于集合 $\mathcal{V}(\Gamma)$ 的有限性, 可以把问题 (7.19) 等价地转化为一个大规模混合整数规划问题.

命题 7.4 不确定性环境下多阶段资源回收系统选址问题 (7.19) 等价于如下混合整数线性规划问题:

$$\begin{aligned} \max_{\boldsymbol{x}} \quad & \sum_{t \in [T]} \left[-\beta^{t-1} \boldsymbol{c}^{\top} \boldsymbol{x}_{t-1} \right] + \gamma \\ \text{s.t.} \quad & \gamma \in \mathcal{K}(\boldsymbol{x}, \boldsymbol{\xi}), \; \boldsymbol{\xi} \in \mathcal{V}(\Gamma), \\ & \boldsymbol{x} \in \mathcal{X}, \end{aligned} \tag{7.22}$$

其中, $\mathcal{K}(\boldsymbol{x}, \boldsymbol{\xi})$ 通过式 (7.21) 定义.

上述问题 (7.22) 为被称为问题 (7.19) 的完全主问题 (complete master problem). 该问题的规模取决于集合 $\mathcal{V}(\Gamma)$ 中点的个数. 此外, 记 ω 为集合 $\mathcal{V}(\Gamma)$ 中点的索引 (index), 即 $\omega = 1, 2, \cdots, |\mathcal{V}(\Gamma)|$, 其中, $|\mathcal{V}(\Gamma)|$ 为集合 $\mathcal{V}(\Gamma)$ 中点的总个数. 放松约束 $\gamma \in \mathcal{K}(\boldsymbol{x}, \boldsymbol{\xi})$, $\boldsymbol{\xi} \in \mathcal{V}(\Gamma)$ 为给定的点 $\omega = 1, \cdots, k, k \leqslant |\mathcal{V}(\Gamma)|$, 可以构建如下松弛主问题 (relaxed master problem):

$$\begin{aligned} \mathrm{UB}_k := \max_{\boldsymbol{x}} \quad & \sum_{t \in [T]} \left[-\beta^{t-1} \boldsymbol{c}^{\top} \boldsymbol{x}_{t-1} \right] + \gamma \\ \text{s.t.} \quad & \gamma \in \mathcal{K}(\boldsymbol{x}, \boldsymbol{\xi}^{\omega}), \; \omega = 1, \cdots, k, \\ & x \in \mathcal{X}, \end{aligned} \tag{7.23}$$

其中, $k \leqslant |\mathcal{V}(\Gamma)|$. 与完全主问题 (7.22) 相比, 松弛主问题 (7.23) 的规模较小, 并给出问题 (7.19) 的一个上界. 同时, 随着新的割平面 $\mathcal{K}(\boldsymbol{x}, \boldsymbol{\xi}^{\omega})$ 在迭代过程中加入

到松弛主问题 (7.23) 中, 上界 UB_k 会逐渐减小. 此外, 给定松弛主问题 (7.23) 的最优选址决策 $\hat{\boldsymbol{x}}$, 通过求解给定 $\hat{\boldsymbol{x}}$ 的问题 (7.14), 我们可以得到问题 (7.19) 的一个上界:

$$\mathrm{LB}(\hat{\boldsymbol{x}}) := \mathcal{Z}_{\mathrm{NPV}}(\hat{\boldsymbol{x}}) = \max_{(\boldsymbol{y},\boldsymbol{z})\in\mathcal{Y}(\hat{\boldsymbol{x}},\boldsymbol{\xi})} \phi_{\mathrm{NPV}}(\hat{\boldsymbol{x}},\boldsymbol{y},\boldsymbol{z}), \tag{7.24}$$

利用命题 7.2, 上述问题的最优值可以通过求解其等价混合整数线性规划问题 (7.15) 来获得. 当 $\mathrm{LB}(\hat{\boldsymbol{x}}) = \mathrm{UB}_k$ 时, 我们得到问题 (7.19) 的最优值和最优解. 特别地, 因为集合 $\mathcal{V}(\Gamma)$ 是有限的, 所以算法在有限步迭代后收敛.

割平面 $\mathcal{K}(\boldsymbol{x},\boldsymbol{\xi}^{\omega})$ 的选择是影响算法计算效率的关键因素. 为了提高割平面的质量, 给定松弛主问题 (7.23) 的解 $\boldsymbol{x}$, 求解问题 (7.24) 的等价问题 (7.15) 可得解 $(\bar{\boldsymbol{\pi}}^+,\bar{\boldsymbol{\pi}}^-)$. 根据命题 7.3, 可以基于上述解构造一个极值情况下原料产量 $\boldsymbol{\xi}^{\dagger}\in\mathcal{V}(\Gamma)$:

$$\xi_{it}^{\dagger}(\bar{\boldsymbol{\pi}}^+,\bar{\boldsymbol{\pi}}^-) := \widehat{\xi}_{i0} + \sum_{\tau\in[t]} \mathcal{F}_{i\tau} + \left[\sum_{\tau\in[t]} \left(\bar{\pi}_{i\tau}^+ - \bar{\pi}_{i\tau}^-\right)\mathcal{S}_{i\tau}\right], \quad \forall\, i\in[I], t\in[T], \tag{7.25}$$

因此, 基于点 $\boldsymbol{\xi}^{\dagger}(\bar{\boldsymbol{\pi}}^+,\bar{\boldsymbol{\pi}}^-)$ 可以构建一个高质量 (或有效) 割平面 $\mathcal{K}(\boldsymbol{x},\boldsymbol{\xi}^{\dagger}(\bar{\boldsymbol{\pi}}^+,\bar{\boldsymbol{\pi}}^-))$.

现在, 我们总结求解多阶段鲁棒资源回收设施选址模型 (7.19) 的割平面算法.

算法 4　割平面算法

输入: 模型参数, 最优性间隙容差 (optimality gap tolerance) ϵ.

初始化: $\mathrm{UB}\leftarrow+\infty, \mathrm{LB}\leftarrow-\infty$, 找到一些集合 $\mathcal{V}(\Gamma)$ 中的点 $\boldsymbol{\xi}^{\omega}$.

1. 求解问题 (7.23) 得到最优值 (上界) UB_N 和最优解 $\boldsymbol{x}_N$, 其中, N 为迭代次数.
2. $\mathrm{UB}\leftarrow\mathrm{UB}_N$, $\boldsymbol{x}_{LB}\leftarrow\boldsymbol{x}_N$.
3. 求解问题 $\mathrm{LB}(\boldsymbol{x}_{LB})$ 得到最优值 (下界)LB_N 和最优解 $(\bar{\boldsymbol{\pi}}^+,\bar{\boldsymbol{\pi}}^-)$.
4. 利用式 (7.25) 构建 $\boldsymbol{\xi}^{\dagger}(\bar{\boldsymbol{\pi}}^+,\bar{\boldsymbol{\pi}}^-)$ 并获得一个新割平面 $\mathcal{K}(\boldsymbol{x},\boldsymbol{\xi}^{\dagger}(\bar{\boldsymbol{\pi}}^+,\bar{\boldsymbol{\pi}}^-))$ 加入问题 (7.23).
5. $\mathrm{LB}\leftarrow\max\{\mathrm{LB},\mathrm{LB}_N\}$.
6. 如果 $\dfrac{\mathrm{UB}-\mathrm{LB}}{\mathrm{LB}}\times 100\% \leqslant \epsilon$, 则 $\boldsymbol{x}^{\star}\leftarrow\boldsymbol{x}_{LB}$ 且算法停止; 否则, 返回第 1 步.

输出: 最优选址决策 $\boldsymbol{x}^{\star}$.

7.4.2　基于原料产量固定比例决策规则的多阶段资源回收设施选址模型

本节介绍一种求解多阶段鲁棒资源回收设施选址问题 (7.19) 的启发式方法. 虽然此方法求得的解不一定是最优解, 但是该方法针对大规模问题 (7.19) 的计算效率高. 记 p_{tij} 是从居民区 i 在第 t 阶段运送到系统回收设施 j 的原料产量的占此阶段居民区 i 总原料产量的比例, q_{ti} 是居民区 i 在第 t 阶段的原料残留量占总原料产量比例. 此外, 记 $(\boldsymbol{p},\boldsymbol{q})$ 为原料产量比例的全部参数, 特别地, 满足 $\sum_{j\in[J]} p_{tij}+q_{ti}=1,\ \forall\ i\in[I],t\in[T]$. 给定原料产量 $\boldsymbol{\xi}$ 和原料产量分配比例 $(\boldsymbol{p},\boldsymbol{q})$, 原料分配决策 $(\boldsymbol{y},\boldsymbol{z})$ 属于如下集合:

$$\boldsymbol{\Lambda}(\boldsymbol{p},\boldsymbol{q},\boldsymbol{\xi}):=\Big\{(\boldsymbol{y},\boldsymbol{z})\ \Big|\ y_{ijt}=p_{ijt}\xi_{it},\ z_{it}=q_{it}\xi_{it},\ \forall t\in[T],i\in[I],j\in[J]\Big\}. \tag{7.26}$$

在上述集合中, 虽然决策 $(\boldsymbol{y},\boldsymbol{z})$ 依赖于不确定原料产量 $\boldsymbol{\xi}$, 但是其与 $\boldsymbol{\xi}$ 为线性关系且比例系数为 $(\boldsymbol{p},\boldsymbol{q})$. 基于集合 $\boldsymbol{\Lambda}(\boldsymbol{p},\boldsymbol{q},\boldsymbol{\xi})$, 我们构建如下多阶段鲁棒资源回收设施选址模型:

$$\begin{aligned}\max_{\boldsymbol{x},\boldsymbol{p},\boldsymbol{q}}\quad & \min_{\boldsymbol{\xi}\in\mathcal{U}^{\Gamma}}\ \max_{(\boldsymbol{y},\boldsymbol{z})\in\mathcal{Y}(\boldsymbol{x},\boldsymbol{\xi})\cap\boldsymbol{\Lambda}(\boldsymbol{p},\boldsymbol{q},\boldsymbol{\xi})} \phi(\boldsymbol{x},\boldsymbol{y},\boldsymbol{z})\\ \text{s.t.}\quad & \sum_{j\in[J]} p_{ijt}+q_{it}=1,\forall\ i\in[I],t\in[T],\\ & p_{ijt},q_{it}\geqslant 0,\forall\ i\in[I],j\in[J],t\in[T],\\ & \boldsymbol{x}\in\mathcal{X},\end{aligned} \tag{7.27}$$

其中, $\mathcal{Y}(\boldsymbol{x},\boldsymbol{\xi})$ 是给定原料产量 $\boldsymbol{\xi}$ 下决策 $(\boldsymbol{y},\boldsymbol{z})$ 的可行集, 其由式 (7.13) 给出. 虽然上述模型利用原料产量比例 $(\boldsymbol{p},\boldsymbol{q})$ 限制决策 $(\boldsymbol{y},\boldsymbol{z})$ 的可行范围, 但是上述模型的计算效率高. 特别地, 它可以被转化为一个规模适中的混合整数线性规划问题.

命题 7.5 模型 (7.27) 等价于如下混合整数线性规划问题:

$$\begin{aligned}\max_{\boldsymbol{x},\boldsymbol{p},\boldsymbol{q}}\ & \sum_{i\in[I]}\sum_{t\in[T]}\left[\widehat{\xi}_{i0}+\sum_{\tau\in[t]}\mathcal{F}_{i\tau}\right]d_{it}^{o}\\ & +\sum_{t\in[T]}\left[\Gamma_t^{\mathrm{Z}}b_t^{o}-\beta^{t-1}\boldsymbol{c}^{\top}\boldsymbol{x}_{t-1}\right]+\sum_{i\in[I]}\Gamma_i^{\mathrm{T}}\varphi_i^{o}+\sum_{i\in[I]}\sum_{t\in[T]}\left[\psi_{it}^{o}-\nu_{it}^{o}\right]\\ \text{s.t.}\ & \beta^{t}\left[\sum_{j\in[J]}r_{ij}p_{ijt}-c_Dq_{it}\right]-d_{it}^{o}=0,\forall\ i\in[I],t\in[T],\\ & b_t^{o}+\varphi_i^{o}+\gamma_{it}^{o}+\varpi_{it}^{o}=0,\forall\ i\in[I],t\in[T],\\ & \psi_{it}^{o}+\nu_{it}^{o}+\varpi_{it}^{o}-\gamma_{it}^{o}-\sum_{\tau\in[t:T]}d_{i\tau}^{o}\mathcal{S}_{i\tau}=0,\forall\ i\in[I],t\in[T],\\ & \sum_{i\in[I]}\left[\widehat{\xi}_{i0}+\sum_{\tau\in[t]}\mathcal{F}_{i\tau}\right]p_{tij}+\sum_{\varsigma\in[T]}\Gamma_{\varsigma}^{\mathrm{Z}}b_{\varsigma}^{tj}+\sum_{i\in[I]}\Gamma_i^{\mathrm{T}}\varphi_i^{tj}\end{aligned}$$

$$
\begin{aligned}
&+\sum_{\varsigma\in[T]}\sum_{i\in[I]}\left[\psi_{i\varsigma}^{tj}-\nu_{i\varsigma}^{tj}\right]\leqslant\sum_{\tau\in[0:t-1]}x_{j\tau}s_j,\forall\, j\in[J],t\in[T],\\
&\sum_{j\in[J]}p_{ijt}+q_{it}=1,\forall\, i\in[I],t\in[T],\\
&b_{\varsigma}^{tj}+\varphi_i^{tj}+\gamma_{i\varsigma}^{tj}+\varpi_{i\varsigma}^{tj}=0,\forall\, i\in[I],\varsigma\in[T],j\in[J],t\in[T],\\
&\psi_{i\varsigma}^{tj}+\nu_{i\varsigma}^{tj}+\varpi_{i\varsigma}^{tj}-\gamma_{i\varsigma}^{tj}=0,\forall\, i\in[I],j\in[J],t\in[T],\varsigma\in[t+1:T],\\
&\psi_{i\varsigma}^{tj}+\nu_{i\varsigma}^{tj}+\varpi_{i\varsigma}^{tj}-\gamma_{i\varsigma}^{tj}-p_{tij}\mathcal{S}_{it}=0,\forall\, i\in[I],j\in[J],t\in[T],\varsigma\in[1:t],\\
&\boldsymbol{d}^o\in\mathbb{R}^{I\times T},\boldsymbol{b}^o\in\mathbb{R}_-^T,\boldsymbol{\varphi}^o\in\mathbb{R}_-^I,\boldsymbol{\gamma}^o,\boldsymbol{\varpi}^o,\boldsymbol{\nu}^o\in\mathbb{R}_+^{I\times T},\\
&\boldsymbol{\psi}^o\in\mathbb{R}_-^{I\times T},\boldsymbol{q}\in\mathbb{R}_+^{I\times J\times T},\boldsymbol{q}\in\mathbb{R}_+^{I\times T},\boldsymbol{x}\in\mathcal{X},\\
&\boldsymbol{b}^{tj}\in\mathbb{R}_+^T,\boldsymbol{\varphi}^{tj}\in\mathbb{R}_+^I,\boldsymbol{\gamma}^{tj},\boldsymbol{\varpi}^{tj},\boldsymbol{\nu}^{tj}\in\mathbb{R}_-^{I\times T},\boldsymbol{\psi}^{tj}\in\mathbb{R}_+^{I\times T},\forall\, j\in[J],t\in[T],
\end{aligned}
\tag{7.28}
$$

其中, $\boldsymbol{d}^o,\boldsymbol{b}^o,\boldsymbol{\varphi}^o,\boldsymbol{\gamma}^o,\boldsymbol{\varpi}^o,\boldsymbol{\nu}^o,\boldsymbol{\psi}^o,\boldsymbol{b}^{tj},\boldsymbol{\varphi}^{tj},\boldsymbol{\gamma}^{tj},\boldsymbol{\varpi}^{tj},\boldsymbol{\nu}^{tj},\boldsymbol{\psi}^{tj},\forall\, j\in[J],t\in[T]$ 都为辅助变量.

上述问题 (7.28) 的最优值是问题 (7.19) 的下界. 因此, 问题 (7.27) 的最优解可以作为割平面算法的一个起始点.

7.5 数值实验

本节基于一个废弃物数据, 设计数值验证实验模型与算法的性能. 考虑一个具有 $I=10$ 个居民区, $J=20$ 个资源回收设施备选地址以及 $T=8$ 个决策阶段的多阶段资源回收设施选址问题, 其中使用新加坡 10 年的废弃物数据来估计原料产量增量预测模型的参数. 我们使用指数加权移动平均预测模型 (例 13) 和平均绝对误差 (例 15) 来构架不确定原料产量预测集 $\mathcal{U}^{\Gamma}$. 本节数值实验主要包括两个部分: ① 给定选址决策 $\boldsymbol{x}$ 分析预测误差预算对极值情况原料产量的影响; ② 资源回收系统选址决策分析.

7.5.1 预测误差预算对极值情况下原料产量的影响

在给定选址决策 $\boldsymbol{x}$ 下, 本节分析预测误差预算对极值情况下原料产量的影响. 我们给定一个选址决策 $\boldsymbol{x}$ (表 7.1), 并利用问题 (7.15) 和式 (7.16) 来计算此决策对应的不同误差预算 Γ 与不同居民区 $i\in[I]$ 的极值原料产量 $\xi_{it}^{\dagger}$, 与二元变量 $\pi_{it}^{+},\pi_{it}^{-}$, $i\in[I],t\in[T]$. 其次, 我们设预算误差 $\Gamma_i^{\mathrm{T}}\equiv\Gamma^{\mathrm{T}}$, $\forall i\in[I]$, 且 $\Gamma_t^{\mathrm{I}}\equiv\Gamma^{\mathrm{I}}$, $\forall t\in[T]$, 并令误差预算 $(\Gamma^{\mathrm{Z}},\Gamma^{\mathrm{T}})$ 的变化范围从 $(5,4)$ 到 $(10,8)$. 此外,

我们还计算不同预算水平下系统的净现值保证水平.

表 7.1 资源回收设施选址决策

决策阶段	$t=1$	$t=2$	$t=3$	$t=4$	$t=5$	$t=6$	$t=7$	$t=8$
每阶段新增选址 $\boldsymbol{x}$	3, 5, 11, 14, 17-18	2	–	10	1	4	20	–

表 7.2 给出了不同预测误差预算下极值原料产量与系统净现值保证水平, 其中, 居民区 4 为原料产量较高地区, 居民区 8 为原料产量较低地区. 结果表明: ① 当预测误差预算较低时, 低于名义原料产量预测值的极值情况下原料产量 $\boldsymbol{\xi}^{\dagger}$ 是影响系统利润保证水平的重要因素, 例如, 当预算值从 $(5,4)$ 变为 $(7,5)$ 时, $\boldsymbol{\pi}^{+}$ 的所有元素都为 0, 且 $\boldsymbol{\pi}^{-}$ 的部分元素为 1, 这说明选址决策对应的极值原料产量小于名义原料产量预测值; ② 当预测误差预算较高时, 高于名义原料产量预测值的极值水平对系统利润保证水平的影响最大, 例如, 当预算值达到 $(8,6)$ 时, $\boldsymbol{\pi}^{-}$ 的所有元素都为 0, 且 $\boldsymbol{\pi}^{+}$ 的部分元素为 1, 这表明在较高的预测误差预算水平下, 选址决策对应的极值原料产量大于名义原料产量预测值.

表 7.2 不同预测误差预算下极值原料产量与系统净现值保证水平

(Γ^{Z},Γ^{T})	净现值保证水平	$\boldsymbol{\xi}_4^{\dagger}$(居民区 4)		$\boldsymbol{\xi}_8^{\dagger}$(居民区 8)	
		$\boldsymbol{pi}_4^{+}$	$\boldsymbol{\pi}_4^{-}$	$\boldsymbol{\pi}_8^{+}$	$\boldsymbol{\pi}_8^{-}$
$(10,8)$	1.6562×10^{8}	$(1,1,1,1,1,1,1,1)$	$(0,0,0,0,0,0,0,0)$	$(1,1,1,1,1,1,1,1)$	$(0,0,0,0,0,0,0,0)$
$(9,7)$	1.6899×10^{8}	$(1,1,1,1,1,1,1,0)$	$(0,0,0,0,0,0,0,0)$	$(0,0,0,0,0,0,1,1)$	$(0,0,0,0,0,0,0,0)$
$(8,6)$	1.7012×10^{8}	$(1,1,1,1,1,1,0,0)$	$(0,0,0,0,0,0,0,0)$	$(0,0,0,0,0,0,1,1)$	$(0,0,0,0,0,0,0,0)$
$(7,5)$	1.7118×10^{8}	$(0,0,0,0,0,0,0,0)$	$(1,1,1,1,0,1,0,0)$	$(0,0,0,0,0,0,0,0)$	$(0,0,0,0,1,1,1,1)$
$(6,4)$	1.7346×10^{8}	$(0,0,0,0,0,0,0,0)$	$(1,1,1,0,1,0,0,0)$	$(0,0,0,0,0,0,0,0)$	$(0,0,0,0,1,1,1,1)$
$(5,4)$	1.7444×10^{8}	$(0,0,0,0,0,0,0,0)$	$(0,0,0,0,1,1,1,1)$	$(0,0,0,0,0,0,0,0)$	$(0,0,0,0,0,1,1,1)$

7.5.2 资源回收系统选址决策分析

本节, 我们比较利用割平面精确求解算法 (算法 4) 得到的模型 (7.18) 的最优选址决策与利用原料产量固定比例决策规则得到的模型 (7.28) 的最优选址决策. 为了表述简洁, 我们记原料产量固定比例决策规则为 “$(\boldsymbol{p},\boldsymbol{q})$-决策规则”.

表 7.3 给出了不同预测误差预算下两种计算方法对应的最优选址决策与设施容量. 结果表明: ① 当预测误差预算从 $(10,8)$ 减少至 $(5,4)$ 时, 两个模型对应的最优设施容量都在减少, 且最优选址个数不增或减少. 这说明当未来原料产量的扰动减小 (或预测精度提升) 时, 决策者通过减少系统设施容量来节约成本. ② 割平面算法对应的最优选址在初始决策阶段的容量比 $(\boldsymbol{p},\boldsymbol{q})$-决策规则对应的最优容量低, 且在之后的决策阶段中逐渐大量增加系统的容量. 此外, 割平面算法对应的最优选址的总容量比 $(\boldsymbol{p},\boldsymbol{q})$-决策规则对应的总容量低. 在大多数情况下, 割平面

算法对应的最优决策与 $(\boldsymbol{p},\boldsymbol{q})$-决策规则对应的最优决策相比, 要么使用较少的决策阶段来扩充系统容量 (例如, 在 $(\Gamma^{\mathrm{Z}},\Gamma^{\mathrm{T}})=(10,8)$ 或 $(8,6)$ 时, 割平面算法对应的最优决策在 4 个决策阶段扩展系统容量, 而 $(\boldsymbol{p},\boldsymbol{q})$-决策规则在 5 个决策阶段扩展容量), 要么在后期决策阶段拓展容量 (例如, 在 $(\Gamma^{\mathrm{Z}},\Gamma^{\mathrm{T}})=(8,6)$ 时, 割平面算法对应的最优决策阶段是 1-3-5-7, 而 $(\boldsymbol{p},\boldsymbol{q})$-决策规则的是 1-3-4-5-6, 此外, 在 $(\Gamma^{\mathrm{Z}},\Gamma^{\mathrm{T}})=(5,4)$ 时, 两者对应的最优决策阶段分别是 1-3-5-8 和 1-3-5-6). 这表明割平面算法或精确求解方法在减少系统容量投资和固定成本方面的有效性.

表 7.3　不同预测误差预算下两种计算方法对应的最优选址决策与设施容量

求解方法	$(\Gamma^{\mathrm{Z}},\Gamma^{\mathrm{T}})$	每个决策阶段选择的地址及设施容量							
		$t=1$	$t=2$	$t=3$	$t=4$	$t=5$	$t=6$	$t=7$	$t=8$
割平面算法	$(10,8)$	$2,3,6,11,14,17\text{-}18$	–	5	–	20	–	8	–
		5408830	–	669650	–	1511400	–	928870	–
	$(8,6)$	$2,3,6,11,14,17\text{-}18$	–	5	–	20	–	1	–
		5408830	–	669650	–	1511400	–	720260	–
	$(5,4)$	$2,3,6,11,14,17\text{-}18$	–	5	–	20	–	–	1
		5408830	–	669650	–	1511400	–	–	720260
$(\boldsymbol{p},\boldsymbol{q})$-决策规则	$(10,8)$	$2,3,5,11,14,17\text{-}18$	–	8	–	1	20	–	13
		5658200	–	928870	–	720260	1511400	–	933860
	$(8,6)$	$2,3,5,11,14,17\text{-}18$	–	6	1	13	20	–	–
		5658200	–	420280	720260	928870	1511400	–	–
	$(5,4)$	$2,3,5,11,14,17\text{-}18$	–	8	–	13	20	–	–
		5658200	–	928870	–	720260	1511400	–	–

表 7.4　不同预测误差预算下两种计算方法的净现值保证水平

$(\Gamma^{\mathrm{Z}},\Gamma^{\mathrm{T}})$	割平面算法	$(\boldsymbol{p},\boldsymbol{q})$-决策规则	净现值保证水平的增长比例
$(10,8)$	16.89×10^7	16.79×10^7	0.59%
$8,6$	17.21×10^7	16.97×10^7	1.39%
$5,4$	17.57×10^7	17.51×10^7	0.34%

最后, 将 $(\boldsymbol{p},\boldsymbol{q})$-决策规则对应的最优选址代入模型 (7.15) 中, 可以得到其对应的净现值保证水平. 表 7.4 给出了不同预测误差预算下两种计算方法的净现值保证水平, 以及割平面算法对应的最优决策的净现值保证水平相比于 $(\boldsymbol{p},\boldsymbol{q})$-决策规则对应的净现值保证水平的相对增加比例. 结果显示: $(\boldsymbol{p},\boldsymbol{q})$-决策规则对应的净现值保证水平与割平面算法对应的净现值保证水平十分相近, 这表明使用 $(\boldsymbol{p},\boldsymbol{q})$-决策规则作为割平面算法的替代方法来求解多阶段鲁棒资源回收系统设施选址模型 (7.19) 的可行性.

7.6 本章小结

本章介绍了多阶段鲁棒资源回收系统设施选址模型的构建与求解方法. 该模型具有如下特点: 首先, 资源回收设施选址决策具有鲁棒性, 即可抵抗原料产量不确定性的影响. 因为在实际情况中不确定原料产量的真实情况无法完全符合模型的假设, 所以任何不确定环境下的决策都是次优的. 然而, 本章模型的最优选址决策可以保证真实的利润不低于模型的名义利润 (模型的最优目标值); 其次, 本章模型具有良好的数据驱动性, 且模型的构建仅需要适量的不确定原料产量信息. 因为模型预测不确定集的构建仅使用历史原料产量数据, 并且不依赖于不确定参数的概率分布假设, 所以本章所介绍的模型不会使决策者在数据收集和参数校准等工作上耗费大量的时间和资金, 这表明模型具有很强的实用价值; 最后, 本章提出了求解模型的割平面精确算法, 在每次迭代时只需求解一个规模适中的混合整数线性规划问题, 这表明本章所介绍的模型具有高效计算性 (tractability).

关于不确定环境下资源回收设施选址问题研究, 读者可以参考更多相关文献, 例如, Šomplák 等 (2013) 考虑了不确定性下的焚烧设施选址优化问题. 该问题在热需求和能源价格等参数不确定的情况下, 确定最佳的垃圾处理设施容量. Wang 等 (2012) 开发了一个废弃物管理系统规划模型, 其使用区间值三角模糊集来描述成本系数, 并使用离散随机变量来描述废弃物产生量. Yeomans (2007) 采用了一种进化仿真优化方法来研究固体废弃物管理问题. 其所提出模型的目标是最小化不确定性造成的负面影响, 并通过在进化仿真优化方法中结合罚函数与灰色规划来求解含有显著不确定性的固体废弃物管理问题. 更多相关研究可参见 (Sun et al., 2012; Tan et al., 2010; Xu et al., 2009).

第 8 章 时间序列分布不确定集与多阶段选址优化

第 7 章介绍了鲁棒性多阶段选址优化问题. 本章主要介绍具有商品需求分布不确定性的多阶段 (或多期) 分布鲁棒枢纽选址问题. 结合时间序列 (time-series) 预测模型, 本章构建基于 Wasserstein 距离的嵌套型分布不确定集 (nested ambiguity set), 并构建一个预算驱动型 (budget-driven) 多阶段枢纽选址模型. 在枢纽容量充足的情况下, 该模型本质上在优化各阶段参考概率分布 (reference probability distribution) 下的夏普比率 (sharpe ratio). 此外, 容量充足型枢纽选址模型可以通过二分法 (bisection search algorithm) 高效求解, 其中每次迭代求解一个混合整数锥优化问题 (mixed-integer conic program). 在枢纽容量限制的情况下, 该模型可以通过拓展线性决策规则 (extended linear decision rule) 方法求得近似解. 最后, 设计数值实验来验证模型性能. 本章的主要结构如下: 8.1 节介绍多阶段枢纽选址问题的背景; 8.2 节介绍嵌套型分布不确定集以及容量充足条件下成本预算驱动型多阶段枢纽选址模型的构建; 8.3 节介绍容量充足型枢纽选址模型的高效计算形式; 8.4 节介绍多阶段容量限制型枢纽选址模型; 最后, 8.5 节介绍实验及结果分析.

8.1 背 景

本节主要介绍多阶段枢纽选址问题的背景, 其中, 该问题每阶段的商品需求具有不确定性, 并且随着时间 (阶段) 推移决策者依次观测到商品需求的实现值. 在观察到商品需求实现值前, 决策者需要在初始阶段确定供应链网络的最优枢纽选址决策, 在每个阶段观测到需求实现值后确定商品的最优运输路线. 多阶段枢纽选址问题在优化建模方面有两大挑战: ① 路径依赖型需求不确定性 (path-dependent uncertainty of demand), 以及 ② 各阶段运营成本波动的风险. 首先, 多阶段枢纽选址问题的选址决策实施周期长, 这导致未来各阶段商品需求是动态的且不确定的. 因此, 商品需求具有路径依赖不确定性 (path-dependent uncertainty). 忽视问题中商品需求的不确定性将导致仓库选址、供应链枢纽选址、货运运输以及其他多阶段供应链优化问题的决策失误 (Snyder, 2006; Unnikrishnan et al., 2009). 因此, 本章重点考虑多阶段枢纽选址问题的路径依赖型需求不确定性. 其次, 供应链网络的建设需要大量的前期投资, 因此运营期内需要保持相对稳定的现金流. 然而, 高度不确定的商品需求造成各阶段间运营成本大幅度波动, 这导致供应链企业易面临财务风险，甚至会造成企业破产. 例如，因现金流中断而延期数十年的柏林勃兰登堡机场项目 (Kostka and Fiedler, 2016). 因此，降低各阶段运营成本波动风险是保证供应链网络平稳运行的一个重要因素.

8.2 多阶段容量充足型枢纽选址模型

本节介绍容量充足条件下多阶段枢纽选址模型. 首先，介绍确定性多阶段枢纽选址模型. 其次，构建预算驱动型多阶段枢纽选址模型. 最后，基于时间序列预测模型，构建嵌套型 Wasserstein 分布不确定集.

8.2.1 确定性多阶段枢纽选址模型

考虑有 I 个备选枢纽地址，K 种商品以及 T 个决策阶段的多阶段枢纽选址问题. 我们记 u_{kt} 为在第 t 阶段第 k 种商品的需求且向量 $\boldsymbol{u}_t=(u_{kt})_{k\in[K]}, t\in[T]$. 记 v_{ijkt} 为在第 t 阶段通过枢纽 i 和 j 运输商品 k 的单位成本. 记 c_{it} 为在第 t 阶段第 i 个枢纽备选地址建设枢纽的固定成本. 该问题有两个决策变量: x_{it} 为 0-1 二元选址决策变量，当且仅当在 t 阶段第 i 个枢纽备选地址建造枢纽时，$x_{it}=1$. 此外，记第 t 阶段的选址决策向量 $\boldsymbol{x}_t=(x_{it})_{i\in[I]}$ 且 $\boldsymbol{X}_t=(\boldsymbol{x}_1,\cdots,\boldsymbol{x}_t)=(x_{is})_{i\in[I],s\leqslant t}$ 为从第 1 阶段到第 t 阶段的选址决策向量. 特别地，假设一个枢纽一旦建成 (或被选择)，在之后的阶段此枢纽将一直保持运行状态. z_{ijkt} 为运输决策，其为在第 t 阶段经一个枢纽 (当 $i=j$ 时) 或两个枢纽 i 和 j 运输第 k 种商品的量. 给定选址决策 $\boldsymbol{X}_t$ 和商品需求 $\boldsymbol{u}_t$，第 t 阶段的商品运输问题 $\mathrm{Q}_t(\boldsymbol{X}_t,\boldsymbol{u}_t)$ 为

$$
\begin{aligned}
\mathrm{Q}_t(\boldsymbol{X}_t,\boldsymbol{u}_t):=\min\quad & \sum_{k\in[K]}\sum_{i,j\in[I]} v_{ijkt}z_{ijk}\\
\text{s.t.}\quad & \sum_{i,j\in[I]} z_{ijk}=u_{kt},\ \forall k\in[K],\\
& \sum_{j\in[I]} z_{ijk}+\sum_{j\in[I]\setminus\{i\}} z_{jik}\leqslant\sum_{s\in[t]} x_{is}u_{kt},\ \forall k\in[K],\ i\in[I],\\
& z_{ijk}\geqslant 0,\ \forall k\in[K],\ i,j\in[I],
\end{aligned}
$$

其中，第一组约束保证每种商品的需求都被满足，第二组约束保证商品不会通过未选择的枢纽点运输. 容量充足型枢纽选址模型的运输成本可表示为关于需求 $\boldsymbol{u}_t$ 的线性函数 (Hamacher et al., 2004) 如下:

$$
\mathrm{Q}_t(\boldsymbol{X}_t,\boldsymbol{u}_t)=\sum_{k\in[K]} u_{kt}\cdot f_{kt}(\boldsymbol{X}_t),\quad \forall t\in[T],
$$

这里, $f_{kt}(\boldsymbol{X}_t)$ 为选址决策 $\boldsymbol{X}_t$ 下商品 k 在第 t 阶段每单位运输成本, 其定义如下:

$$\begin{aligned} f_{kt}(\boldsymbol{X}_t) := \min \quad & \sum_{i,j\in[I]} v_{ijkt}y_{ij} \\ \text{s.t.} \quad & \sum_{i,j\in[I]} y_{ij} = 1, \\ & \sum_{j\in[I]} y_{ij} + \sum_{j\in[I]\setminus\{i\}} y_{ji} \leqslant \sum_{s\in[t]} x_{is},\ \forall i\in[I], \\ & y_{ij} \geqslant 0,\ \forall i,j\in[I], \end{aligned}$$

其中, y_{ij} 为通过枢纽 i 和 j 运送商品 k 的量占该商品总量的比例. 给定商品需求 $\boldsymbol{u}_t, t\in[T]$, 我们构建如下多阶段容量充足型枢纽选址问题如下:

$$\begin{aligned} \min \quad & \sum_{t\in[T]} \tau_t \\ \text{s.t.} \quad & \boldsymbol{c}_t^\top \boldsymbol{x}_t + \mathrm{Q}_t(\boldsymbol{X}_t, \boldsymbol{u}_t) \leqslant \tau_t,\ \forall t\in[T], \\ & \boldsymbol{X}\in\mathcal{X},\ \boldsymbol{\tau}\geqslant \boldsymbol{0}, \end{aligned} \tag{8.1}$$

其中, 上述问题的目标是最小化 T 个阶段的枢纽建造成本和商品运输成本. 对于每个 $t\in[T]$, 记 $\boldsymbol{c}_t=(c_{it})_{i\in[I]}$, $\boldsymbol{\tau}=(\tau_1,\cdots,\tau_T)$, τ_t 为第 t 阶段总成本的提升 (epigraphical) 决策变量, $\mathcal{X}$ 是选址决策 $\boldsymbol{X}=(\boldsymbol{x}_1,\cdots,\boldsymbol{x}_T)=(x_{it})_{i\in[I],t\in[T]}$ 的可行集. 特别地, 我们假设枢纽设施被选择后, 则一直保持开放, 即 $\mathcal{X}=\{\boldsymbol{x}\in\{0,1\}^{I\times T}|\sum_{t\in[T]}x_{it}\leqslant 1,\ \forall i\in[I]\}$.

8.2.2　预算驱动型多阶段枢纽选址模型

为了保证供应链运营的可持续性, 决策者需要控制各阶段的成本. 因此, 每个阶段的成本预算 τ_t 会被预先设定. 此外, 决策者还需要考虑商品需求的不确定性. 假设真实商品需求的概率分布 $\mathbb{P}$ 属于一个分布不确定集 $\mathcal{F}(\theta)$, 其中, θ 表示分布不确定集的大小, 即当 $\theta_1\leqslant\theta_2$ 时, $\mathcal{F}(\theta_1)\subseteq\mathcal{F}(\theta_2)$. 给定成本预算 $\boldsymbol{\tau}$, 在不违反成本预算的前提下, 我们通过最大化分布不确定集的大小 θ 来获得一个能够对抗商品需求分布不确定性的选址决策 $\boldsymbol{X}$. 特别地, 我们构建如下预算驱动型枢纽选址模型:

$$\begin{aligned} \max \quad & \theta \\ \text{s.t.} \quad & \mathrm{Z}_t(\boldsymbol{X},\theta) = \sup_{\mathbb{P}\in\mathcal{F}(\theta)} \mathbb{E}_{\mathbb{P}}[\boldsymbol{c}_t^\top\boldsymbol{x}_t + \mathrm{Q}_t(\boldsymbol{X}_t,\tilde{\boldsymbol{u}}_t)] \leqslant \tau_t,\ \forall t\in[T], \\ & \boldsymbol{X}\in\mathcal{X},\theta\geqslant 0, \end{aligned} \tag{8.2}$$

其中, $\mathrm{Z}_t(\boldsymbol{X},\theta)$ 为极值概率分布下第 t 阶段的期望总成本. 模型 (8.2) 的最优解 $\theta^\star$ 反映最优选址决策 $\boldsymbol{X}^\star$ 的鲁棒性水平. 此外, 商品需求 $\tilde{\boldsymbol{u}}_1,\cdots,\tilde{\boldsymbol{u}}_T$ 随决策阶段依

次被观测到或实现. 因此, 将商品需求看作具有概率分布不确定的时间序列 (time series) 是合理的.

8.2.3 基于时间序列预测模型的嵌套型分布不确定集

假设有一组商品需求历史样本 $\hat{\boldsymbol{u}}_{-1},\cdots,\hat{\boldsymbol{u}}_{1-N}$, 其中, 对于每个 $n\in[N]$, $\hat{\boldsymbol{u}}_{-n}=(u_{-nk})_{k\in[K]}$ 表示商品在决策前 n 个阶段的历史需求. 基于历史商品需求数据, 可以得到需求预测模型:

$$\tilde{\boldsymbol{u}}_t=\boldsymbol{u}_t^0+\boldsymbol{A}_t^1\tilde{\boldsymbol{u}}_1+\boldsymbol{A}_t^2\tilde{\boldsymbol{u}}_2+\cdots+\boldsymbol{A}_t^{t-1}\tilde{\boldsymbol{u}}_{t-1}+\tilde{\boldsymbol{\xi}}_t\quad\forall t\in[T],\tag{8.3}$$

其中, $\boldsymbol{u}_t^0$ 是截距 (intercept), $\boldsymbol{A}_t^1,\cdots,\boldsymbol{A}_t^{t-1}$ 表示 $\boldsymbol{u}_t$ 与之前各阶段需求相关关系的系数矩阵, $\tilde{\boldsymbol{\xi}}_t$ 是估计的随机噪声 (estimated random noise, ERN). 接下来, 我们给出几个常用的时间序列预测模型.

例 17 (p 阶向量自回归模型) 利用多元最小二乘估计 (multivariate least squares estimation), 可以构建如下 p 阶向量自回归 (p^{th}-order vector autoregression) 模型 VAR(p):

$$\tilde{\boldsymbol{u}}_t=\boldsymbol{\kappa}+\boldsymbol{M}_1\tilde{\boldsymbol{u}}_{t-1}+\cdots+\boldsymbol{M}_p\tilde{\boldsymbol{u}}_{t-p}+\tilde{\boldsymbol{\xi}}_t,$$

其中, $(\boldsymbol{\kappa},\boldsymbol{M}_1,\cdots,\boldsymbol{M}_p)=\boldsymbol{U}\boldsymbol{Z}^\top(\boldsymbol{Z}\boldsymbol{Z}^\top)^{-1}$ 且 $\boldsymbol{U}=(\hat{\boldsymbol{u}}_{-1},\cdots,\hat{\boldsymbol{u}}_{-N})$, 以及 $\boldsymbol{Z}=(\boldsymbol{Z}_{-1},\cdots,\boldsymbol{Z}_{-N})$ 且 $\boldsymbol{Z}_{-n}=(1,\hat{\boldsymbol{u}}_{-n-1},\cdots,\hat{\boldsymbol{u}}_{-n-p})^\top\ \forall n\in[N]$. 我们将上述 VAR($p$) 模型转化为模型 (8.3) 的形式:

$$\tilde{\boldsymbol{u}}_t=\boldsymbol{u}_t^0+\boldsymbol{A}_t^1\tilde{\boldsymbol{u}}_1+\boldsymbol{A}_t^2\tilde{\boldsymbol{u}}_2+\cdots+\boldsymbol{A}_t^{t-1}\tilde{\boldsymbol{u}}_{t-1}+\tilde{\boldsymbol{\xi}}_t\quad\forall t\in[T],$$

其中, $\boldsymbol{u}_t^0=\boldsymbol{\kappa}+\sum_{s\in[p-t+1]}\boldsymbol{M}_{t+s}\hat{\boldsymbol{u}}_{-s}$ 且对每个 $s\in[t-1]$, $\boldsymbol{A}_t^s=\mathbb{I}[s\geqslant t-p]\cdot\boldsymbol{M}_{t-s}$, $\mathbb{I}[\cdot]$ 为示性函数.

例 18 ((p,q) 阶向量自回归移动平均模型) 给定参数 p 和 q, (p,q) 阶向量自回归移动平均 (vector autoregressive moving average) 模型 VARMA (p,q) 为

$$\tilde{\boldsymbol{u}}_t=\boldsymbol{\kappa}+\boldsymbol{M}_1\tilde{\boldsymbol{u}}_{t-1}+\cdots+\boldsymbol{M}_p\tilde{\boldsymbol{u}}_{t-p}+\tilde{\boldsymbol{\xi}}_t+\boldsymbol{L}_1\tilde{\boldsymbol{\xi}}_{t-1}+\cdots+\boldsymbol{L}_q\tilde{\boldsymbol{\xi}}_{t-q}.$$

上述 VARMA (p,q) 模型可以转化为模型 (8.3) 的形式, 其中, 参数

$$\begin{aligned}
\boldsymbol{u}_t^0&=\boldsymbol{\kappa}+\sum_{s\in[p-t+1]}\boldsymbol{M}_{t+s}\hat{\boldsymbol{u}}_{-s}+\sum_{s\in[q-t+1]}\boldsymbol{L}_{t+s}\hat{\boldsymbol{\xi}}_{-s}-\sum_{s\in[t-1]}\boldsymbol{L}_s\boldsymbol{u}_{t-s}^0,\\
\boldsymbol{A}_t^s&=\sum_{\substack{j+i_1+i_2+\cdots=t-s\\1\leqslant i_1\leqslant i_2\leqslant\cdots\leqslant q,\ 1\leqslant j}}\boldsymbol{L}_{i_1}\boldsymbol{L}_{i_2}\cdots(\boldsymbol{M}_j+\boldsymbol{L}_j),\qquad\forall s=1,\cdots,t,
\end{aligned}$$

这里, 索引 $i_1,i_2,\cdots$ 的数量可以为零, 根据历史需求数据可计算出历史估计噪声 $\hat{\boldsymbol{\xi}}_{-1},\cdots,\hat{\boldsymbol{\xi}}_{-N}$.

基于时间序列模型 (8.3), 我们对各个阶段随机噪声的分布不确定性进行建模. 假设每个阶段 $t \in [T]$, 随机噪声 $\tilde{\boldsymbol{\xi}}_t$ 的真实概率分布为 $\mathbb{Q}_t$. 一般情况下, 虽然决策者很难获得真实概率分布的信息, 但是可以利用历史数据构建一个与真实概率分布相似的参考概率分布 (reference probability distribution) 或经验概率分布 $\hat{\mathbb{Q}}_t$. 我们利用 1-型 Wasserstein 距离来度量概率分布 $\mathbb{Q}_t$ 和 $\hat{\mathbb{Q}}_t$ 之间的距离 (Mohajerin Esfahani and Kuhn, 2018; Gao and Kleywegt, 2023):

$$
\begin{aligned}
& d(\mathbb{Q}_t, \hat{\mathbb{Q}}_t) \\
=& \inf_{\Pi}\left\{\mathbb{E}_{\Pi}[\|\tilde{\boldsymbol{\xi}}_t - \tilde{\boldsymbol{\xi}}_t'\|] : \Pi \in \mathcal{P}(\mathbb{R}^K \times \mathbb{R}^K)\text{且其边际概率分布分别为 } \mathbb{Q}_t \text{ 和 } \hat{\mathbb{Q}}_t\right\},
\end{aligned}
$$

其中, $\tilde{\boldsymbol{\xi}}_t \sim \mathbb{Q}_t$, $\tilde{\boldsymbol{\xi}}_t' \sim \hat{\mathbb{Q}}_t$, $\mathcal{P}(\mathbb{R}^K \times \mathbb{R}^K)$ 表示支撑集为 $\mathbb{R}^K \times \mathbb{R}^K$ 的所有联合概率分布的集合, 且 $\|\cdot\|$ 为范数, 例如, L_ℓ 范数 $\|\cdot\|_\ell$ 和 Mahalanobis 范数. 基于 Wasserstein 距离 $d(\mathbb{Q}_t, \hat{\mathbb{Q}}_t)$, 我们构建如下关于随机噪声 $\hat{\boldsymbol{\xi}}_t$ 的 Wasserstein 分布不确定集:

$$
\mathcal{B}_t(\theta) = \{\mathbb{Q} \in \mathcal{P}(\mathbb{R}^K) \mid d(\mathbb{Q}, \hat{\mathbb{Q}}_t) \leqslant \theta\}, \tag{8.4}
$$

其中, $\mathcal{B}_t(0) = \{\hat{\mathbb{Q}}_t\}$, 因为 $d(\mathbb{Q}_t, \hat{\mathbb{Q}}_t) = 0$ 当且仅当 $\mathbb{Q}_t = \hat{\mathbb{Q}}_t$. 上述分布不确定集 (8.4) 是与参考概率分布 $\hat{\mathbb{Q}}_t$ 之间的 1-型 Wasserstein 距离不超过 θ 的所有概率分布集合.

接下来, 基于随机噪声分布不确定集 (8.4), 我们用嵌套的方式 (nested fashion) 对需求 $\tilde{\boldsymbol{u}}_1, \cdots, \tilde{\boldsymbol{u}}_T$ 的分布不确定性进行建模. 根据时间序列模型 (8.3), 可知第 $t=1$ 阶段不确定商品需求 $\tilde{\boldsymbol{u}}_1$ 的概率分布 $\mathbb{P}_1$ 由常数 $\boldsymbol{u}_1^0$ 和随机噪声 $\tilde{\boldsymbol{\xi}}_1$ 决定. 因此, 我们定义如下关于 $\tilde{\boldsymbol{u}}_1$ 的分布不确定集:

$$
\mathcal{F}_1(\theta) = \left\{\mathbb{P}_1 \in \mathcal{P}(\mathbb{R}_+^K) \;\middle|\; \begin{array}{l} \tilde{\boldsymbol{u}}_1 \sim \mathbb{P}_1,\ \tilde{\boldsymbol{\xi}}_1 \sim \mathbb{Q}_1, \\ \tilde{\boldsymbol{u}}_1 = \boldsymbol{u}_1^0 + \tilde{\boldsymbol{\xi}}_1, \\ \mathbb{Q}_1 \in \mathcal{B}_1(\theta) \end{array}\right\}.
$$

对于其余 $t=2,\cdots,T$ 的决策阶段, 给定随机需求 $(\tilde{\boldsymbol{u}}_1, \cdots, \tilde{\boldsymbol{u}}_{t-1})$ 的实现值 $\boldsymbol{U}_{t-1} = (\boldsymbol{u}_1, \cdots, \boldsymbol{u}_{t-1})$, 第 t 阶段商品需求 $\tilde{\boldsymbol{u}}_t$ 的条件概率分布不确定集 $\mathcal{F}_{t|\boldsymbol{U}_{t-1}}$ 定义如下:

$$
\mathcal{F}_{t|\boldsymbol{U}_{t-1}}(\theta) = \left\{\mathbb{P}_{t|\boldsymbol{U}_{t-1}} \in \mathcal{P}(\mathbb{R}_+^K) \;\middle|\; \begin{array}{l} \tilde{\boldsymbol{u}}_t \sim \mathbb{P}_{t|\boldsymbol{U}_{t-1}},\ \tilde{\boldsymbol{\xi}}_t \sim \mathbb{Q}_t, \\ \tilde{\boldsymbol{u}}_t = \boldsymbol{u}_t^0 + \boldsymbol{A}_t^1\boldsymbol{u}_1 + \cdots + \boldsymbol{A}_t^{t-1}\boldsymbol{u}_{t-1} + \tilde{\boldsymbol{\xi}}_t, \\ \mathbb{Q}_t \in \mathcal{B}_t(\theta) \end{array}\right\},
$$

其中, $\mathbb{P}_{t|\boldsymbol{U}_{t-1}}$ 是给定 $\boldsymbol{U}_{t-1}$ 下的条件概率分布. 最后, 基于上述分布不确定集, 我

们构建如下不确定需求路径 $(\tilde{\boldsymbol{u}}_1,\cdots,\tilde{\boldsymbol{u}}_T)$ 的嵌套型分布不确定集:

$$\mathcal{F}(\theta)=\left\{\mathbb{P}\in\mathcal{P}\left(\mathbb{R}_{+}^{K\times T}\right)\ \middle|\ \begin{array}{l}(\tilde{\boldsymbol{u}}_1,\cdots,\tilde{\boldsymbol{u}}_T)\sim\mathbb{P},\\ \tilde{\boldsymbol{u}}_1\sim\mathbb{P}_1,\ \tilde{\boldsymbol{u}}_{t|\boldsymbol{U}_{t-1}}\sim\mathbb{P}_{t|\boldsymbol{U}_{t-1}},\ \forall t=2,\cdots,T,\\ \mathbb{P}_1\in\mathcal{F}_1(\theta),\ \mathbb{P}_{t|\boldsymbol{U}_{t-1}}\in\mathcal{F}_{t|\boldsymbol{U}_{t-1}}(\theta),\ \forall t=2,\cdots,T\end{array}\right\}. \tag{8.5}$$

下面介绍几个关于随机噪声的常用参考概率分布 $\hat{\mathbb{Q}}_t, t\in[T]$.

例 19 (高斯分布) 在 VARMA(p,q) 模型中, 通常假设随机噪声 $\tilde{\boldsymbol{\xi}}$ 服从均值为零且协方差矩阵为 $\boldsymbol{\Sigma}$ 的高斯分布 (Gaussian distribution) (Lütkepohl, 2005). 在此假设下, 每个阶段 $t\in[T]$ 的参考概率分布为 $\hat{\mathbb{Q}}_t=\mathcal{N}(\boldsymbol{0},\boldsymbol{\Sigma})$.

例 20 (经验分布) 利用历史需求样本和时间序列模型, 可以计算出历史随机噪声样本 $\hat{\boldsymbol{\xi}}_{-n}, n\in[N]$, 例如, 对于 VAR(1) 模型, 可通过 $\hat{\boldsymbol{\xi}}_{-n}=\hat{\boldsymbol{u}}_{-n}-\boldsymbol{\kappa}-\boldsymbol{M}_1\hat{\boldsymbol{u}}_{-n-1}$ 计算出随机噪声样本. 则参考概率分布可以定义为经验概率分布 (empirical probability distribution) $\hat{\mathbb{Q}}=\dfrac{1}{N}\sum_{n\in[N]}\delta_{\hat{\boldsymbol{\xi}}_{-n}}$, 其中, $\delta_{\hat{\boldsymbol{\xi}}_{-n}}$ 是 Dirac 测度, 表示点 $\hat{\boldsymbol{\xi}}_{-n}$ 处的概率为 1.

例 21 (一般分布) 我们还可将一般的概率分布 $\hat{\mathbb{Q}}_t$ 看作每个阶段 $t\in[T]$ 的分布不确定集 $\mathcal{B}_t(\theta)$ 中的参考概率分布. 通常基于历史样本数据构造一个近似离散概率分布来表示参考概率分布 $\hat{\mathbb{Q}}_t$.

8.3 模型的高效计算形式与鲁棒性水平分析

本节介绍容量充足条件下预算驱动型多阶段枢纽选址模型的高效计算形式, 并分析成本预算对模型最优鲁棒性水平的影响.

8.3.1 模型的高效计算形式

基于嵌套型分布不确定集 (8.5), 我们可以将模型 (8.2) 在极值概率分布下每阶段期望总成本 $\mathrm{Z}_t(\boldsymbol{X},\theta)$ 转化为如下嵌套形式.

命题 8.1 给定嵌套型分布不确定集 (8.5), 对于每个阶段 $t\in[T]$, 我们有

$$\begin{aligned}&\mathrm{Z}_t(\boldsymbol{X},\theta)\\ =&\sup_{\mathbb{P}_1\in\mathcal{F}_1(\theta)}\mathbb{E}_{\mathbb{P}_1}\left[\sup_{\mathbb{P}_2\in\mathcal{F}_{2|\tilde{\boldsymbol{U}}_1}(\theta)}\mathbb{E}_{\mathbb{P}_2}\left[\cdots\sup_{\mathbb{P}_t\in\mathcal{F}_{t|\tilde{\boldsymbol{U}}_{t-1}}(\theta)}\mathbb{E}_{\mathbb{P}_t}\left[\boldsymbol{c}_t^{\top}\boldsymbol{x}_t+\mathrm{Q}_t(\boldsymbol{X}_t,\tilde{\boldsymbol{u}}_t)\middle|\tilde{\boldsymbol{U}}_{t-1}\right]\cdots\middle|\tilde{\boldsymbol{U}}_1\right]\right].\end{aligned} \tag{8.6}$$

上述嵌套形式 (8.6) 给出各阶段需求不确定性之间的关系, 并且一般情况下这种嵌套结构会造成多阶段优化问题求解上的维度灾难问题 (Shapiro et al., 2021).

由于多阶段容量充足型枢纽选址问题的目标函数具有线性结构, 即对所有阶段 $t \in [T]$, $\mathrm{Q}_t(\boldsymbol{X}_t, \boldsymbol{u}_t) = \sum_{k\in[K]} u_{kt} f_{kt}(\boldsymbol{X}_t)$, 并且嵌套型分布不确定集也具有线性结构, 我们可以根据每个 $\tilde{\boldsymbol{\xi}}_t$ 重新构造式 (8.6) 的各项以得到 $\mathrm{Z}_t(\boldsymbol{X}, \theta)$ 的标准分布鲁棒优化等价形式 (regular distributionally robust optimization reformulation).

命题 8.2　给定嵌套型分布不确定集 (8.5), 对于每个阶段 $t \in [T]$, 我们有

$$\mathrm{Z}_t(\boldsymbol{X}, \theta) = \sup_{\mathbb{Q}_{[t]} \in \mathcal{B}_{[t]}(\theta)} \mathbb{E}_{\mathbb{Q}_{[t]}} \left[\boldsymbol{c}_t^\top \boldsymbol{x}_t + \boldsymbol{\eta}_t^\top \boldsymbol{f}_t(\boldsymbol{X}_t) + \sum_{s\in[t]} [\boldsymbol{B}_t^s \tilde{\boldsymbol{\xi}}_s]^\top \boldsymbol{f}_t(\boldsymbol{X}_t) \right]. \tag{8.7}$$

其中, $\boldsymbol{f}_t(\cdot) = (f_{kt}(\cdot))_{k\in[K]}$, $\mathbb{Q}_{[t]} = \bigotimes_{s\in[t]} \mathbb{Q}_s$, $\mathcal{B}_{[t]}(\theta) = \bigotimes_{s\in[t]} \mathcal{B}_s(\theta)$, 向量 $\boldsymbol{\eta}_t$ 和矩阵 $\boldsymbol{B}_t^s$ 是时间序列模型 (8.3) 参数, 分别为

$$\boldsymbol{\eta}_t = \boldsymbol{u}_t^0 + \sum_{s\in[t-1]} \boldsymbol{A}_t^s \boldsymbol{\eta}_s \quad 且 \quad \boldsymbol{B}_t^s = \sum_{s<i_1<i_2<\cdots<i_l<t-1} \boldsymbol{A}_{i_1}^s \boldsymbol{A}_{i_2}^{i_1} \cdots \boldsymbol{A}_t^{i_l}, \quad \forall s = 1, \cdots, t-1,$$

此外, $\boldsymbol{B}_t^t$ 是一个 K 维单位矩阵.

特别地, 对于每个阶段 $t \in [T]$, 式 (8.7) 关于 $\tilde{\boldsymbol{\xi}}_t$ 是线性的, 这使得我们能够使用正则化 (Mohajerin Esfahani and Kuhn, 2018; Gao et al., 2024) 将模型 (8.2) 转化为如下形式.

定理 8.1　预算驱动型多阶段容量充足型枢纽选址模型 (8.2) 可以等价地转化为如下问题:

$$\max_{\boldsymbol{X}\in\mathcal{X}} \min_{t\in[T]} \gamma_t(\boldsymbol{X}), \quad 其中 \quad \gamma_t(\boldsymbol{X}) = \frac{\left(\tau_t - \mathbb{E}_{\hat{\mathbb{Q}}_{[t]}} \left[\boldsymbol{c}_t^\top \boldsymbol{x}_t + \tilde{\boldsymbol{u}}_t^\top \boldsymbol{f}_t(\boldsymbol{X}_t) \right] \right)^+}{\nabla_\theta \mathrm{Z}_t(\boldsymbol{X}, \theta)}, \tag{8.8}$$

这里, $\hat{\mathbb{Q}}_{[t]} = \bigotimes_{s\in[t]} \hat{\mathbb{Q}}_s$, 且 $\nabla_\theta \mathrm{Z}_t(\boldsymbol{X}, \theta) = \sum_{s\in[t]} \left\| [\boldsymbol{B}_t^s]^\top \boldsymbol{f}_t(\boldsymbol{X}_t) \right\|_*$ 是 $\mathrm{Z}_t(\boldsymbol{X}, \theta)$ 在点 θ 处的梯度.

定理 8.1 表明: 对每个阶段 $t \in [T]$, 模型 (8.8) 目标函数中的项 $\gamma_t(\boldsymbol{X})$ 是一个类似夏普比率的决策标准, 其中, 分子是参考概率分布 $\hat{\mathbb{Q}}_{[t]}$ 下的期望成本预算剩余量, 分母是极值概率分布下阶段 t 的期望总成本 $\mathrm{Z}_t(\boldsymbol{X}, \theta)$ 关于分布不确定集大小 θ 的敏感度. 此外, 最优鲁棒性水平 $\theta^\star$ 是通过优化 T 个阶段中最坏情况下的夏普比率来获得.

推论 8.1　假设对于每个 $t \in [T]$, $\hat{\mathbb{Q}}_t = \mathcal{N}(\boldsymbol{0}, \boldsymbol{\Sigma})$ 且 Wasserstein 分布不确定集 $\mathcal{B}_t(\theta)$ 中的范数 $\|\cdot\|$ 为基于估计协方差矩阵 $\boldsymbol{\Sigma}$ 的 Mahalanobis 范数, 即

$\|\boldsymbol{b}\| = \|\boldsymbol{\Sigma}^{-1/2}\boldsymbol{b}\|_2$. 对每个 $t \in [T]$, 我们有

$$\gamma_t(\boldsymbol{X}) = \frac{\left(\tau_t - \mathbb{E}_{\hat{\mathbb{Q}}_{[t]}}\left[\boldsymbol{c}_t^\top \boldsymbol{x}_t + \tilde{\boldsymbol{u}}_t^\top \boldsymbol{f}_t(\boldsymbol{X}_t)\right]\right)^+}{\sum_{s\in[t]} \sqrt{\mathbb{V}_{\hat{\mathbb{Q}}_s}[\boldsymbol{B}_t^s \tilde{\boldsymbol{\xi}}_s]^\top \boldsymbol{f}_t(\boldsymbol{X}_t)]}},$$

其中, $\mathbb{V}_{\hat{\mathbb{Q}}_t}[\cdot]$ 是经验分布 $\hat{\mathbb{Q}}_t$ 的方差. 特别地, 在单阶段问题中 (即 $T = 1$), 我们有

$$\gamma_1(\boldsymbol{X}) = \frac{\left(\tau_1 - \mathbb{E}_{\hat{\mathbb{Q}}_1}\left[\boldsymbol{c}_1^\top \boldsymbol{x}_1 + \tilde{\boldsymbol{u}}_1^\top \boldsymbol{f}_1(\boldsymbol{x}_1)\right]\right)^+}{\sqrt{\mathbb{V}_{\hat{\mathbb{Q}}_1}[\boldsymbol{c}_1^\top \boldsymbol{x}_1 + \tilde{\boldsymbol{u}}_1^\top \boldsymbol{f}_1(\boldsymbol{x}_1)]}}.$$

此外, 给定任意选址决策 $\boldsymbol{X}$, 模型 (8.2) 的鲁棒性水平可以通过下式来计算:

$$\theta^\star(\boldsymbol{X}) = \min_{t\in[T]} \gamma_t(\boldsymbol{X}) = \min_{t\in[T]} \left\{ \frac{\left(\tau_t - \mathbb{E}_{\hat{\mathbb{Q}}_{[t]}}\left[\boldsymbol{c}_t^\top \boldsymbol{x}_t + \tilde{\boldsymbol{u}}_t^\top \boldsymbol{f}_t(\boldsymbol{X}_t)\right]\right)^+}{\sum_{s\in[t]} \|[\boldsymbol{B}_t^s]^\top \boldsymbol{f}_t(\boldsymbol{X}_t)\|_*} \right\}.$$

进一步, 可以利用二分法 (bisection search algorithm) 在可行集 $\mathcal{X}$ 中寻找最优鲁棒性水平的选址决策, 其中每次迭代需要验证一个混合整数锥约束系统 (mixed-integer conic constraint system) 的可行性.

定理 8.2 给定嵌套型分布不确定集 (8.5), 模型 (8.2) 的可行集等价于如下混合整数锥约束系统:

$$\begin{cases} \boldsymbol{c}_t^\top \boldsymbol{x}_t + \boldsymbol{\eta}_t^\top \boldsymbol{h}_t + \sum\limits_{s\in[t]} \boldsymbol{h}_t^\top \boldsymbol{B}_t^s \bar{\boldsymbol{\xi}}_s + \theta\lambda_t^s \leqslant \tau_t,\ \forall t \in [T], \\ \|[\boldsymbol{B}_t^s]^\top \boldsymbol{h}_t\|_* \leqslant \lambda_t^s,\ \forall t \in [T],\ s \in [t], \\ \sum\limits_{i,j\in[I]} v_{ijkt} y_{ijkt} \leqslant h_{kt},\ \forall t \in [T],\ k \in [K], \\ \sum\limits_{i,j\in[I]} y_{ijkt} = 1,\ \forall t \in [T],\ k \in [K], \\ \sum\limits_{j\in[I]} y_{ijkt} + \sum\limits_{j\in[I]\setminus\{i\}} y_{jikt} \leqslant \sum\limits_{s\in[t]} x_{is},\ \forall t \in [T],\ k \in [K],\ i \in [I], \\ \lambda_t^s \in \mathbb{R}_+,\ y_{ijkt} \in \mathbb{R}_+,\ \boldsymbol{h}_t \in \mathbb{R}_+^K,\ \forall t \in [T],\ k \in [K],\ i, j \in [I],\ s \in [t], \\ \boldsymbol{X} \in \mathcal{X},\ \theta \geqslant 0, \end{cases} \tag{8.9}$$

其中, 对于每个 $t \in [T]$, 有 $\bar{\boldsymbol{\xi}}_t = \mathbb{E}_{\hat{\mathbb{Q}}_t}[\tilde{\boldsymbol{\xi}}_t]$.

现在, 我们总结求解模型 (8.2) 的二分法算法如下.

算法 5 二分法求解模型 (8.2) 的最优鲁棒性水平 $\theta^\star$

输入: 模型参数, 最优误差间隙 ϵ.
初始化: $\underline{\theta} \leftarrow 0$, $\bar{\theta} \leftarrow M$, M 为一个足够大的数.
1. 计算 $\theta = (\underline{\theta} + \bar{\theta})/2$.
2. 验证混合整数锥约束系统 (8.9) 的可行性.
3. 如果约束 (8.9) 是可行的, 则令 $\underline{\theta} \leftarrow \theta$; 否则, $\bar{\theta} \leftarrow \theta$.
4. 如果 $\bar{\theta} - \underline{\theta} \leqslant \varepsilon$, 则令 $\theta^\star = (\underline{\theta} + \bar{\theta})/2$, 且算法停止; 否则, 返回第 1 步.

输出: 模型最优鲁棒性水平 $\theta^\star$.

8.3.2 成本预算对模型最优鲁棒性水平的影响

本小节探讨成本预算值 $\boldsymbol{\tau}$ 对模型 (8.2) 最优鲁棒性水平 $\theta^\star(\boldsymbol{\tau})$ 的影响. 首先, 定义如下最小夏普比率对应的阶段集合:

$$\mathcal{T}^\star(\boldsymbol{\tau}, \boldsymbol{X}) = \arg\min_{t\in[T]} \frac{\left(\tau_t - \mathbb{E}_{\hat{\mathbb{Q}}_{[t]}}\left[\boldsymbol{c}_t^\top \boldsymbol{x}_t + \tilde{\boldsymbol{u}}_t^\top \boldsymbol{f}_t(\boldsymbol{X}_t)\right]\right)^+}{\sum_{s\in[t]} \|[\boldsymbol{B}_t^s]^\top \boldsymbol{f}_t(\boldsymbol{X}_t)\|_*}.$$

接下来, 分别考虑上述集合的元素个数 $|\mathcal{T}^\star(\boldsymbol{\tau}, \boldsymbol{X})|$ 为 1 或大于等于 2 两种情况下, 成本预算对模型最优鲁棒性水平的影响.

定理 8.3 假设给定成本预算 $\boldsymbol{\tau}$ 下, 模型 (8.2) 具有唯一最优选址决策 $\boldsymbol{X}^\star$. 如果 $\mathcal{T}^\star(\boldsymbol{\tau}, \boldsymbol{X}^\star) = \{t^\star\}$, 那么

$$\nabla_{\tau_{t^\star}} \theta^\star(\boldsymbol{\tau}) = \frac{1}{\sum_{s\in[t^\star]} \|[\boldsymbol{B}_{t^\star}^s]^\top \boldsymbol{f}_{t^\star}(\boldsymbol{X}_{t^\star}^\star)\|_*},$$

且对于每个 $t \in [T] \setminus \{t^\star\}$, 有 $\nabla_{\tau_t} \theta^\star(\boldsymbol{\tau}) = 0$. 如果 $|\mathcal{T}^\star(\boldsymbol{\tau}, \boldsymbol{X}^\star)| \geqslant 2$, 那么

$$\nabla_{\tau_t}^- \theta^\star(\boldsymbol{\tau}) = \frac{1}{\sum_{s\in[t]} \|[\boldsymbol{B}_t^s]^\top \boldsymbol{f}_t(\boldsymbol{X}_t^\star)\|_*} \quad \text{且} \quad \nabla_{\tau_t}^+ \theta^\star(\boldsymbol{\tau}) = 0 \quad \forall t \in \mathcal{T}^\star(\boldsymbol{\tau}, \boldsymbol{X}^\star);$$

并且对于每个 $t \in [T] \setminus \mathcal{T}^\star(\boldsymbol{\tau}, \boldsymbol{X}^\star)$, 都有 $\nabla_{\tau_t} \theta^\star(\boldsymbol{\tau}) = 0$.

定理 8.3 表明: 如果只有一个阶段的成本预算 $\boldsymbol{\tau}$ 与最优鲁棒性水平 $\theta^\star(\boldsymbol{\tau})$ 相关, 则 $\theta^\star(\boldsymbol{\tau})$ 仅受此阶段成本预算增加或减少的影响. 此外, 如果有多个阶段的成本预算 $\boldsymbol{\tau}$ 与最优鲁棒性水平 $\theta^\star(\boldsymbol{\tau})$ 相关, 则 $\theta^\star(\boldsymbol{\tau})$ 仅受这些阶段成本预算减少的影响. 因为, 对于每个 $t \in [T]$, 有 $\nabla_\theta \mathrm{Z}_t(\boldsymbol{X}, \theta) = \sum_{s\in[t]} \left\|[\boldsymbol{B}_t^s]^\top \boldsymbol{f}_t(\boldsymbol{X}_t)\right\|_*$, 所以最优鲁棒性水平对成本预算变化的敏感度以及极值概率分布下期望成本对鲁棒性水平的敏感度有等式联系 $[\nabla_\theta \mathrm{Z}_1(\boldsymbol{X}, \theta)), \cdots, \nabla_\theta \mathrm{Z}_T(\boldsymbol{X}, \theta)]^\top \nabla_{\boldsymbol{\tau}}^- \theta^\star(\boldsymbol{\tau}) = |\mathcal{T}^\star(\boldsymbol{\tau}, \boldsymbol{X})|$.

8.4 多阶段容量限制型枢纽选址模型

本节在容量限制条件下构建基于嵌套型分布不确定集的预算驱动型多阶段枢纽选址模型, 并使用扩展线性决策规则方法 (extended linear decision rule approach) 来近似求解模型.

8.4.1 容量限制条件下预算驱动型多阶段枢纽选址模型

假设每个枢纽 i 在阶段 t 内总运输商品量的上限为 Γ_{it}. 在第 t 阶段, 任何需求未被满足的商品 k 的都会产生惩罚成本, 其单位惩罚成本为 w_{kt}. 第 $t \in [T]$ 阶段的总成本由如下问题表示:

$$
\begin{aligned}
&H_t(\boldsymbol{X}_t, \boldsymbol{u}_t) \\
= \min \quad & \sum_{k\in[K]}\sum_{i,j\in[I]} v_{ijkt}z_{ijkt} + \sum_{k\in[K]} w_{kt}\left(u_{kt} - \sum_{i,j\in[I]} z_{ijkt}\right) \\
\text{s.t.} \quad & \sum_{i,j\in[I]} z_{ijkt} \leqslant u_{kt}, \ \forall t \in [T],\ k \in [K], \\
& \sum_{k\in[K]}\left(\sum_{j\in[I]} z_{ijkt} + \sum_{j\in[I]\setminus\{i\}} z_{jikt}\right) \leqslant \sum_{s\in[t]} x_{is}\Gamma_{it}, \ \forall t \in [T],\ i \in [I], \\
& z_{ijkt} \geqslant 0, \ \forall t \in [T],\ k \in [K],\ i,j \in [I],
\end{aligned}
$$

其中, 目标函数包括运输成本和未满足需求的惩罚成本, 第二组约束确保商品的总运输量不超过枢纽的容量上限. 特别地, 由于第二组约束不再能根据商品类型 $k \in [K]$ 分解, 因此容量限制型枢纽选址问题 $H_t(\boldsymbol{X}_t, \boldsymbol{u}_t)$ 不具有容量充足型枢纽选址模型关于 $\boldsymbol{u}_t$ 的线性结构.

基于问题 $H_t(\boldsymbol{X}_t, \boldsymbol{u}_t)$, 可构建如下预算驱动型多阶段容量限制型枢纽选址模型:

$$
\begin{aligned}
\max \quad & \theta \\
\text{s.t.} \quad & Y_t(\boldsymbol{X}, \theta) = \sup_{\mathbb{P}\in\mathcal{F}(\theta)} \mathbb{E}_{\mathbb{P}}\left[\boldsymbol{c}_t^\top \boldsymbol{x}_t + H_t(\boldsymbol{X}_t, \tilde{\boldsymbol{u}}_t)\right] \leqslant \tau_t, \ \forall t \in [T], \\
& \boldsymbol{X} \in \mathcal{X}, \ \theta \geqslant 0.
\end{aligned} \tag{8.10}
$$

与命题 8.1 的结果类似, 给定选址决策 $\boldsymbol{X}$, 第 t 阶段极值概率分布下期望成本可以等价地转化为如下嵌套形式:

$$
Y_t(\boldsymbol{X}, \theta) = \sup_{\mathbb{P}_1\in\mathcal{F}_1(\theta)} \mathbb{E}_{\mathbb{P}_1}\Bigg[\sup_{\mathbb{P}_2\in\mathcal{F}_{2|\tilde{\boldsymbol{U}}_1}(\theta)} \mathbb{E}_{\mathbb{P}_2}\Bigg[\cdots \sup_{\mathbb{P}_t\in\mathcal{F}_{t|\tilde{\boldsymbol{U}}_{t-1}}(\theta)} \mathbb{E}_{\mathbb{P}_t}\Big[\boldsymbol{c}_t^\top \boldsymbol{x}_t
$$

$$+ H_t(\boldsymbol{X}_t, \tilde{\boldsymbol{u}}_t)\bigg|\tilde{\boldsymbol{U}}_{t-1}\bigg]\cdots\bigg|\tilde{\boldsymbol{U}}_1\bigg]\bigg].$$

由于函数 $H_t(\boldsymbol{X}_t, \tilde{\boldsymbol{u}}_t)$ 不是关于不确定需求 $\tilde{\boldsymbol{u}}_t$ 的线性函数, 并且对于每个阶段 t, 运输决策 $\boldsymbol{z}_t$ 的最优值需通过求解一个包含不确定需求的参数化问题获得, 且该问题是关于依阶段顺序实现的不确定需求 $\tilde{\boldsymbol{u}}_s$, $s \in [t]$ 的函数. 因此, 容量限制条件下预算驱动型多阶段枢纽选址模型 (8.10) 具有典型多阶段分布鲁棒优化问题的嵌套结构, 其难以被精确求解 (Shapiro et al., 2021). 接下来, 我们介绍一种可以近似求解此模型的决策规则近似方法.

为了表述简洁, 对于每个 $t \in [T]$, 我们定义如下符号:

$$\bar{\boldsymbol{v}}_t = (v_{ijkt} - w_{kt})_{i,j\in[I],k\in[K]}, \quad \bar{\boldsymbol{v}}_t^\top \boldsymbol{z}_t = \sum_{i,j\in[I]}\sum_{k\in[K]}(v_{ijkt} - w_{kt})z_{ijkt},$$

$$\boldsymbol{1}^\top \boldsymbol{z}_t = \left(\sum_{i,j\in[I]} z_{ijkt}\right)_{k\in[K]},$$

$$\boldsymbol{D}\boldsymbol{z}_t = \sum_{k\in[K]}\left(\sum_{j\in[I]} z_{ijkt} + \sum_{j\in[I]\setminus\{i\}} z_{jikt}\right)_{i\in[I]}, \quad 且 \quad \boldsymbol{\Gamma}_t\boldsymbol{X}_t = \left(\Gamma_{it}\sum_{s\in[t]} x_{is}\right)_{i\in[I]},$$

其中, $\boldsymbol{1} \in \mathbb{R}^{I^2}, \boldsymbol{D} \in \mathbb{R}^{I\times I^2K}$ 和 $\boldsymbol{\Gamma}_t \in \mathbb{R}^{I\times It}$ 是相应的系数向量或矩阵. 利用上述符号, 我们可以将问题 $H_t(\boldsymbol{X}_t, \boldsymbol{u}_t)$ 表示为如下形式:

$$\begin{aligned} H_t(\boldsymbol{X}_t, \boldsymbol{u}_t) = \min \quad & \bar{\boldsymbol{v}}_t^\top \boldsymbol{z}_t + \boldsymbol{w}_t^\top \boldsymbol{u}_t \\ \text{s.t.} \quad & \boldsymbol{1}^\top \boldsymbol{z}_t \leqslant \boldsymbol{u}_t,\ \boldsymbol{D}\boldsymbol{z}_t \leqslant \boldsymbol{\Gamma}_t\boldsymbol{X}_t,\ \boldsymbol{z}_t \geqslant \boldsymbol{0}, \end{aligned} \qquad \forall t \in [T].$$

基于上述形式, 我们可以得到问题 (8.10) 对应的决策规则近似形式.

命题 8.3 给定问题 (8.10) 的一个可行选址决策 $\boldsymbol{X} \in \mathcal{X}$, 基于嵌套型分布不确定集 (8.5), 极值概率分布下的期望总成本 $Y_t(\boldsymbol{X}, \theta)$ 等价于如下问题:

$$\begin{aligned} Y_t(\boldsymbol{X}, \theta) = \min \quad & \boldsymbol{c}_t^\top \boldsymbol{x}_t + \sup_{\mathbb{Q}_{[t]}\in\mathcal{B}_{[t]}(\theta)} \mathbb{E}_{\mathbb{Q}_{[t]}}\left[\bar{\boldsymbol{v}}_t^\top \boldsymbol{z}_t(\tilde{\boldsymbol{\xi}}_{[t]}) + \boldsymbol{w}_t^\top\left(\boldsymbol{\eta}_t + \sum_{s\in[t]}\boldsymbol{B}_t^s\tilde{\boldsymbol{\xi}}_s\right)\right] \\ \text{s.t.} \quad & \boldsymbol{1}^\top \boldsymbol{z}_t(\tilde{\boldsymbol{\xi}}_{[t]}) \leqslant \boldsymbol{\eta}_t + \sum_{s\in[t]}\boldsymbol{B}_t^s\tilde{\boldsymbol{\xi}}_s, \\ & \boldsymbol{D}\boldsymbol{z}_t(\tilde{\boldsymbol{\xi}}_{[t]}) \leqslant \boldsymbol{\Gamma}_t\boldsymbol{X}_t, \qquad \mathbb{Q}_{[t]}\text{-a.s.} \quad \forall \mathbb{Q}_{[t]} \in \mathcal{B}_{[t]}(\theta), \\ & \boldsymbol{z}_t(\tilde{\boldsymbol{\xi}}_{[t]}) \geqslant \boldsymbol{0}, \\ & \boldsymbol{z}_t(\cdot) \in \mathcal{R}^{Kt,I^2K}, \end{aligned}$$

其中, $\tilde{\boldsymbol{\xi}}_{[t]} = (\tilde{\boldsymbol{\xi}}_1, \cdots, \tilde{\boldsymbol{\xi}}_t) = (\xi_{ks})_{k\in[K],s\in[t]}$, 且 $\mathcal{R}^{Kt,I^2K}$ 是从 $\mathbb{R}^{Kt}$ 到 $\mathbb{R}^{I^2K}$ 的可测函数空间.

8.4.2 拓展的线性决策规则近似形式

本小节利用扩展的线性决策规则 (extended linear decision rule, ELDR) 来近似求解模型 (8.10). 扩展的线性决策规则方法在两个方面扩展了经典的线性决策规则方法: 首先, 通过引入辅助变量来提升 (lifts) 每个阶段随机噪声对应的 Wasserstein 分布不确定集, 这提高了决策规则的灵活性, 使其具有更紧的逼近形式; 其次, 扩展的线性决策规则方法利用了自适应鲁棒优化的二次决策规则 (quadratic decision rule) 思想 (de Ruiter, 2018), 引入提升变量来处理决策规则中的二次项. 基于上述辅助变量和提升变量, 扩展的线性决策规则方法具有比经典线性决策规则方法更紧的逼近形式 (Bertsimas et al., 2019; Chen et al., 2020; Saif and Delage, 2021).

记集合 Ξ_t 为阶段 $t \in [T]$ 随机噪声支撑集, 并假设 $\Xi_t, t \in [T]$ 有界且具有高效计算的锥表示形式 (tractable conic representable), 并且集合至少存在一个内点. 此外, 利用例 21 中的参考概率分布 $\hat{\mathbb{Q}}_t = \dfrac{1}{N}\sum_{n\in[N]} \delta_{\hat{\boldsymbol{\xi}}_t^n}$ 作为每个 Wasserstein 球 $\mathcal{B}_t(\theta), t \in [T]$ 中的参考概率分布. 令 $\boldsymbol{\xi}_t^{(m)}$ 表示将 $\boldsymbol{\xi}_t$ 中的每个元素作 m 次方后构成的向量, 即 $\boldsymbol{\xi}_t^{(m)} = (\xi_{1t}^m, \cdots, \xi_{Kt}^m)$. 接下来, 定义与原始 Wasserstein 球 $\mathcal{B}_t(\theta)$ 对应的分布不确定集 $\mathcal{G}_t(\theta)$:

$$
\mathcal{G}_t(\theta) = \left\{ \mathbb{Q} \in \mathcal{P}(\mathbb{R}^{2K+1}) \;\middle|\; \begin{array}{l} (\tilde{\boldsymbol{\xi}}_t, \tilde{\boldsymbol{\zeta}}_t, \tilde{\pi}_t) \sim \mathbb{Q},\ \tilde{\boldsymbol{\xi}}_t' \sim \hat{\mathbb{Q}}_t, \\ \dfrac{1}{N}\displaystyle\sum_{n\in[N]} \mathbb{E}_{\mathbb{Q}}\left[\tilde{\pi}_t \,\middle|\, \tilde{\boldsymbol{\xi}}_t' = \hat{\boldsymbol{\xi}}_t^n\right] \leqslant \theta, \\ \mathbb{Q}\left[(\tilde{\boldsymbol{\xi}}_t, \tilde{\boldsymbol{\zeta}}_t, \tilde{\pi}_t) \in \mathcal{W}_{tn} \,\middle|\, \tilde{\boldsymbol{\xi}}_t' = \hat{\boldsymbol{\xi}}_t^n\right] = 1,\ \forall n \in [N] \end{array} \right\},
$$

其中, $\tilde{\pi}_t$ 和 $\tilde{\boldsymbol{\zeta}}_t$ 分别为辅助变量和提升变量, 每个 $\mathcal{W}_{tn}$ 为给定 $\tilde{\boldsymbol{\xi}}_t' = \hat{\boldsymbol{\xi}}_t^n$ 的条件支撑集, 它定义为

$$
\mathcal{W}_{tn} = \left\{ (\boldsymbol{\xi}_t, \boldsymbol{\zeta}_t, \pi_t) \in \mathbb{R}^{2K+1} \;\middle|\; \boldsymbol{\xi}_t^{(2)} \leqslant \boldsymbol{\zeta}_t,\ \|\boldsymbol{\xi}_t - \hat{\boldsymbol{\xi}}_t^n\|_\ell \leqslant \pi_t,\ \boldsymbol{\xi}_t \in \Xi_t \right\},\ \forall n \in [N]. \tag{8.11}
$$

分布不确定集 $\mathcal{G}_t(\theta)$ 包含随机变量 $(\tilde{\boldsymbol{\xi}}_t, \tilde{\boldsymbol{\zeta}}_t, \tilde{\pi}_t)$ 的所有联合分布. 我们定义基于 $(\boldsymbol{\xi}_{[t]}, \boldsymbol{\zeta}_{[t]}, \pi_{[t]})$ 的拓展的线性决策规则集合:

$$
\mathcal{L}_t = \left\{ \boldsymbol{z}_t(\cdot) \in \mathcal{R}^{(2K+1)t, I^2K} \;\middle|\; \begin{array}{l} \exists \boldsymbol{z}_0,\ \boldsymbol{Y}_s \in \mathbb{R}^{I^2K},\ \boldsymbol{Z}_s,\ \boldsymbol{W}_s \in \mathbb{R}^{K\times I^2K},\ \forall s \in [t]: \\ \boldsymbol{z}_t\left(\boldsymbol{\xi}_{[t]}, \boldsymbol{\zeta}_{[t]}, \pi_{[t]}\right) = \boldsymbol{z}_0 + \sum_{s\in[t]} \boldsymbol{Z}_s\boldsymbol{\xi}_s + \boldsymbol{W}_s\boldsymbol{\zeta}_s + \boldsymbol{Y}_s\pi_s \end{array} \right\}. \tag{8.12}
$$

利用 $\mathcal{G}_t(\theta)$ 和 $\mathcal{L}_t$, 可以构建问题 $Y_t(\boldsymbol{X},\theta)$ 对应的拓展的线性决策规则近似形式 $\bar{Y}_t(\boldsymbol{X},\theta)$:

$$
\begin{aligned}
& \bar{Y}_t(\boldsymbol{X},\theta) \\
= \min \quad & \boldsymbol{c}_t^\top \boldsymbol{x}_t \\
& + \sup\nolimits_{\mathbb{Q}_{[t]}\in\mathcal{G}_{[t]}(\theta)} \mathbb{E}_{\mathbb{Q}_{[t]}}\left[\bar{\boldsymbol{v}}_t^\top \boldsymbol{z}_t(\tilde{\boldsymbol{\xi}}_{[t]},\tilde{\boldsymbol{\zeta}}_{[t]},\tilde{\pi}_{[t]}) + \boldsymbol{w}_t^\top\left(\boldsymbol{\eta}_t + \sum_{s\in[t]}\boldsymbol{B}_t^s\tilde{\boldsymbol{\xi}}_s\right)\right] \\
\text{s.t.} \quad & \boldsymbol{1}^\top \boldsymbol{z}_t(\tilde{\boldsymbol{\xi}}_{[t]},\tilde{\boldsymbol{\zeta}}_{[t]},\tilde{\pi}_{[t]}) \leqslant \boldsymbol{\eta}_t + \sum_{s\in[t]}\boldsymbol{B}_t^s\boldsymbol{\xi}_s, \\
& \boldsymbol{D}\boldsymbol{z}_t(\tilde{\boldsymbol{\xi}}_{[t]},\tilde{\boldsymbol{\zeta}}_{[t]},\tilde{\pi}_{[t]}) \leqslant \boldsymbol{\Gamma}_t\boldsymbol{X}_t, \qquad \mathbb{Q}_{[t]}\text{-a.s.} \quad \forall\mathbb{Q}_{[t]}\in\mathcal{G}_{[t]}(\theta) \\
& \boldsymbol{z}_t(\tilde{\boldsymbol{\xi}}_{[t]},\tilde{\boldsymbol{\zeta}}_{[t]},\tilde{\pi}_{[t]}) \geqslant \boldsymbol{0}, \\
& \boldsymbol{z}_t(\cdot) \in \mathcal{L}_t,
\end{aligned}
$$

其中, $\mathcal{G}_{[t]}(\theta) = \bigotimes_{s\in[t]}\mathcal{G}_s(\theta)$. 最后, 将 $\bar{Y}_t(\boldsymbol{X},\theta)$ 代入模型 (8.10) 得到预算驱动型多阶段容量限制型枢纽选址模型的近似形式:

$$
\begin{aligned}
\max \quad & \theta \\
\text{s.t.} \quad & \bar{Y}_t(\boldsymbol{X},\theta) \leqslant \tau_t, \ \forall t\in[T], \\
& \boldsymbol{X}\in\mathcal{X}, \ \theta \geqslant 0.
\end{aligned}
$$

同样地, 上述模型可以通过二分法求解, 并且在每次迭代中, 可行性问题等价于一个混合整数锥优化问题.

定理8.4 给定任意选址决策 $\boldsymbol{X}\in\mathcal{X}$, 基于拓展线性决策规则的问题 $\bar{Y}_t(\boldsymbol{X},\theta)$ 等价于如下问题:

$$
\begin{aligned}
\min \quad & \boldsymbol{c}_t^\top\boldsymbol{x}_t + \boldsymbol{w}_t^\top\boldsymbol{\eta}_t + \bar{\boldsymbol{v}}_t^\top\boldsymbol{z}_0 + \sum_{s\in[t]}\left(\theta\lambda_s + \frac{1}{N}\sum_{n\in[N]}\sigma_s^n\right) \\
\text{s.t.} \quad & (\bar{\boldsymbol{v}}_t^\top\boldsymbol{Z}_s + \boldsymbol{w}_t^\top\boldsymbol{B}_t^s)\boldsymbol{\xi}_s + \bar{\boldsymbol{v}}_t^\top\boldsymbol{W}_s\boldsymbol{\zeta}_s + (\bar{\boldsymbol{v}}_t^\top\boldsymbol{Y}_s - \lambda_s)\pi_s \leqslant \sigma_s^n, \\
& \sum_{s\in[t]}(\boldsymbol{1}^\top\boldsymbol{Z}_s - \boldsymbol{B}_t^s)\boldsymbol{\xi}_s + \boldsymbol{1}^\top\boldsymbol{W}_s\boldsymbol{\zeta}_s + \boldsymbol{1}^\top\boldsymbol{Y}_s\pi_s \leqslant \boldsymbol{\eta}_t - \boldsymbol{1}^\top\boldsymbol{z}_0, \quad \forall(\boldsymbol{\xi}_s,\boldsymbol{\zeta}_s,\pi_s)\in\mathcal{W}_{sn}, \\
& \sum_{s\in[t]}\boldsymbol{D}\left(\boldsymbol{Z}_s\boldsymbol{\xi}_s + \boldsymbol{W}_s\boldsymbol{\zeta}_s + \boldsymbol{Y}_s\pi_s\right) \leqslant \boldsymbol{\Gamma}_t\boldsymbol{X}_t - \boldsymbol{D}\boldsymbol{z}_0, \\
& \sum_{s\in[t]} -\boldsymbol{Z}_s\boldsymbol{\xi}_s - \boldsymbol{W}_s\boldsymbol{\zeta}_s - \boldsymbol{Y}_s\pi_s \leqslant \boldsymbol{z}_0, \\
& \lambda_s\in\mathbb{R}_+, \ \sigma_s^n\in\mathbb{R}, \ \boldsymbol{z}_0, \ \boldsymbol{Y}_s\in\mathbb{R}^{I^2K}, \ \boldsymbol{Z}_s, \ \boldsymbol{W}_s\in\mathbb{R}^{K\times I^2K}, \ \forall n\in[N], \ s\in[t],
\end{aligned}
$$

其中, 对每个 $s\in[t]$ 和 $n\in[N]$, $\mathcal{W}_{sn}$ 由式 (8.11) 定义.

特别地, 经典的线性决策规则集合为

$$\mathcal{L}_t^0 = \left\{ \boldsymbol{z}_t(\cdot) \in \mathcal{R}^{Kt, I^2K} \;\middle|\; \exists \boldsymbol{z}_0 \in \mathbb{R}^{I^2K}, \right.$$
$$\left. \boldsymbol{Z}_s \in \mathbb{R}^{K \times I^2K}\ \forall s \in [t]:\ \boldsymbol{z}_t(\boldsymbol{\xi}_{[t]}) = \boldsymbol{z}_0 + \sum_{s\in[t]} \boldsymbol{Z}_s \boldsymbol{\xi}_s \right\}.$$

记利用经典线性决策规则 $\mathcal{L}_t^0$ 近似的问题 $Y_t(\boldsymbol{X}, \theta)$ 为 $\bar{Y}_t^0(\boldsymbol{X}, \theta)$. 我们有如下命题:

命题 8.4 给定任意选址决策 $\boldsymbol{X} \in \mathcal{X}$, 我们有

$$Y_t(\boldsymbol{X}, \theta) \leqslant \bar{Y}_t(\boldsymbol{X}, \theta) \leqslant \bar{Y}_t^0(\boldsymbol{X}, \theta), \quad \forall t \in [T].$$

8.5 数值实验

本章同样利用美国民用航空委员会 (Civil Aeronautics Board, CAB) 数据, 设计数值实验验证模与算法的性能. 特别地, 我们考虑一个具有 $K = 100$ 个商品, $I = 15$ 个备选枢纽地址, $T = 5, 10, 15$ 个阶段的枢纽选址问题. 实验包括两个部分: 首先, 分析成本预算对枢纽选址决策的影响; 其次, 验证在分布不确定集中考虑时间序列模型的价值.

我们利用 VAR(1) 模型生成商品需求历史样本 $\hat{\boldsymbol{u}}_{-1}, \hat{\boldsymbol{u}}_{-2}, \cdots, \hat{\boldsymbol{u}}_{-N}$. 基于该样本, 我们构建时间序列模型 (8.3), 并利用此模型预测未来商品需求, 再结合 VAR(1) 模型生成 10000 条未来需求路径 $(\boldsymbol{u}_1, \cdots, \boldsymbol{u}_T)$ 测试样本. 图 8.1 给出了前两种商品的时间序列模型预测结果, 可以看出前两种商品的需求在未来 10 个阶段内逐渐增加, 表明需求不确定性的时间传导性 (前一阶段的需求不确定性会累积到下一阶段). 最后, 我们利用此时间序列模型构建嵌套型分布不确定集 (8.5). 为了表述简洁, 我们记预算驱动型多阶段枢纽选址模型为 BDO, 并记分布鲁棒枢纽选址模型为 DRO.

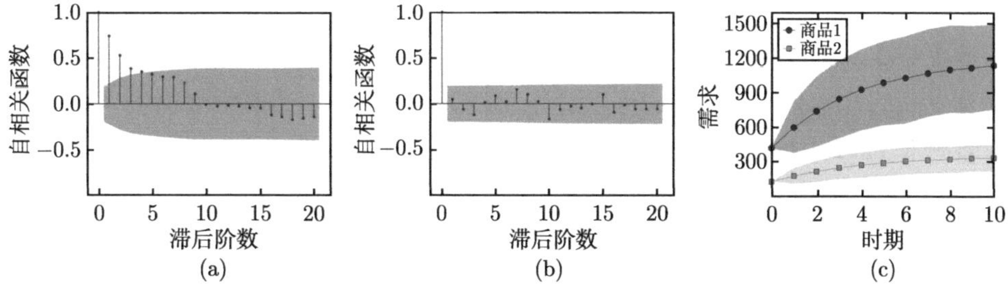

图 8.1 时间序列模型的预测结果. 图 (a) 为第一种商品历史需求的自相关函数, 图 (b) 为对应的历史随机噪声自相关函数. 图 (c) 为时间序列模型的商品需求预测值, 其中实线是前两种商品的预测需求路径. 各图中阴影区域为置信度为 95% 的置信区间

8.5.1 成本预算对枢纽选址决策的影响

假设每阶段的成本预算 τ_t 随阶段以 $r \in \{-2\%, 0\%, 2\%\}$ 的速率增长, 即 $\tau_t = (1+r)^{t-1}\tau_1 \quad \forall t \in [T]$. 特别地, 我们记模型 (8.2) 的总成本预算为 $\sum_{t\in[T]} \tau_t = \rho\tau_0$. 其中, $\rho \geqslant 1$ 代表预算水平, τ_0 为经验分布下最优期望成本, 即

$$\tau_0 \;=\; \min_{\boldsymbol{X}\in\mathcal{X}} \; \mathbb{E}_{\hat{\mathbb{Q}}_{[T]}}\left[\sum_{t\in[T]} \boldsymbol{c}_t^\top \boldsymbol{x}_t + \tilde{\boldsymbol{u}}_t^\top \boldsymbol{f}_t(\mathbf{X}_t)\right].$$

我们同时考虑基于嵌套型分布不确定集的分布鲁棒枢纽选址 (DRO) 模型, 该模型优化极值概率分布下期望总成本. 此外, 我们采用五折交叉验证方法决定模型中的超参数, 如模型 (8.2) 的参数 ρ 以及 DRO 模型的参数 θ.

表 8.1 给出两个模型在不同阶段中的最优选址决策. 结果表明: 首先, 两种模型都倾向于在第一阶段选择更多的枢纽设施, 这是由于第一阶段建造的枢纽可以服务以后所有阶段的商品运输. 其次, 相比于 DRO 模型, BDO 模型倾向于将枢纽建设计划推迟到后续决策阶段, 以规避第一阶段总成本超过成本预算的风险.

表 8.1 最优枢纽选址决策

模型	r	各个阶段的最优枢纽选址决策					总计
		$t=1$	$t=2$	$t=3$	$t=4$	$t=5\sim 10$	
BDO	−2%	{2,3}	{1}	{15}	{12}	—	5
	0%	{2,3}	{1}	{15}	{12}	—	5
	2%	{2,3}	{4,15}	{12,14}	—	—	6
DRO	—	{2,3,4,12,15}	—	—	—	—	5

给定两个模型最优选址决策, 我们可以分析选址决策的样本外表现. 图 8.2 给出了两个模型最优选址决策对应的归一化阶段性成本及其极差. 结果表明: ① BDO 模型最优选址决策对应的第一阶段的成本显著低于 DRO 模型对应的成本, 并且具有更加平稳的阶段性成本; ② 在不同的预算增长率下, BDO 模型对应的阶段成本波动均小于 DRO 模型对应的成本波动, 这说明 BDO 模型的最优选址决策在决策期内具有更平稳的运营成本; ③ BDO 模型相比于 DRO 模型需要有更高的样本外总成本, 这可以看作 BDO 模型为了控制阶段成本波动所需付出的额外 (机会) 成本.

8.5.2 时间序列模型的价值

本节验证在分布不确定集中考虑时间序列模型的价值. 为了比较模型的性能, 我们考虑基于一般 Wasserstein 分布不确定集的预算驱动型多阶段枢纽选址模型 (记为“BDO_W”), 其中, 分布不确定集将不确定需求路径作为一个随机向量 $\tilde{\mathbf{U}} = (\tilde{\boldsymbol{u}}_1, \cdots, \tilde{\boldsymbol{u}}_T)$. 特别地, 基于不确定需求 $\tilde{\mathbf{U}}$ 构建 1-型 Wasserstein 球

$\mathcal{F}_{\mathrm{W}}(\theta) = \{\mathbb{P} \in \mathcal{P}(\mathbb{R}_+^{K\times T}) \mid d(\mathbb{P}, \hat{\mathbb{P}}) \leqslant \theta\}$. 其中, 利用历史需求样本来构建经验概率分布 $\hat{\mathbb{P}} = \dfrac{1}{N}\sum_{n\in[N]}\delta_{\hat{\mathbf{U}}^n}$. 基于该分布不确定集, $\mathrm{BDO_W}$ 模型可以表示为

$$\begin{aligned} \max \quad & \theta \\ \text{s.t.} \quad & \sup_{\mathbb{P}\in\mathcal{F}_{\mathrm{W}}(\theta)} \mathbb{E}_{\mathbb{P}}[\boldsymbol{c}_t^\top \boldsymbol{x}_t + \mathrm{Q}_t(\mathbf{X}_t, \tilde{\boldsymbol{u}}_t)] \leqslant \tau_t, \ \forall t \in [T], \\ & \mathbf{X} \in \mathcal{X}, \ \theta \geqslant 0, \end{aligned} \tag{8.13}$$

因此, BDO 模型与 $\mathrm{BDO_W}$ 模型的唯一区别是前者将时间序列模型 (8.3) 加入分布不确定集中.

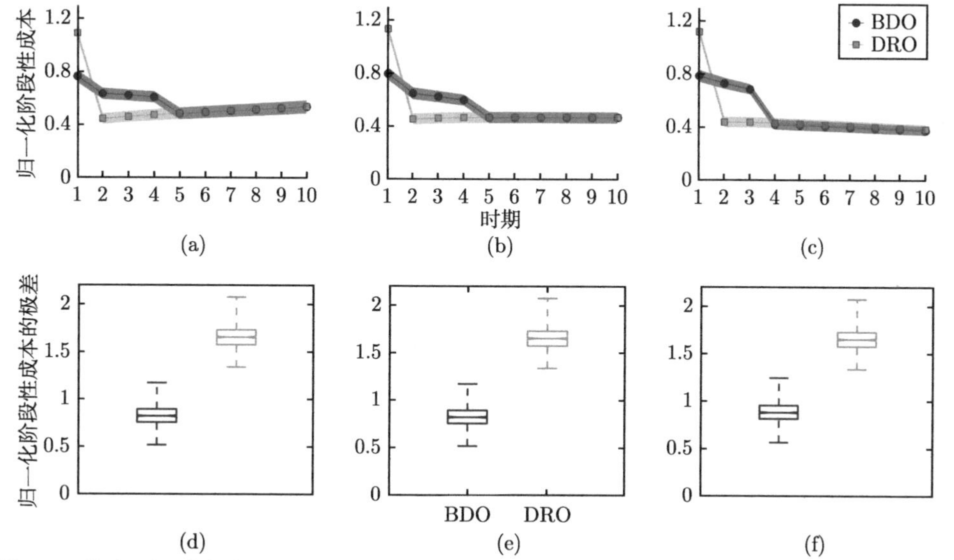

图 8.2 给定不同增长率 $r = -2\%$ (a), $r = 0\%$ (b) 及 $r = 2\%$ (c) 下, 归一化阶段性成本 (第一行) 及相应的归一化阶段性成本极差的箱线图 (第二行). 在第一行中, 每个归一化阶段内成本为该成本除以该阶段成本预算, 阴影部分表示 $5\% \sim 95\%$ 的置信区间

图 8.3 给出了不同归一化预算水平 ρ 下 BDO 模型与 $\mathrm{BDO_W}$ 模型的样本外满足预算概率. 结果表明: ① BDO 模型在不同预算水平 ρ 下满足成本预算约束概率都要高于 $\mathrm{BDO_W}$ 模型, 如 $\rho \in [1.04, 1.08]$, 对所有 $T \in \{5, 10, 15\}$ 的情况, BDO 模型均存在最优选址决策满足阶段性成本预算约束的场景, 而 $\mathrm{BDO_W}$ 模型不存在选址决策能满足阶段性预算约束的场景. 这说明包含时间序列模型信息的 BDO 模型能够更好提供满足阶段性成本预算的选址方案, 以满足阶段性预算约束. ② 当成本预算水平较大时 ($\rho \geqslant 1.12$), 对于所有 T 的情况, BDO 模型满足阶

段性预算的概率也高于 $\mathrm{BDO_W}$. 这些结果验证了在 BDO 模型中加入时间序列预测模型信息的价值.

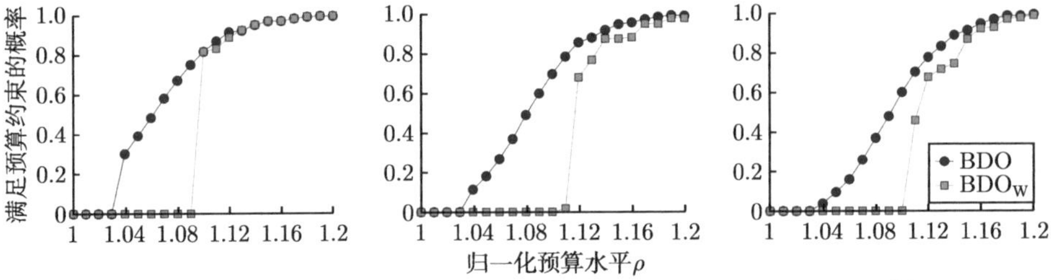

图 8.3　满足预算约束的概率: $T = 5$ (a), $T = 10$ (b) 及 $T = 15$ (c). 给定特定的预算水平, 概率为 0 代表模型预算约束无法满足

8.6　本 章 小 结

本章围绕多阶段枢纽选址问题, 介绍了一种基于嵌套型分布不确定集的预算驱动型多阶段枢纽选址模型, 并给出了模型的高效计算形式. 本章介绍的预算驱动型模型保证了供应链网络在多个阶段内成本的稳定性, 并且在容量充足的情况下给出了模型的经济解释: 模型本质上是在优化各阶段中最坏情况下的夏普比率.

关于不确定性环境下多阶段选址优化问题与基于残差的分布鲁棒优化方法的研究, 读者可以参考更多相关文献, 例如, Correia 等 (2018) 研究了基于给定需求分布的随机容量有限型多阶段枢纽选址问题. Yu 等 (2021) 研究了一个随机多阶段仓库选址问题. Wang 等 (2021a) 基于不确定需求的均值和支撑集信息提出了目标 (预算) 驱动的多阶段选址优化模型, 并开发 Benders 分解算法来求解该模型. 基于 ∞ 型 Wasserstein 分布不确定集, Wang 等 (2021b) 研究了人道主义物流中的多阶段仓库选址问题. Dou 和 Anitescu (2019) 提出了一个单阶段分布鲁棒模型, 其样本数据来自 VAR(1) 过程. 针对可利用的协变量信息, Kannan 等 (2020) 研究了基于回归模型残差的 Wasserstein 分布鲁棒优化建模框架. Perakis 等 (2022) 提出了一种信息-分割分布不确定集, 来刻画加式需求模型中的残差分布, 并将其研究的生产定价优化问题转化为混合整数线性规划问题. Qi 等 (2022) 研究了协变量固定下的分布鲁棒条件分位回归模型.

附录 A　证明及其他相关内容

A.1　第 3 章的证明

引理 3.1 的证明

因为总量模型 (3.1) 和分数模型 (3.2) 等价, 所以给定分布不确定集 $\mathcal{F}$ 和决策 $\boldsymbol{x}$, 我们有

$$\mathbb{P}\left[\mathrm{Q_u}(\boldsymbol{x},\tilde{\boldsymbol{u}},\tilde{\boldsymbol{v}})=\mathrm{F_u}(\boldsymbol{x},\tilde{\boldsymbol{u}},\tilde{\boldsymbol{v}})\right]=1,\quad \forall\mathbb{P}\in\mathcal{F}.$$

因此, 引理 3.1 的结果成立.

定理 3.1 的证明

为了证明定理 3.1, 我们只需要证明, 对于任意给定的 $\boldsymbol{x}$, 下述等式成立, 即

$$\sup_{\mathbb{P}\in\mathcal{F}}\mathbb{E}_{\mathbb{P}}[\mathrm{F_u}(\boldsymbol{x},\tilde{\boldsymbol{u}},\tilde{\boldsymbol{v}})]=\max_{(\boldsymbol{u},\boldsymbol{v})\in\mathcal{W}}\mathrm{F_u}(\boldsymbol{x},\boldsymbol{u},\boldsymbol{v}).\tag{A.1}$$

首先, 对于任意的 $(\boldsymbol{u}^\star,\boldsymbol{v}^\star)\in\mathcal{W}$, 构造一个 Dirac 分布 $\mathbb{P}^\star=\delta_{(\boldsymbol{u}^\star,\boldsymbol{v}^\star)}$, 此分布在点 $(\boldsymbol{u}^\star,\boldsymbol{v}^\star)$ 处的概率为 1 且满足 $\mathbb{P}^\star\in\mathcal{F}$. 因为随机变量 $\tilde{\boldsymbol{u}}$ 和 $\tilde{\boldsymbol{v}}$ 相互独立, 我们可以得到等式:

$$\mathbb{E}_{\mathbb{P}^\star}[\mathrm{F_u}(\boldsymbol{x},\tilde{\boldsymbol{u}},\tilde{\boldsymbol{v}})]=\sum_{k\in[K]}u_k^\star\left[\min_{\boldsymbol{y}_k\in\mathcal{Y}(\boldsymbol{x})}\sum_{i,j\in[N]}\boldsymbol{d}_{ijk}^\top\boldsymbol{v}_k^\star y_{ijk}\right].$$

因此, 可以得到如下不等式:

$$\sup_{\mathbb{P}\in\mathcal{F}}\mathbb{E}_{\mathbb{P}}[\mathrm{F_u}(\boldsymbol{x},\tilde{\boldsymbol{u}},\tilde{\boldsymbol{v}})]\geqslant\max_{(\boldsymbol{u},\boldsymbol{v})\in\mathcal{W}}\mathrm{F_u}(\boldsymbol{x},\boldsymbol{u},\boldsymbol{v}).\tag{A.2}$$

其次, 给定任意概率分布 $\mathbb{P}^\star\in\mathcal{F}$, 定义 $(\boldsymbol{u}^\star,\boldsymbol{v}^\star)=\mathbb{E}_{\mathbb{P}^\star}[(\tilde{\boldsymbol{u}},\tilde{\boldsymbol{v}})]$, 其满足 $(\boldsymbol{u}^\star,\boldsymbol{v}^\star)\in\mathcal{W}$. 因为对于任意的 $k\in[K]$ 和 $\boldsymbol{u}^\star\geqslant\boldsymbol{0}$,

$$\min_{\boldsymbol{y}_k\in\mathcal{Y}(\boldsymbol{x})}\sum_{i,j\in[N]}\boldsymbol{d}_{ijk}^\top\boldsymbol{v}_k y_{ijk}$$

是关于 $\boldsymbol{v}_k$ 的凹函数. 利用 Jensen 不等式, 可以得到如下不等式:

$$\mathbb{E}_{\mathbb{P}^\star}[\mathrm{F}_{\mathrm{u}}(\boldsymbol{x}, \tilde{\boldsymbol{u}}, \tilde{\boldsymbol{v}})] \leqslant \sum_{k\in[K]} u_k^\star \left[\min_{\boldsymbol{y}_k \in \mathcal{Y}(\boldsymbol{x})} \sum_{i,j\in[N]} \boldsymbol{d}_{ijk}^\top \boldsymbol{v}_k^\star y_{ijk} \right].$$

因此, 可以得到不等式:

$$\sup_{\mathbb{P}\in\mathcal{F}} \mathbb{E}_{\mathbb{P}}[\mathrm{F}_{\mathrm{u}}(\boldsymbol{x}, \tilde{\boldsymbol{u}}, \tilde{\boldsymbol{v}})] \leqslant \max_{(\boldsymbol{u},\boldsymbol{v})\in\mathcal{W}} \mathrm{F}_{\mathrm{u}}(\boldsymbol{x}, \boldsymbol{u}, \boldsymbol{v}),$$

结合第一个不等式 (A.2), 证明了等式 (A.1) 成立. 因此, 定理 3.1 的结果成立.

定理 3.2 的证明

利用定理 3.1 和不确定集 $\mathcal{W}$ 的约束依赖性, 给定一个选址决策 $\boldsymbol{x}$, 可以将极值概率分布下第二阶段问题转化为如下形式:

$$\begin{aligned} \max_{(\boldsymbol{u},\boldsymbol{v})\in\mathcal{W}} \mathrm{F}_{\mathrm{u}}(\boldsymbol{x}, \boldsymbol{u}, \boldsymbol{v}) &= \max_{\boldsymbol{u}\in\bar{\mathcal{U}}, \boldsymbol{v}_k\in\bar{\mathcal{V}}_k, k\in[K]} \sum_{k\in[K]} u_k \left[\min_{\boldsymbol{y}_k\in\mathcal{Y}(\boldsymbol{x})} \sum_{i,j\in[N]} \boldsymbol{d}_{ijk}^\top \boldsymbol{v}_k y_{ijk} \right] \\ &= \max_{\boldsymbol{u}\in\bar{\mathcal{U}}} \sum_{k\in[K]} u_k \left[\max_{\boldsymbol{v}_k\in\bar{\mathcal{V}}_k} \min_{\boldsymbol{y}_k\in\mathcal{Y}(\boldsymbol{x})} \sum_{i,j\in[N]} \boldsymbol{d}_{ijk}^\top \boldsymbol{v}_k y_{ijk} \right]. \end{aligned}$$

因为集合 $\mathcal{Y}(\boldsymbol{x})$ 和 $\bar{\mathcal{V}}_k$ 都是有界多面体集, 利用经典的最小最大化定理 (classical minimax theorem) (Sion, 1958), 可将问题

$$\max_{\boldsymbol{v}_k\in\bar{\mathcal{V}}_k} \min_{\boldsymbol{y}_k\in\mathcal{Y}(\boldsymbol{x})} \sum_{i,j\in[N]} \boldsymbol{d}_{ijk}^\top \boldsymbol{v}_k y_{ijk}$$

中的最大化问题和最小化问题进行等价交换, 进而得到问题 (3.10). 接下来, 对于任意的 $k \in [K]$, 引入一个辅助变量 λ_k 将优化问题 (3.10) 转化为如下最小化问题:

$$\min \left\{ \lambda_k \;\middle|\; \lambda_k \geqslant \max_{\boldsymbol{v}_k\in\bar{\mathcal{V}}_k} \sum_{i,j\in[N]} \boldsymbol{d}_{ijk}^\top \boldsymbol{v}_k y_{ijk},\ \boldsymbol{y}_k \in \mathcal{Y}(\boldsymbol{x}) \right\},$$

其中, 上述问题中同时最小化 $(\lambda_k, \boldsymbol{y}_k)$ 等价于在问题 (3.10) 中优化 $\boldsymbol{y}_k$. 因此, 通过引入辅助变量 λ_k, $k \in [K]$, 可以等价地将问题 (3.8) 转化为如下问题:

$$\begin{aligned} \min \quad & \left\{ \boldsymbol{c}^\top \boldsymbol{x} + \max_{\boldsymbol{u}\in\bar{\mathcal{U}}} \sum_{k\in[K]} u_k \min_{\lambda_k} \lambda_k \right\} \\ \text{s.t.} \quad & \lambda_k \geqslant \max_{\boldsymbol{v}_k\in\bar{\mathcal{V}}_k} \sum_{i,j\in[N]} \boldsymbol{d}_{ijk}^\top \boldsymbol{v}_k y_{ijk},\ \forall k \in [K], \\ & \boldsymbol{y}_k \in \mathcal{Y}(\boldsymbol{x}),\ \forall k \in [K], \\ & \boldsymbol{x} \in \{0,1\}^N. \end{aligned}$$

因为, 任意的 $\boldsymbol{u}\in\bar{\mathcal{U}}$ 都是非负的, 所以可以将目标函数中的 $\min_{\lambda_k}\lambda_k$ 等价地替换为 λ_k, 进而得到定理 3.2 的结果.

定理 3.3 的证明

利用定理 3.2, 可以将问题 (3.8) 等价地转化为如下问题:

$$\begin{aligned}\min\quad & \left\{\boldsymbol{c}^{\top}\boldsymbol{x}+\max_{\boldsymbol{u}\in\bar{\mathcal{U}}}\sum_{k\in[K]}u_k\lambda_k\right\}\\ \text{s.t.}\quad & \lambda_k\geqslant\min_{\boldsymbol{y}_k\in\mathcal{Y}(\boldsymbol{x})}\max_{\boldsymbol{v}_k\in\bar{\mathcal{V}}_k}\sum_{i,j\in[N]}\boldsymbol{d}_{ijk}^{\top}\boldsymbol{v}_k y_{ijk},\ \forall k\in[K],\\ & \boldsymbol{x}\in\{0,1\}^N.\end{aligned}\tag{A.3}$$

对于任意的 $k\in[K]$, 对上述问题的约束中最大最小化问题

$$\min_{\boldsymbol{y}_k\in\mathcal{Y}(\boldsymbol{x})}\max_{\boldsymbol{v}_k\in\bar{\mathcal{V}}_k}\sum_{i,j\in[N]}\boldsymbol{d}_{ijk}^{\top}\boldsymbol{v}_k y_{ijk}$$

的最大化问题做对偶, 可以得到如下等价的线性规划问题:

$$\begin{aligned}\min\quad & \overline{\boldsymbol{\phi}}_k^{\top}\boldsymbol{v}_k^{+}-\underline{\boldsymbol{\phi}}_k^{\top}\boldsymbol{v}_k^{-}+\boldsymbol{\gamma}_k^{\top}\boldsymbol{b}_k+\sum_{\ell\in[L_k]}(\eta_{k\ell}\varepsilon_{k\ell}-\boldsymbol{\psi}_{k\ell}^{\top}\boldsymbol{C}_{k\ell}\bar{\boldsymbol{v}}_k)\\ \text{s.t.}\quad & \overline{\boldsymbol{\phi}}_k-\underline{\boldsymbol{\phi}}_k+\boldsymbol{B}_k^{\top}\boldsymbol{\gamma}_k-\sum_{\ell\in[L_k]}\boldsymbol{C}_{k\ell}^{\top}\boldsymbol{\psi}_{k\ell}-\sum_{i,j\in[N]}\boldsymbol{d}_{ijk}y_{ijk}\geqslant\boldsymbol{0},\\ & \eta_{k\ell}\in\mathbb{R},\ \boldsymbol{\psi}_{k\ell}\in\mathbb{R}_+^{N_{k\ell}},\ \eta_{k\ell}\geqslant\|\boldsymbol{\psi}_{k\ell}\|_{\infty},\ \forall\ell\in[L_k],\\ & \overline{\boldsymbol{\phi}}_k,\underline{\boldsymbol{\phi}}_k\in\mathbb{R}_+^3,\ \boldsymbol{\gamma}_k\in\mathbb{R}_+^{M_k},\end{aligned}$$

其中, 与不确定集 $\bar{\mathcal{V}}_k$ 的约束 $\|\boldsymbol{C}_{k\ell}(\boldsymbol{v}_k-\bar{\boldsymbol{v}}_k)\|_1\leqslant\varepsilon_{k\ell}$ 对应的对偶变量 $\eta_{k\ell}$ 和 $\boldsymbol{\psi}_{k\ell}$ 满足 $\eta_{k\ell}\geqslant\|\boldsymbol{\psi}_{k\ell}\|_{\infty}$. 用上述对偶问题替换原最大化问题可以得到等价的线性规划问题 (3.12), 以求解最优运输决策 $\boldsymbol{y}^{\star}$. 此外, 利用上述线性规划问题, 可以得到问题 (A.3) 中第一组约束的如下等价形式:

$$\left\{\begin{aligned}& \lambda_k\geqslant\overline{\boldsymbol{\phi}}_k^{\top}\boldsymbol{u}_k^{+}-\underline{\boldsymbol{\phi}}_k^{\top}\boldsymbol{u}_k^{-}+\boldsymbol{\gamma}_k^{\top}\boldsymbol{b}_k+\sum_{\ell\in[L_k]}(\eta_{k\ell}\varepsilon_{k\ell}-\boldsymbol{\psi}_k^{\top}\boldsymbol{C}_{k\ell}\bar{\boldsymbol{v}}_k),\ \forall k\in[K],\\ & \overline{\boldsymbol{\phi}}_k-\underline{\boldsymbol{\phi}}_k+\boldsymbol{B}_k^{\top}\boldsymbol{\gamma}_k-\sum_{\ell\in[L_k]}\boldsymbol{C}_{k\ell}^{\top}\boldsymbol{\psi}_{k\ell}-\sum_{i,j\in[N]}\boldsymbol{d}_{ijk}y_{ijk}\geqslant\boldsymbol{0},\ \forall k\in[K],\\ & \eta_{k\ell}\in\mathbb{R},\ \boldsymbol{\psi}_{k\ell}\in\mathbb{R}_+^{N_{k\ell}},\ \eta_{k\ell}\geqslant\|\boldsymbol{\psi}_{k\ell}\|_{\infty},\ \forall k\in[K],\ \ell\in[L_k],\\ & \overline{\boldsymbol{\phi}}_k,\underline{\boldsymbol{\phi}}_k\in\mathbb{R}_+^3,\ \boldsymbol{\gamma}_k\in\mathbb{R}_+^{M_k},\ \boldsymbol{y}_k\in\mathcal{Y}(\boldsymbol{x}),\ \forall k\in[K].\end{aligned}\right.$$

最后, 对问题 (A.3) 目标函数中最大化问题做对偶, 可以得到定理 3.3 中的混合整数线性规划问题 (3.13).

定理 3.4 的证明

首先, 定义 $G(\boldsymbol{y},\boldsymbol{u},\boldsymbol{v})=\sum_{i,j\in[N]}\sum_{k\in[K]}(\boldsymbol{d}_{ijk}^{\top}\boldsymbol{v}_k)u_k y_{ijk}$. 其次, 给定选址决策 $\boldsymbol{x}$, 考虑如下极值概率分布下第二阶段问题:

$$\begin{aligned}\sup_{\mathbb{P}}\quad & \mathbb{E}_{\mathbb{P}}\left[\min_{\boldsymbol{y}_k\in\mathcal{Y}(\boldsymbol{x})} G(\boldsymbol{y},\tilde{\boldsymbol{u}},\tilde{\boldsymbol{v}})\right]\\ \text{s.t.}\quad & \mathbb{E}_{\mathbb{P}}[\tilde{\boldsymbol{u}}]\in[\boldsymbol{u}^-,\boldsymbol{u}^+],\\ & \mathbb{E}_{\mathbb{P}}[\tilde{\boldsymbol{v}}_k]\in[\boldsymbol{v}_k^-,\boldsymbol{v}_k^+],\ \forall k\in[K],\\ & \mathbb{E}_{\mathbb{P}}[\|\boldsymbol{C}_l(\tilde{\boldsymbol{v}}-\bar{\boldsymbol{v}})\|_1]\leqslant\varepsilon_l,\ \forall l\in[L],\\ & \mathbb{P}[\tilde{\boldsymbol{u}}\in\mathcal{U}_{\text{CH}},\tilde{\boldsymbol{v}}\in\mathcal{V}]=1.\end{aligned}$$

利用强对偶定理 (Bertsimas et al., 2019), 可以得到上述问题的对偶问题如下:

$$\begin{aligned}\inf\quad & \alpha+\overline{\boldsymbol{\beta}}^{\top}\boldsymbol{u}^+-\underline{\boldsymbol{\beta}}^{\top}\boldsymbol{u}^-+\sum_{k\in[K]}(\overline{\boldsymbol{\gamma}}_k^{\top}\boldsymbol{v}_k^+-\underline{\boldsymbol{\gamma}}_k^{\top}\boldsymbol{v}_k^-)+\sum_{l\in[L]}\rho_l\varepsilon_l\\ \text{s.t.}\quad & \alpha+(\overline{\boldsymbol{\beta}}-\underline{\boldsymbol{\beta}})^{\top}\boldsymbol{u}+\sum_{k\in[K]}(\overline{\boldsymbol{\gamma}}_k-\underline{\boldsymbol{\gamma}}_k)^{\top}\boldsymbol{v}_k\\ &\quad +\sum_{l\in[L]}\rho_l\|\boldsymbol{C}_l(\boldsymbol{v}-\bar{\boldsymbol{v}})\|_1\geqslant\min_{\boldsymbol{y}_k\in\mathcal{Y}(\boldsymbol{x})}G(\boldsymbol{y},\boldsymbol{u},\boldsymbol{v}),\ \forall\boldsymbol{u}\in\mathcal{U}_{\text{CH}},\ \boldsymbol{v}\in\mathcal{V},\\ & \alpha\in\mathbb{R},\ \overline{\boldsymbol{\beta}},\underline{\boldsymbol{\beta}}\geqslant\mathbb{R}_+^K,\ \overline{\boldsymbol{\gamma}}_k,\underline{\boldsymbol{\gamma}}_k\in\mathbb{R}_+^3,\ \rho_l\geqslant 0,\ \forall k\in[K],\ l\in[L].\end{aligned}\tag{A.4}$$

对于任意给定的 $\boldsymbol{v}\in\mathcal{V}$, 上述问题的第一组约束等价于如下不等式:

$$\begin{aligned}&\alpha+\sum_{k\in[K]}(\overline{\boldsymbol{\gamma}}_k-\underline{\boldsymbol{\gamma}}_k)^{\top}\boldsymbol{v}_k+\sum_{l\in[L]}\rho_l\|\boldsymbol{C}_l(\boldsymbol{v}-\bar{\boldsymbol{v}})\|_1\\ \geqslant&\max_{\boldsymbol{u}\in\mathcal{U}_{\text{CH}}}\min_{\boldsymbol{y}_k\in\mathcal{Y}(\boldsymbol{x})}\left\{G(\boldsymbol{y},\boldsymbol{u},\boldsymbol{v})-(\overline{\boldsymbol{\beta}}-\underline{\boldsymbol{\beta}})^{\top}\boldsymbol{u}\right\}.\end{aligned}$$

因为函数 $G(\boldsymbol{y},\boldsymbol{u},\boldsymbol{v})-(\overline{\boldsymbol{\beta}}-\underline{\boldsymbol{\beta}})^{\top}\boldsymbol{u}$ 是关于变量 $\boldsymbol{y}$ 和 $\boldsymbol{u}$ 的线性函数, 且集合 $\mathcal{U}_{\text{CH}}$ 和 $\mathcal{Y}(\boldsymbol{x})$ 是有界多面体集. 利用最小最大化定理 (Sion, 1958), 可以将上述不等式转化为如下不等式:

$$\begin{aligned}&\alpha+\sum_{k\in[K]}(\overline{\boldsymbol{\gamma}}_k-\underline{\boldsymbol{\gamma}}_k)^{\top}\boldsymbol{v}_k+\sum_{l\in[L]}\rho_l\|\boldsymbol{C}_l(\boldsymbol{v}-\bar{\boldsymbol{v}})\|_1\\ \geqslant&\min_{\boldsymbol{y}_k\in\mathcal{Y}(\boldsymbol{x})}\max_{\boldsymbol{u}\in\mathcal{U}_{\text{CH}}}\left\{G(\boldsymbol{y},\boldsymbol{u},\boldsymbol{v})-(\overline{\boldsymbol{\beta}}-\underline{\boldsymbol{\beta}})^{\top}\boldsymbol{u}\right\},\end{aligned}$$

其中, 约束右端项中最大化问题的目标函数是关于变量 $\boldsymbol{u}$ 线性凸函数. 因此, 最大化问题的最优解在可行集 $\mathcal{U}_{\text{CH}}$ 的极值点处取到, 即最优解在点集 $\boldsymbol{u}^s$, $s\in[S]$ 中.

利用此性质, 可以将上述约束等价地转化为如下不等式:

$$\begin{aligned}&\alpha+\sum_{k\in[K]}(\overline{\gamma}_k-\underline{\gamma}_k)^\top \boldsymbol{v}_k+\sum_{l\in[L]}\rho_l\|\boldsymbol{C}_l(\boldsymbol{v}-\bar{\boldsymbol{v}})\|_1\\ \geqslant&\min_{\boldsymbol{y}_k\in\mathcal{Y}(\boldsymbol{x})}\left\{G(\boldsymbol{y},\boldsymbol{u}^s,\boldsymbol{v})-(\overline{\boldsymbol{\beta}}-\underline{\boldsymbol{\beta}})^\top\boldsymbol{u}^s\right\},\quad \forall s\in[S].\end{aligned}$$

利用上述不等式, 可以将问题 (A.4) 的第一组约束等价地转化为如下不等式:

$$\begin{aligned}&\alpha+(\overline{\boldsymbol{\beta}}-\underline{\boldsymbol{\beta}})^\top\boldsymbol{u}^s\\ \geqslant&\max_{\boldsymbol{v}\in\mathcal{V}}\min_{\boldsymbol{y}_k\in\mathcal{Y}(\boldsymbol{x})}\left\{G(\boldsymbol{y},\boldsymbol{u}^s,\boldsymbol{v})-\sum_{k\in[K]}(\overline{\gamma}_k-\underline{\gamma}_k)^\top\boldsymbol{v}_k-\sum_{l\in[L]}\rho_l\|\boldsymbol{C}_l(\boldsymbol{v}-\bar{\boldsymbol{v}})\|_1\right\},\quad\forall s\in[S],\end{aligned}$$

再次利用最小最大化定理, 可以得到上述不等式的等价形式:

$$\begin{aligned}&\alpha+(\overline{\boldsymbol{\beta}}-\underline{\boldsymbol{\beta}})^\top\boldsymbol{u}^s\\ \geqslant&\min_{\boldsymbol{y}_k\in\mathcal{Y}(\boldsymbol{x})}\max_{\boldsymbol{v}\in\mathcal{V}}\left\{G(\boldsymbol{y},\boldsymbol{u}^s,\boldsymbol{v})-\sum_{k\in[K]}(\overline{\gamma}_k-\underline{\gamma}_k)^\top\boldsymbol{v}_k-\sum_{l\in[L]}\rho_l\|\boldsymbol{C}_l(\boldsymbol{v}-\bar{\boldsymbol{v}})\|_1\right\},\quad\forall s\in[S].\end{aligned}$$

对上述约束右端项中关于变量 $\boldsymbol{v}$ 的最大化问题做对偶, 且将对偶问题与外层关于变量 $\boldsymbol{y}$ 的最大化问题进行合并, 可以得到问题 (A.4) 的等价混合整数线性规划形式 (3.15). 因此, 得到定理 3.4 的结果.

命题 3.1 的证明

根据问题 (3.17) 的约束, 我们可得 $\mathbb{P}_v^\dagger\in\mathcal{F}^\dagger$. 接下来, 我们只需要证明问题 (3.17) 的最优值 $\Omega^\star$ 等于问题 (3.15) 的最优值减去 $\boldsymbol{c}^\top\boldsymbol{x}^\star$ 的差值. 问题 (3.17) 是一个线性规划问题, 利用强对偶定理, 可以得到问题 (3.17) 的对偶问题如下:

$$\begin{aligned}\min\quad&\alpha+\overline{\boldsymbol{\beta}}^\top\boldsymbol{u}^+-\underline{\boldsymbol{\beta}}^\top\boldsymbol{u}^-+\sum_{k\in[K]}(\overline{\gamma}_k^\top\boldsymbol{v}_k^+-\underline{\gamma}_k^\top\boldsymbol{v}_k^-)+\sum_{l\in[L]}\rho_l\varepsilon_l\\ \text{s.t.}\quad&\alpha+(\overline{\boldsymbol{\beta}}-\underline{\boldsymbol{\beta}})^\top\boldsymbol{u}^s+\sum_{k\in[K]}(\overline{\gamma}_k-\underline{\gamma}_k)^\top\boldsymbol{v}_k^s\\ &\quad+\sum_{l\in[L]}\rho_l\|\boldsymbol{C}_l(\boldsymbol{v}^s-\bar{\boldsymbol{v}})\|_1\geqslant\sum_{i,j\in[N]}\sum_{k\in[K]}\boldsymbol{d}_{ijk}^\top\boldsymbol{v}_k^s u_k^s y_{ijk}^{s*},\quad\forall s\in[S]\\ &\alpha\in\mathbb{R},\ \overline{\boldsymbol{\beta}},\underline{\boldsymbol{\beta}}\geqslant\boldsymbol{0},\ \overline{\gamma}_k,\underline{\gamma}_k\geqslant\boldsymbol{0},\ \forall k\in[K],\quad\rho_l\geqslant0,\ \forall l\in[L].\end{aligned}$$

因为 $(\boldsymbol{y}^{s\star})_{s\in[S]}$ 是问题 (3.15) 的最优解, 所以利用 $(\boldsymbol{v}^s)_{s\in[S]}$ 的表达式 (3.16) 可以将上述问题的第一组约束等价地转化为如下形式:

$$\begin{cases} \quad \alpha + (\overline{\boldsymbol{\beta}} - \underline{\boldsymbol{\beta}})^\top \boldsymbol{u}^s \\ \geqslant \max\limits_{\boldsymbol{v}\in\mathcal{V}} \left\{ \sum\limits_{i,j\in[N]} \sum\limits_{k\in[K]} \boldsymbol{d}_{ijk}^\top \boldsymbol{v}_k u_k^s y_{ijk} - \sum\limits_{k\in[K]} (\overline{\boldsymbol{\gamma}}_k - \underline{\boldsymbol{\gamma}}_k)^\top \boldsymbol{v}_k - \sum\limits_{l\in[L]} \rho_l \|\boldsymbol{C}_l(\boldsymbol{v} - \bar{\boldsymbol{v}})\|_1 \right\} \\ \boldsymbol{y}^s \in \mathcal{Y}(\boldsymbol{x}). \end{cases}$$

将上述约束右端的最大化问题等价地替换为其对偶问题, 可以得到问题 (3.17) 等价于给定选址决策 $\boldsymbol{x}^\star$ 下问题 (3.15). 因此 $\Omega^\star$ 等于问题 (3.15) 的最优值减去 $\boldsymbol{c}^\top \boldsymbol{x}^\star$ 的差值.

定理 3.5 的证明

给定 $\boldsymbol{x}$ 和 $(\boldsymbol{u}, \boldsymbol{v})$, 第二阶段问题 $\mathrm{Q_c}(\boldsymbol{x}, \boldsymbol{u}, \boldsymbol{v})$ 是一个线性规划问题. 利用强对偶定理, 可以推出问题 $\mathrm{Q_c}(\boldsymbol{x}, \boldsymbol{u}, \boldsymbol{v})$ 的对偶问题如下:

$$\mathrm{Q_c}(\boldsymbol{x}, \boldsymbol{u}, \boldsymbol{v}) = \max_{(\boldsymbol{r},\boldsymbol{t})\in\mathcal{H}(\boldsymbol{v})} \left\{ (\boldsymbol{r} + \boldsymbol{h})^\top \boldsymbol{u} + \sum_{i\in[N]} x_i q_i t_i \right\},$$

其中, 对偶变量的可行集为 $\mathcal{H}(\boldsymbol{v}) = \{(\boldsymbol{r}, \boldsymbol{t}) \in \mathbb{R}_-^K \times \mathbb{R}_-^N \mid r_k + t_i + t_j \cdot \mathbb{I}[j \neq i] \leqslant \boldsymbol{d}_{ijk}^\top \boldsymbol{v}_k - h_k \ \forall i, j \in [N],\ k \in [K]\}$, 且符号 $\mathbb{I}[\cdot]$ 为示性函数. 利用强对偶定理, 可以得出极值概率分布下第二阶段问题 $\sup_{\mathbb{P}\in\mathcal{F}^\dagger} \mathbb{E}_{\mathbb{P}}[\mathrm{Q_c}(\boldsymbol{x}, \tilde{\boldsymbol{u}}, \tilde{\boldsymbol{v}})]$ 等价于如下最小化问题:

$$\begin{aligned} \inf \quad & \alpha + \overline{\boldsymbol{\beta}}^\top \boldsymbol{u}^+ - \underline{\boldsymbol{\beta}}^\top \boldsymbol{u}^- + \sum_{k\in[K]} (\overline{\boldsymbol{\gamma}}_k^\top \boldsymbol{v}_k^+ - \underline{\boldsymbol{\gamma}}_k^\top \boldsymbol{v}_k^-) + \sum_{l\in[L]} \rho_l \varepsilon_l \\ \text{s.t.} \quad & \alpha + (\overline{\boldsymbol{\beta}} - \underline{\boldsymbol{\beta}})^\top \boldsymbol{u} + \sum_{k\in[K]} (\overline{\boldsymbol{\gamma}}_k - \underline{\boldsymbol{\gamma}}_k)^\top \boldsymbol{v}_k \\ & \quad + \sum_{l\in[L]} \rho_l \|\boldsymbol{C}_l(\boldsymbol{v} - \bar{\boldsymbol{v}})\|_1 \geqslant \mathrm{Q_c}(\boldsymbol{x}, \boldsymbol{u}, \boldsymbol{v}),\ \forall \boldsymbol{u} \in \mathcal{U}_{\mathrm{CH}},\ \boldsymbol{v} \in \mathcal{V}, \\ & \alpha \in \mathbb{R},\ \overline{\boldsymbol{\beta}}, \underline{\boldsymbol{\beta}} \in \mathbb{R}_+^K,\ \overline{\boldsymbol{\gamma}}_k, \underline{\boldsymbol{\gamma}}_k \in \mathbb{R}_+^3, \quad \forall k \in [K],\ \boldsymbol{\rho} \in \mathbb{R}_+^L, \end{aligned} \tag{A.5}$$

其中, 上述问题的第一组约束可以等价地表示为

$$\begin{aligned} & \alpha + \sum_{k\in[K]} (\overline{\boldsymbol{\gamma}}_k - \underline{\boldsymbol{\gamma}}_k)^\top \boldsymbol{v}_k + \sum_{l\in[L]} \rho_l \|\boldsymbol{C}_l(\boldsymbol{v} - \bar{\boldsymbol{v}})\|_1 \\ \geqslant & \max_{\boldsymbol{u}\in\mathcal{U}_{\mathrm{CH}}} \{\mathrm{Q_c}(\boldsymbol{x}, \boldsymbol{u}, \boldsymbol{v}) - (\overline{\boldsymbol{\beta}} - \underline{\boldsymbol{\beta}})^\top \boldsymbol{u}\}, \quad \forall \boldsymbol{v} \in \mathcal{V}. \end{aligned}$$

给定任意 $\boldsymbol{v}\in\mathcal{V}$, 根据问题 $\mathrm{Q_c}(\boldsymbol{x},\boldsymbol{u},\boldsymbol{v})$ 的对偶问题, 可知 $\mathrm{Q_c}(\boldsymbol{x},\boldsymbol{u},\boldsymbol{v})-(\overline{\boldsymbol{\beta}}-\underline{\boldsymbol{\beta}})^\top\boldsymbol{u}$ 是关于变量 $\boldsymbol{u}$ 的凸函数, 且 $\mathcal{U}_{\mathrm{CH}}$ 是一个多面体集. 因此, 上述不等式右端项最大化问题的最优值在集合 $\mathcal{U}_{\mathrm{CH}}$ 的极值点 $\boldsymbol{u}^s, s\in[S]$ 处取到, 这使得我们可以将上述不等式转化为如下等价形式:

$$\begin{aligned}&\alpha+\sum_{k\in[K]}(\overline{\boldsymbol{\gamma}}_k-\underline{\boldsymbol{\gamma}}_k)^\top\boldsymbol{v}_k+\sum_{l\in[L]}\rho_l\|\boldsymbol{C}_l(\boldsymbol{v}-\bar{\boldsymbol{v}})\|_1\\&\geqslant\mathrm{Q_c}(\boldsymbol{x},\boldsymbol{u}^s,\boldsymbol{v})-(\overline{\boldsymbol{\beta}}-\underline{\boldsymbol{\beta}})^\top\boldsymbol{u}^s,\quad\forall\boldsymbol{v}\in\mathcal{V},s\in[S].\end{aligned}$$

利用对偶问题 $\mathrm{Q_c}(\boldsymbol{x},\boldsymbol{u},\boldsymbol{v})$ 的原问题 $\mathrm{Q_u}(\boldsymbol{x},\boldsymbol{u},\boldsymbol{v})$, 可以得到如下不等式:

$$\begin{aligned}&\alpha+\sum_{k\in[K]}(\overline{\boldsymbol{\gamma}}_k-\underline{\boldsymbol{\gamma}}_k)^\top\boldsymbol{v}_k+\sum_{l\in[L]}\rho_l\|\boldsymbol{C}_l(\boldsymbol{v}-\bar{\boldsymbol{v}})\|_1\\&\geqslant\min_{\boldsymbol{z}\in\mathcal{Z}_c(\boldsymbol{x},\boldsymbol{u}^s)}\left\{\sum_{k\in[K]}\sum_{i,j\in[N]}(\boldsymbol{d}_{ijk}^\top\boldsymbol{v}_k-h_k)z_{ijk}+\boldsymbol{h}^\top\boldsymbol{u}^s\right\}\\&\quad-(\overline{\boldsymbol{\beta}}-\underline{\boldsymbol{\beta}})^\top\boldsymbol{u}^s,\quad\forall\boldsymbol{v}\in\mathcal{V},s\in[S].\end{aligned}$$

对于任意 $s\in[S]$, 可以将上述不等式转化为如下形式:

$$\begin{aligned}&\alpha+(\overline{\boldsymbol{\beta}}-\underline{\boldsymbol{\beta}}-\boldsymbol{h})^\top\boldsymbol{u}^s\\&\geqslant\max_{\boldsymbol{v}\in\mathcal{V}}\left\{\min_{\boldsymbol{z}\in\mathcal{Z}_c(\boldsymbol{x},\boldsymbol{u}^s)}\left\{\sum_{k\in[K]}\sum_{i,j\in[N]}(\boldsymbol{d}_{ijk}^\top\boldsymbol{v}_k-h_k)z_{ijk}\right\}\right.\\&\quad\left.-\sum_{k\in[K]}(\overline{\boldsymbol{\gamma}}_k-\underline{\boldsymbol{\gamma}}_k)^\top\boldsymbol{v}_k-\sum_{l\in[L]}\rho_l\|\boldsymbol{C}_l(\boldsymbol{v}-\bar{\boldsymbol{v}})\|_1\right\}.\end{aligned}$$

接下来, 利用最小最大化定理可以得到如下等价不等式组:

$$\left\{\begin{aligned}&\alpha+(\overline{\boldsymbol{\beta}}-\underline{\boldsymbol{\beta}}-\boldsymbol{h})^\top\boldsymbol{u}^s\geqslant\max_{\boldsymbol{v}\in\mathcal{V}}\left\{\sum_{k\in[K]}\sum_{i,j\in[N]}(\boldsymbol{d}_{ijk}^\top\boldsymbol{v}_k-h_k)z_{ijk}^s\right.\\&\qquad\qquad\left.-\sum_{k\in[K]}(\overline{\boldsymbol{\gamma}}_k-\underline{\boldsymbol{\gamma}}_k)^\top\boldsymbol{v}_k-\sum_{l\in[L]}\rho_l\|\boldsymbol{C}_l(\boldsymbol{v}-\bar{\boldsymbol{v}})\|_1\right\},\\&\boldsymbol{z}^s\in\mathcal{Z}_c(\boldsymbol{x},\boldsymbol{u}^s).\end{aligned}\right.$$

最后, 将上述约束中关于变量 $\boldsymbol{v}$ 的最大化问题用其对偶问题替换, 并将上述不等式组代入问题 (A.5), 则可以得到定理 3.5 中的混合整数线性规划问题 (3.20).

A.2 第 4 章的证明及其他相关内容

本节将提供第 4 章中引理、命题和定理的证明以及其他相关内容.

A.2.1 第 4 章的证明

引理 4.1 的证明

首先, 当产能充足时, 即条件 (4.10) 成立时, 我们可以推出

$$x_{ij} \leqslant \tilde{d}_j y_i, \forall i \in [I], j \in [J] \Longrightarrow \sum_{j\in[J]} x_{ij} \leqslant q_i y_i, \forall i \in [I] \ \ \mathbb{P}\text{-几乎处处成立}, \ \forall \mathbb{P} \in \mathcal{F}.$$

也就是说, 我们有如下等式成立

$$\mathbb{P}\left[\mathcal{X}(\boldsymbol{y}, \tilde{\boldsymbol{d}}) \supseteq \mathcal{X}_{\mathrm{u}}(\boldsymbol{y}, \tilde{\boldsymbol{d}})\right] = 1, \quad \forall \mathbb{P} \in \mathcal{F}.$$

其次, 对于任意的 $i \in [I]$, 如果 $y_i = 0$, 我们有

$$\sum_{j\in[J]} x_{ij} \leqslant q_i y_i = 0 \Longrightarrow x_{ij} \leqslant \tilde{d}_j y_i = 0, \forall j \in [J] \ \ \mathbb{P}\text{-几乎处处成立}, \ \forall \mathbb{P} \in \mathcal{F}.$$

如果 $y_i = 1$, 有如下不等式成立:

$$\begin{aligned} x_{ij} &\leqslant \min\left\{\sum_{i\in I} x_{ij}, \sum_{j\in[J]} x_{ij}\right\} \\ &\leqslant \min\left\{\tilde{d}_j, q_i y_i\right\} = \tilde{d}_j, \ \forall j \in [J] \ \ \mathbb{P}\text{-几乎处处成立}, \ \forall \mathbb{P} \in \mathcal{F}. \end{aligned}$$

结合上述两种 y_i 的取值情况, 我们可以推出

$$\mathbb{P}\left[\mathcal{X}(\boldsymbol{y}, \tilde{\boldsymbol{d}}) \subseteq \mathcal{X}_{\mathrm{u}}(\boldsymbol{y}, \tilde{\boldsymbol{d}})\right] = 1, \quad \forall \mathbb{P} \in \mathcal{F}.$$

因此, 引理 4.1 的结果成立.

引理 4.2 的证明

给定状态 $\tilde{k}$ 的概率分布 $\mathbb{P}[\tilde{k} = k] = p_k, \forall k \in [K]$, 我们用 $\mathbb{P}_k$ 表示给定 $\tilde{k} = k$ $(k \in [K])$ 下需求 $\tilde{\boldsymbol{d}}$ 的条件概率分布, 进而可以将任意概率分布 $\mathbb{P} \in \mathcal{F}$ 及其期望 $\mathbb{E}_{\mathbb{P}}$ 分别分解为 $\{\mathbb{P}_k, k \in [K]\}$ 和 $\{\mathbb{E}_{\mathbb{P}_k}, k \in [K]\}$. 因此, 给定任一可行的选址决

策 $\boldsymbol{y}$, 我们可以使用全期望公式将极值概率分布下第二阶段问题 (4.13) 转化为如下形式:

$$\sup_{\mathbb{P}\in\mathcal{F}(\boldsymbol{\Theta})} \mathbb{E}_{\mathbb{P}}\left[\mathrm{Q}(\boldsymbol{y},\tilde{\boldsymbol{d}})\right] = \sum_{k\in[K]} p_k \sup_{\mathbb{P}\in\mathcal{F}_k} \mathbb{E}_{\mathbb{P}}\left[\mathrm{Q}(\boldsymbol{y},\tilde{\boldsymbol{d}})\right], \tag{A.6}$$

其中, $\mathcal{F}_k$ 是状态 $k\in[K]$ 下不确定需求 $\tilde{\boldsymbol{d}}$ 条件概率分布的分布不确定集:

$$\mathcal{F}_k = \left\{ \mathbb{P}\in\mathcal{P}(\mathbb{R}^J) \;\middle|\; \begin{array}{l} \tilde{\boldsymbol{d}}\sim\mathbb{P}, \\ \displaystyle\int_{\tilde{\boldsymbol{d}}\in\mathcal{D}_k} \tilde{\boldsymbol{d}}\,\mathrm{d}\mathbb{P}_k(\tilde{\boldsymbol{d}}) \leqslant \boldsymbol{d}_k^+, \\ \displaystyle\int_{\tilde{\boldsymbol{d}}\in\mathcal{D}_k} \tilde{\boldsymbol{d}}\,\mathrm{d}\mathbb{P}_k(\tilde{\boldsymbol{d}}) \geqslant \boldsymbol{d}_k^-, \\ \displaystyle\int_{\tilde{\boldsymbol{d}}\in\mathcal{D}_k} \boldsymbol{g}_k(\tilde{\boldsymbol{d}})\,\mathrm{d}\mathbb{P}_k(\tilde{\boldsymbol{d}}) \leqslant \boldsymbol{\sigma}_k, \\ \displaystyle\int_{\tilde{\boldsymbol{d}}\in\mathcal{D}_k} \mathrm{d}\mathbb{P}_k(\tilde{\boldsymbol{d}}) = 1 \end{array} \right\}.$$

接下来, 我们利用强对偶定理, 推导每个状态 $k\in K$ 下条件极值概率分布下第二阶段期望成本

$$\sup_{\mathbb{P}\in\mathcal{F}_k} \mathbb{E}_{\mathbb{P}}\left[\mathrm{Q}(\boldsymbol{y},\tilde{\boldsymbol{d}})\right] \tag{A.7}$$

的对偶形式. 根据 (Shapiro, 2001) 中的命题 3.4, 为了得到问题 (A.7) 的对偶形式, 我们首先证明对任意 $k\in[K]$,

$$(\boldsymbol{d}_k^+,\boldsymbol{d}_k^-,\boldsymbol{\sigma}_k,1)\in\mathrm{int}[\mathcal{A}(\mathcal{M}^+(\mathcal{D}_k))-\mathcal{C}], \tag{A.8}$$

其中, $\mathcal{M}^+(\mathcal{D}_k)$ 表示所有支撑集为 $\mathcal{D}_k$ 的非负博雷尔 (Borel) 测度集合, $\mathcal{A}(\mathcal{M}^+(\mathcal{D}_k))$ 表示从 $\mathcal{M}^+(\mathcal{D}_k)$ 到 $\mathbb{R}^4$ 的线性映射:

$$\begin{aligned} &\mathcal{A}(\mathcal{M}^+(\mathcal{D}_k)):\mu\in\mathcal{M}^+(\mathcal{D}_k) \\ &\longmapsto \left(\int_{\mathcal{D}_k}\tilde{\boldsymbol{d}}\,\mathrm{d}\mu(\tilde{\boldsymbol{d}}),\int_{\mathcal{D}_k}\tilde{\boldsymbol{d}}\,\mathrm{d}\mu(\tilde{\boldsymbol{d}}),\int_{\mathcal{D}_k}\boldsymbol{g}_k(\tilde{d})\,\mathrm{d}\mu(\tilde{\boldsymbol{d}}),\int_{\mathcal{D}_k}\mathrm{d}\mu(\tilde{\boldsymbol{d}})\right), \end{aligned}$$

并且 $\mathcal{C}$ 是如下所示的一个闭凸锥:

$$\mathcal{C}=\mathbb{R}_-^J\times\mathbb{R}_+^J\times\mathbb{R}_-^{n_k}\times\{0\}.$$

对于每个 $k\in[K]$, 证明式 (A.8) 成立等价于证明点 $(\boldsymbol{d}_k^+,\boldsymbol{d}_k^-,\boldsymbol{\sigma}_k,1)$ 位于如下凸锥

的内部 (interior):

$$\mathcal{V}_k=\left\{(\boldsymbol{b},\boldsymbol{c},\boldsymbol{h},a)\in\mathbb{R}^J\times\mathbb{R}^J\times\mathbb{R}^{n_k}\times\mathbb{R}\ \middle|\ \exists\mu\in\mathcal{M}^+(\mathcal{D}_k)\quad,\ \begin{array}{l}\int_{\mathcal{D}_k}\boldsymbol{d}\ \mathrm{d}\mu(\boldsymbol{d})\leqslant\boldsymbol{b},\\ \int_{\mathcal{D}_k}\boldsymbol{d}\ \mathrm{d}\mu(\boldsymbol{d})\geqslant\boldsymbol{c},\\ \int_{\mathcal{D}_k}\boldsymbol{g}_k(\boldsymbol{d})\ \mathrm{d}\mu(\boldsymbol{d})\leqslant\boldsymbol{h},\\ \int_{\mathcal{D}_k}\mathrm{d}\mu(\boldsymbol{d})=a\end{array}\right\}.$$

记 $\mathcal{B}_\epsilon(\boldsymbol{u})$ 为球心为点 $\boldsymbol{u}$、半径为 $\epsilon>0$ 的闭欧氏球. 根据假设, 存在一个 $\boldsymbol{d}_k^\dagger\in(\boldsymbol{d}_k^-,\boldsymbol{d}_k^+)$ 使得 $\boldsymbol{g}_k(\boldsymbol{d}_k^\dagger)<\boldsymbol{\sigma}_k$ 和 $[\boldsymbol{d}_k^-,\boldsymbol{d}_k^+]\subseteq\mathcal{D}_k$ 成立, 我们总是可以选择一个足够小的 $\epsilon>0$ 使得任意点 $(\boldsymbol{r},\boldsymbol{s},\boldsymbol{t},\lambda)\in\mathcal{B}_\epsilon(\boldsymbol{d}_k^+)\times\mathcal{B}_\epsilon(\boldsymbol{d}_k^-)\times\mathcal{B}_\epsilon(\boldsymbol{\sigma}_k)\times\mathcal{B}_\epsilon(1)$ 都满足 $\boldsymbol{r}\geqslant\boldsymbol{d}_k^\dagger\geqslant\boldsymbol{s}$. 接下来, 我们可以构造一个支撑集为 $\mathcal{D}_k$ 的 Dirac 测度 $\mu^\natural:=\lambda\cdot\delta_{\boldsymbol{d}_k^\dagger/\lambda}$, 其表示在点 $\boldsymbol{d}_k^\dagger/\lambda$ 处的质量 (mass) 为 λ 其余为 0, 并满足

$$\int_{\mathcal{D}_k}\mathrm{d}\mu^\natural(\tilde{\boldsymbol{d}})=\lambda,\quad \int_{\mathcal{D}_k}\tilde{\boldsymbol{d}}\ \mathrm{d}\mu^\natural(\tilde{\boldsymbol{d}})=\boldsymbol{d}_k^\dagger\leqslant\boldsymbol{r},\quad \int_{\mathcal{D}_k}\tilde{\boldsymbol{d}}\ \mathrm{d}\mu^\natural(\tilde{\boldsymbol{d}})=\boldsymbol{d}_k^\dagger\geqslant\boldsymbol{s}.$$

此外, 因为 $\boldsymbol{g}_k(\boldsymbol{d}_k^\dagger)<\boldsymbol{\sigma}_k$ 且 ϵ 足够小, 我们有 $\boldsymbol{g}_k(\boldsymbol{d}_k^\dagger)<\boldsymbol{t}$ 成立. 因为正常凸映射 (proper convex mapping) $\boldsymbol{g}_k(\cdot)$ 是连续的, 所以对于足够小的 ϵ, 测度 $\mu^\natural$ 满足

$$\int_{\mathcal{D}_k}\boldsymbol{g}_k(\tilde{\boldsymbol{d}})\mathrm{d}\mu^\natural(\tilde{\boldsymbol{d}})=\lambda\boldsymbol{g}_k(\boldsymbol{d}_k^\dagger/\lambda)\leqslant\boldsymbol{t}.$$

因此, $(\boldsymbol{r},\boldsymbol{s},\boldsymbol{t},\lambda)\in\mathcal{V}_k$ 且 $(\boldsymbol{d}_k^+,\boldsymbol{d}_k^-,\boldsymbol{\sigma}_k,1)\in\mathrm{int}(\mathcal{V}_k)$. 所以, 根据 (Shapiro, 2001) 中的命题 3.4, 强对偶性成立, 且对于任意 $k\in[K]$, 有如下对偶形式 $\sup_{\mathbb{P}\in\mathcal{F}_k}\mathbb{E}_{\mathbb{P}}\Big[\mathrm{Q}(\boldsymbol{y},\tilde{\boldsymbol{d}})\Big]$:

$$\begin{aligned}\min\ &\alpha_k+\overline{\boldsymbol{\beta}}_k^\top\boldsymbol{d}_k^+-\underline{\boldsymbol{\beta}}_k^\top\boldsymbol{d}_k^-+\boldsymbol{\rho}_k^\top\boldsymbol{\sigma}_k,\\ \text{s.t.}\ &\alpha_k+(\overline{\boldsymbol{\beta}}_k-\underline{\boldsymbol{\beta}}_k)^\top\boldsymbol{d}+\boldsymbol{\rho}_k^\top\boldsymbol{g}_k(\boldsymbol{d})\geqslant\mathrm{Q}(\boldsymbol{y},\boldsymbol{d}),\ \boldsymbol{d}\in\mathcal{D}_k,\\ &\alpha_k\in\mathbb{R},\ \overline{\boldsymbol{\beta}}_k,\underline{\boldsymbol{\beta}}_k\in\mathbb{R}_+^J,\ \boldsymbol{\rho}_k\in\mathbb{R}_+^{n_k},\end{aligned}\tag{A.9}$$

其中, $\overline{\boldsymbol{\beta}}_k,\underline{\boldsymbol{\beta}}_k,\boldsymbol{\rho}_k,\alpha_k$ 为对偶变量.

接下来, 可以将问题 (A.9) 的约束转换为如下形式:

$$\alpha_k+(\overline{\boldsymbol{\beta}}_k-\underline{\boldsymbol{\beta}}_k)^\top\boldsymbol{d}+\boldsymbol{\rho}_k^\top\boldsymbol{w}\geqslant\mathrm{Q}(\boldsymbol{y},\boldsymbol{d}),\quad(\boldsymbol{d},\boldsymbol{w})\in\mathcal{W}_k,$$

其中, $\boldsymbol{w}$ 为辅助变量, $\mathcal{W}_k$ 为状态 $k \in [K]$ 下的提升 (lifted) 不确定集:

$$\mathcal{W}_k = \left\{ (\boldsymbol{d}, \boldsymbol{w}) \in \mathbb{R}^{|J|} \times \mathbb{R}^{n_k} \,\middle|\, \boldsymbol{d} \in \mathcal{D}_k,\ \boldsymbol{g}_k(\boldsymbol{d}) \leqslant \boldsymbol{w} \right\}.$$

因此, 对偶形式 (A.9) 可以等价转换为如下最小化问题:

$$\begin{aligned} \min\quad & \alpha_k + \overline{\boldsymbol{\beta}}_k^\top \boldsymbol{d}_k^+ - \underline{\boldsymbol{\beta}}_k^\top \boldsymbol{d}_k^- + \boldsymbol{\rho}_k^\top \boldsymbol{\sigma}_k, \\ \text{s.t.}\quad & \alpha_k + (\overline{\boldsymbol{\beta}}_k - \underline{\boldsymbol{\beta}}_k)^\top \boldsymbol{d} + \boldsymbol{\rho}_k^\top \boldsymbol{w} \geqslant \mathrm{Q}(\boldsymbol{y}, \boldsymbol{d}),\ (\boldsymbol{d}, \boldsymbol{w}) \in \mathcal{W}_k, \quad \forall k \in [K]. \\ & \alpha_k \in \mathbb{R},\ \overline{\boldsymbol{\beta}}_k, \underline{\boldsymbol{\beta}}_k \in \mathbb{R}_+^J,\ \boldsymbol{\rho}_k \in \mathbb{R}_+^{n_k}, \end{aligned} \tag{A.10}$$

记 $\Pi_k(\boldsymbol{d}_k^+, \boldsymbol{d}_k^-, \boldsymbol{\sigma}_k | \boldsymbol{y})$ 为对偶形式 (A.10), 则有如下结果成立:

$$\sup_{\mathbb{P} \in \mathcal{F}(\boldsymbol{\Theta})} \mathbb{E}_{\mathbb{P}} \left[\mathrm{Q}(\boldsymbol{y}, \tilde{\boldsymbol{d}}) \right] = \sum_{k \in [K]} p_k \sup_{\mathbb{P} \in \mathcal{F}_k} \mathbb{E}_{\mathbb{P}} \left[\mathrm{Q}(\boldsymbol{y}, \tilde{\boldsymbol{d}}) \right] = \sum_{k \in [K]} p_k \Pi_k(\boldsymbol{d}_k^+, \boldsymbol{d}_k^-, \boldsymbol{\sigma}_k | \boldsymbol{y}).$$

引理 4.2 的结果成立.

命题 4.1 的证明

给定选址决策 $\boldsymbol{y}$, 我们只需证明参数 $\boldsymbol{d}^+$ 变动 $\Delta_{\boldsymbol{d}^+} = (\Delta_{\boldsymbol{d}_k^+})_{k \in K} \geqslant 0$ 所导致的极值概率分布下第二阶段期望成本的变化量 $\Delta\Pi(\boldsymbol{d}^+ | \boldsymbol{y})$ 满足

$$\Delta\Pi(\boldsymbol{d}^+ | \boldsymbol{y}) \leqslant \sum_{k \in K} p_k \left(\Delta_{\boldsymbol{d}_k^+}^\top \overline{\boldsymbol{\beta}}_k^\star \right).$$

其他参数变化对应的结果可以类似地证明.

根据引理 4.2 并记 $\left(\alpha_k^\star, \overline{\boldsymbol{\beta}}_k^\star, \underline{\boldsymbol{\beta}}_k^\star, \boldsymbol{\rho}_k^\star\right)_{k \in K}$ 为问题 (4.14) 的最优解, 则有

$$\begin{aligned} \Pi(\boldsymbol{d}^+ + \Delta_{\boldsymbol{d}^+} | \boldsymbol{y}) &= \sum_{k \in K} p_k \Pi_k(\boldsymbol{d}_k^+ + \Delta_{\boldsymbol{d}_k^+} | \boldsymbol{y}) \\ &\leqslant \sum_{k \in [K]} p_k \left(\alpha_k^\star + \left[\boldsymbol{d}_k^+ + \Delta_{\boldsymbol{d}_k^+} \right]^\top \overline{\boldsymbol{\beta}}_k^\star - \underline{\boldsymbol{\beta}}_k^{\star\top} \boldsymbol{d}_k^- + \boldsymbol{\rho}_k^{\star\top} \boldsymbol{\sigma}_k \right) \\ &= \sum_{k \in [K]} p_k \Pi_k(\boldsymbol{d}_k^+ | \boldsymbol{y}) + \sum_{k \in [K]} p_k \left(\Delta_{\boldsymbol{d}_k^+}^\top \overline{\boldsymbol{\beta}}_k^\star \right) \\ &= \Pi(\boldsymbol{d}^+ | \boldsymbol{y}) + \sum_{k \in [K]} p_k \left(\Delta_{\boldsymbol{d}_k^+}^\top \overline{\boldsymbol{\beta}}_k^\star \right), \end{aligned}$$

其中, 不等式成立是因为 $\left(\alpha_k^\star, \overline{\boldsymbol{\beta}}_k^\star, \underline{\boldsymbol{\beta}}_k^\star, \boldsymbol{\rho}_k^\star\right)$ 为问题 $\Pi_k(\boldsymbol{d}_k^+ + \Delta_{\boldsymbol{d}_k^+} | \boldsymbol{y})$ 的一个可行解且 $p_k, \forall k \in [K]$ 是非负的. 类似地, 也可以推导出命题 4.1 的其他结果.

推论 4.1 的证明

对于结果 (1), 注意到

$$
\begin{aligned}
&\Pi^\star(\boldsymbol{p}+\Delta_{\boldsymbol{p}})\\
&=\min_{\boldsymbol{y}\in\mathcal{Y}}\left\{\boldsymbol{f}^\top\boldsymbol{y}+\Pi(\boldsymbol{p}+\Delta_{\boldsymbol{p}}|\boldsymbol{y})\right\}\\
&\leqslant \boldsymbol{f}^\top\boldsymbol{y}^\star+\Pi(\boldsymbol{p}+\Delta_{\boldsymbol{p}}|\boldsymbol{y}^\star)\\
&=\Pi^\star(\boldsymbol{p})+\Delta_{\boldsymbol{p}}^\top\left(\alpha_k^\star(\boldsymbol{y}^\star)+\overline{\boldsymbol{\beta}}_k^\star(\boldsymbol{y}^\star)^\top\boldsymbol{d}_k^+-\underline{\boldsymbol{\beta}}_k^\star(\boldsymbol{y}^\star)^\top\boldsymbol{d}_k^-+\boldsymbol{\rho}_k^\star(\boldsymbol{y}^\star)^\top\boldsymbol{\sigma}_k\right)_{k\in[K]},
\end{aligned}
$$

因此, 结果 (2) 成立.

对于结果 (2), 只需要证明关于 $\Delta_{\boldsymbol{d}^+}$ 的结果, 其他结果的证明是类似的. 我们注意到

$$
\begin{aligned}
\Pi^\star(\boldsymbol{d}^++\Delta_{\boldsymbol{d}^+})&=\min_{\boldsymbol{y}\in\mathcal{Y}}\left\{\boldsymbol{f}^\top\boldsymbol{y}+\Pi(\boldsymbol{d}^++\Delta_{\boldsymbol{d}^+}|\boldsymbol{y})\right\}\\
&\leqslant \boldsymbol{f}^\top\boldsymbol{y}^\star+\Pi(\boldsymbol{d}^++\Delta_{\boldsymbol{d}^+}|\boldsymbol{y}^\star)\\
&\leqslant \boldsymbol{f}^\top\boldsymbol{y}^\star+\Pi(\boldsymbol{d}^+|\boldsymbol{y}^\star)+\sum_{k\in[K]}p_k\Delta_{\boldsymbol{d}_k^+}^\top\overline{\boldsymbol{\beta}}_k^\star(\boldsymbol{y}^\star),
\end{aligned}
$$

其中, 最后一个不等式是命题 4.1 结果的第一个不等式. 因此, 结果 (2) 成立.

命题 4.2 的证明

分三步来证明命题 4.2.

第一步　首先证明对于每个 $k\in[K]$, $\mathcal{K}_k$ 都是一个正常锥 (proper cone).

(1) $\mathcal{K}_k$ 是凸锥. 由 $\mathcal{K}_k$ 的定义和其中关于正常凸映射 $\boldsymbol{g}_k(\cdot)$ 的约束可直接证得.

(2) $\mathcal{K}_k$ 是闭锥 (closed cone). 因为 $\mathcal{K}_k$ 是由闭包定义的, 该结果显然.

(3) $\mathcal{K}_k$ 是实锥 (solid cone, 即 $\mathcal{K}_k$ 包含非空内部). 根据 $\mathcal{K}_k$ 的定义, 总能找到一个点 $(\hat{\boldsymbol{d}},\hat{\boldsymbol{w}})$ 使得 $\boldsymbol{g}_k(\hat{\boldsymbol{d}})<\hat{\boldsymbol{w}}$ 成立. 因此, 可以找到集合 $\mathcal{K}_k$ 中的一个内点 $(\hat{\boldsymbol{d}},\hat{\boldsymbol{w}},1)$, 即 $\mathcal{K}_k$ 存在内点.

(4) $\mathcal{K}_k$ 是尖锥 (pointed cone). 假设存在一个点 $(\boldsymbol{d},\boldsymbol{w},t)\in\mathcal{K}_k$ 使得 $(-\boldsymbol{d},-\boldsymbol{w},-t)\in\mathcal{K}_k$. 如果 $t>0$, 那么 $-t<0$ 且 $(-\boldsymbol{d},-\boldsymbol{w},-t)\notin\mathcal{K}_k$, 这与假设矛盾. 如果 $-t>0$, 那么 $t<0$ 且 $(\boldsymbol{d},\boldsymbol{w},t)\notin\mathcal{K}_k$, 这仍与假设矛盾. 因此 t 必为 0. 因为 $\boldsymbol{d}$ 和 $-\boldsymbol{d}$ 都是非负的, 所以必有 $\boldsymbol{d}=\boldsymbol{0}$. 鉴于 $t=0$, 如果 $\boldsymbol{w}>0$, 则有 $\dfrac{-\boldsymbol{w}}{0}=-\infty$ 且有

$\boldsymbol{g}_k\left(\dfrac{-\boldsymbol{d}}{0}\right) \leqslant -\infty$, 从而与 $\boldsymbol{g}_k(\cdot)$ 为正常凸映射的假设矛盾. 当 $\boldsymbol{w} < 0$ 时也会出现这种矛盾. 因此, $\boldsymbol{w}$ 同样必须为零. 综上所述, 如果 $(-\boldsymbol{d}, -\boldsymbol{w}, -t)$ 和 $(\boldsymbol{d}, \boldsymbol{w}, t)$ 两个点都在集合 $\mathcal{K}_k$ 中, 那么 $(\boldsymbol{d}, \boldsymbol{w}, t) = (\boldsymbol{0}, \boldsymbol{0}, 0)$ 必然成立, 即, $\mathcal{K}_k$ 是尖锥.

第二步 证明极值概率分布 $\mathbb{P}_k^\star \in \mathcal{F}_k, \forall k \in [K]$. 因为点 $(\boldsymbol{\eta}_\star^{e_k}, \zeta_\star^{e_k})_{e_k \in E}$ 是问题 (4.22) 的可行解, 我们有 $\sum_{e_k \in E} \zeta_\star^{e_k} = 1$ 和 $\zeta_\star^{e_k} \in \mathbb{R}_+, \forall e_k \in [E]$ 成立. 因此, 对于每个 $k \in [K]$, 由式 (4.21) 定义的 $\mathbb{P}_k^\star$ 都是一个概率测度. 进一步, 利用问题 (4.22) 中的第二个和第三个约束, 我们可以得到

$$\mathbb{E}_{\mathbb{P}_k^\star}\left[\tilde{\boldsymbol{d}}\right] = \sum_{e_k \in [E]} \zeta_\star^{e_k} \frac{\boldsymbol{\eta}_\star^{e_k}}{\zeta_\star^{e_k}} = \sum_{e_k \in [E]} \boldsymbol{\eta}_\star^{e_k} \leqslant \boldsymbol{d}_k^+,$$

$$\mathbb{E}_{\mathbb{P}_k^\star}\left[\tilde{\boldsymbol{d}}\right] = \sum_{e_k \in E} \zeta_\star^{e_k} \frac{\boldsymbol{\eta}_\star^{e_k}}{\zeta_\star^{e_k}} = \sum_{e_k \in E} \boldsymbol{\eta}_\star^{e_k} \geqslant \boldsymbol{d}_k^-.$$

根据问题 (4.22) 的第五个约束, 有下式成立

$$\boldsymbol{A}_k\left(\frac{\boldsymbol{\eta}_\star^{e_k}}{\zeta_\star^{e_k}}\right) \leqslant \boldsymbol{b}_k, \forall e_k \in [E] \Longleftrightarrow \mathbb{E}_{\mathbb{P}_k^\star}\mathbb{I}_{[\boldsymbol{A}_d\tilde{\boldsymbol{d}} \leqslant \boldsymbol{b}_k]} = 1 \Longleftrightarrow \mathbb{P}_k^\star\left[\tilde{\boldsymbol{d}} \in \mathcal{D}_k\right] = 1,$$

其中, $\mathbb{I}_{[\mathcal{A}]}$ 是关于集合 $\mathcal{A}$ 的示性函数. 接下来, 利用问题 (4.22) 的第四个和第六个约束, 我们可以得到

$$\mathbb{E}_{\mathbb{P}_k^\star}\left[\boldsymbol{g}_k(\tilde{\boldsymbol{d}})\right] = \sum_{e_k \in [E]} \zeta_\star^{e_k} \boldsymbol{g}_k\left(\frac{\boldsymbol{\eta}_\star^{e_k}}{\zeta_\star^{e_k}}\right) \leqslant \sum_{e_k \in [E]} \zeta_\star^{e_k} \frac{\boldsymbol{\tau}^{e_k}}{\zeta_\star^{e_k}} \leqslant \boldsymbol{\sigma}_k.$$

上述推得的四个结果保证了由式 (4.21)-(4.22) 定义的分布 $\mathbb{P}_k^\star$ 是 $\mathcal{F}_k$ 中的一个可行分布, 也就是说, $\mathbb{P}_k^\star \in \mathcal{F}_k$, $\forall k \in [K]$.

第三步 根据假设的 Slater 条件, 给定集合 $[E]$, 对每个 $k \in [K]$, 记

$$\left(\hat{\boldsymbol{\eta}}^{e_k}, \hat{\boldsymbol{\tau}}^{e_k}, \hat{\zeta}^{e_k}\right)_{e_k \in [E]} = \left(\boldsymbol{d}_k^\dagger/E, \boldsymbol{\sigma}_k/E, 1/E\right)_{e_k \in [E]}.$$

容易证明 $(\hat{\boldsymbol{\eta}}^{e_k}, \hat{\boldsymbol{\tau}}^{e_k}, \hat{\zeta}^{e_k})_{e_k \in [E]}$ 是问题 (4.22) 的一个可行解且满足

$$\boldsymbol{g}_k\left(\frac{\hat{\eta}^{e_k}}{\hat{\zeta}^{e_k}}\right) = \boldsymbol{g}_k\left(\frac{\boldsymbol{d}_k^\dagger/E}{1/E}\right) = \boldsymbol{g}_k\left(\boldsymbol{d}_k^\dagger\right) < \boldsymbol{\sigma}_k = \frac{\hat{\boldsymbol{\tau}}^{e_k}}{\hat{\zeta}^{e_k}}.$$

因此, 根据锥强对偶性, 问题 (4.22) 和问题 (4.17) 的最优值相等. 命题 4.2 证毕.

命题 4.3 的证明

给定选址决策 $\boldsymbol{y}$, 用 $\hat{\boldsymbol{y}}$ 表示在决策 $\boldsymbol{y}$ 中选址不变的情况下, 在地址 i 新开设工厂的选址决策. 首先, 证明

$$c^{-}(i)=\sum_{k\in[K]}p_k\sup_{\mathbb{P}\in\mathcal{F}_k}\mathbb{E}_{\mathbb{P}}\left[\mathrm{Q}(\boldsymbol{y},\tilde{\boldsymbol{d}})\right]-\sum_{k\in[K]}p_k\sup_{\mathbb{P}\in\mathcal{F}_k}\mathbb{E}_{\mathbb{P}}\left[\mathrm{Q}(\hat{\boldsymbol{y}},\tilde{\boldsymbol{d}})\right]\geqslant 0. \tag{A.11}$$

也就是说, 在地址 i 开设一个新的工厂会降低极值概率分布下第二阶段期望成本. 我们注意到

$$\mathrm{Q}(\boldsymbol{y},\boldsymbol{d})=\min_{\boldsymbol{x}\in\mathcal{X}(\boldsymbol{y},\boldsymbol{d})}\sum_{i\in[I]}\sum_{j\in[J]}c_{ij}x_{ij}+\sum_{j\in[J]}r_j\left(d_j-\sum_{i\in[I]}x_{ij}\right),$$

并且在选址决策 $\boldsymbol{y}$ 中开设一个新工厂 i (即 $y_i=1$) 会扩大第二阶段可行集, 即 $\mathcal{X}(\boldsymbol{y},\boldsymbol{d})\subseteq\mathcal{X}(\hat{\boldsymbol{y}},\boldsymbol{d})$. 则可以推出 $\mathrm{Q}(\boldsymbol{y},\tilde{\boldsymbol{d}})\geqslant\mathrm{Q}(\hat{\boldsymbol{y}},\tilde{\boldsymbol{d}})$ 几乎处处成立, 进而证明了不等式 (A.11) $c^{-}(i)\geqslant 0$ 成立.

对于一个给定选址决策 $\boldsymbol{y}\in\mathcal{Y}$, 记 $\{(\hat{\boldsymbol{\nu}}^{e_k},\hat{\boldsymbol{\lambda}}^{e_k})_{e_k\in[E]}\}$ 是集合 $\mathcal{H}$ 的极值点集, 并取 $(\boldsymbol{\eta}_\star^{e_k},\zeta_\star^{e_k})_{e_k\in[E]}$ 为问题 $\sup_{\mathbb{P}\in\mathcal{F}_k}\mathbb{E}_{\mathbb{P}}[\mathrm{Q}(\boldsymbol{y},\tilde{\boldsymbol{d}})]$ 的最优解. 根据命题 4.2 的结果, 我们有

$$\begin{aligned}\sup_{\mathbb{P}\in\mathcal{F}_k}\mathbb{E}_{\mathbb{P}}\left[\mathrm{Q}(\hat{\boldsymbol{y}},\tilde{\boldsymbol{d}})\right]\geqslant&\sum_{e_k\in[E]}(\boldsymbol{r}-\hat{\boldsymbol{\nu}}^{e_k})^\top\boldsymbol{\eta}_\star^{e_k}-\sum_{e_k\in[E]}\sum_{i\in[I]}q_i\hat{y}_i\hat{\lambda}_i^{e_k}\zeta_\star^{e_k}\\
=&\sup_{\mathbb{P}\in\mathcal{F}_k}\mathbb{E}_{\mathbb{P}}\left[\mathrm{Q}(\boldsymbol{y},\tilde{\boldsymbol{d}})\right]-\sum_{i\in[I]}\left(q_i\sum_{e^k\in[E]}\hat{\lambda}_i^{e}\zeta_\star^{e_k}\right)(\hat{y}_i-y_i)\\
=&\sup_{\mathbb{P}\in\mathcal{F}_k}\mathbb{E}_{\mathbb{P}}\left[\mathrm{Q}(\boldsymbol{y},\tilde{\boldsymbol{d}})\right]-q_i\sum_{e^k\in[E]}\hat{\lambda}_i^{e}\zeta_\star^{e_k},\end{aligned}$$

其中, 因为 $(\boldsymbol{\eta}_\star^{e_k},\zeta_\star^{e_k})_{e_k\in[E]}$ 是问题 $\sup_{\mathbb{P}\in\mathcal{F}_k}\mathbb{E}_{\mathbb{P}}\left[\mathrm{Q}(\hat{\boldsymbol{y}},\tilde{\boldsymbol{d}})\right]$ 的最优解, 第一个不等式成立. 因此,

$$\begin{aligned}&\sum_{k\in[K]}p_k\left(\sup_{\mathbb{P}\in\mathcal{F}_k}\mathbb{E}_{\mathbb{P}}\left[\mathrm{Q}(\hat{\boldsymbol{y}},\tilde{\boldsymbol{d}})\right]-\sup_{\mathbb{P}\in\mathcal{F}_k}\mathbb{E}_{\mathbb{P}}\left[\mathrm{Q}(\boldsymbol{y},\tilde{\boldsymbol{d}})\right]\right)\\
\geqslant&-\sum_{k\in[K]}\left(p_kq_i\sum_{e^k\in[E]}\hat{\lambda}_i^{e_k}\zeta_\star^{e_k}\right)\end{aligned}$$

$$= -q_i\left(\sum_{k\in[K]}\sum_{e^k\in[E]}\hat{\lambda}_i^{e_k}\zeta_\star^{e_k}p_k\right),$$

或者, 等价地写为

$$c^-(i) \leqslant q_i\left(\sum_{k\in[K]}\sum_{e^k\in E}\hat{\lambda}_i^{e_k}\zeta_\star^{e_k}p_k\right).$$

利用相同的证明逻辑, 可以推出

$$c^+(l) \geqslant q_l\left(\sum_{k\in[K]}\sum_{e^k\in[E]}\hat{\lambda}_i^{e_k}\zeta_\star^{e_k}p_k\right) \geqslant 0,$$

其中, 因为变量 $\hat{\lambda}_i^{e_k}$ 和 $\zeta_\star^{e_k}$ 是非负的, 所以上述第二个不等式成立.

命题 4.4 的证明

利用引理 4.2 中的对偶形式 (4.14), 可以将松弛主问题 (4.33) 转化为如下形式:

$$\begin{aligned}
\min\quad & \boldsymbol{f}^\top\boldsymbol{y} + \sum_{k\in[K]} p_k\left(\alpha_k + \overline{\boldsymbol{\beta}}_k^\top\boldsymbol{d}_k^+ - \underline{\boldsymbol{\beta}}_k^\top\boldsymbol{d}_k^- + \boldsymbol{\rho}_k^\top\boldsymbol{\sigma}_k\right)\\
\text{s.t.}\quad & \alpha_k + (\overline{\boldsymbol{\beta}}_k - \underline{\boldsymbol{\beta}}_k)^\top\boldsymbol{d} + \boldsymbol{\rho}_k^\top\boldsymbol{w} \geqslant \mathrm{Q}^L(\boldsymbol{y},\boldsymbol{d}),\ \forall k\in[K],\ (\boldsymbol{d},\boldsymbol{w})\in\mathcal{W}_k,\\
& \alpha_k\in\mathbb{R},\ \overline{\boldsymbol{\beta}}_k,\underline{\boldsymbol{\beta}}_k\in\mathbb{R}_+^J,\ \boldsymbol{\rho}_k\in\mathbb{R}_+^{n_k},\ \forall k\in[K],\ \boldsymbol{y}\in\mathcal{Y}.
\end{aligned} \tag{A.12}$$

利用下界问题 $\mathrm{Q}^L(\boldsymbol{y},\boldsymbol{d})$ 的对偶形式, 可以将问题 (A.12) 的约束转化为如下形式:

$$\begin{aligned}
&\alpha_k + (\overline{\boldsymbol{\beta}}_k - \underline{\boldsymbol{\beta}}_k)^\top\boldsymbol{d} + \boldsymbol{\rho}_k^\top\boldsymbol{w}\\
\geqslant\ & \boldsymbol{r}^\top\boldsymbol{d} + \max_{e\in[L]}\left\{-\boldsymbol{d}^\top\boldsymbol{\nu}^e - \sum_{i\in[I]} q_i y_i\lambda_i^e\right\},\quad \forall k\in[K],(\boldsymbol{d},\boldsymbol{w})\in\mathcal{W}_k.
\end{aligned}$$

因此, 对于任意给定的 $(\boldsymbol{d},\boldsymbol{w})\in\mathcal{W}_k, k\in[K]$, 有如下结果成立:

$$\alpha_k + (\overline{\boldsymbol{\beta}}_k - \underline{\boldsymbol{\beta}}_k - \boldsymbol{r})^\top\boldsymbol{d} + \boldsymbol{\rho}_k^\top\boldsymbol{w} \geqslant \max_{e\in[L]}\left\{-\boldsymbol{d}^\top\boldsymbol{\nu}^e - \sum_{i\in[I]} q_i y_i\lambda_i^e\right\},\quad k\in[K].$$

给定顶点集的一个子集 $(\boldsymbol{\nu}^{e_k},\boldsymbol{\lambda}^{e_k})_{e_k\in[L]}$, 我们可以得到如下结果:

$$\alpha_k + (\overline{\boldsymbol{\beta}}_k - \underline{\boldsymbol{\beta}}_k - \boldsymbol{r})^\top\boldsymbol{d} + \boldsymbol{\rho}_k^\top\boldsymbol{w} \geqslant -\boldsymbol{d}^\top\boldsymbol{\nu}^{e_k} - \sum_{i\in[I]} q_i y_i\lambda_i^{e_k},\quad \forall e_k\in[L],\ k\in[K].$$

将上述不等式代入到问题 (A.12) 的约束中则有

$$\alpha_k + \sum_{i\in[I]} q_i y_i \hat{\lambda}_i^{e_k}$$
$$\geqslant \max_{(\boldsymbol{d},\boldsymbol{w})\in\mathcal{W}_k} \boldsymbol{d}^\top \left(-\hat{\boldsymbol{\nu}}^{e_k} + \boldsymbol{r} - \overline{\boldsymbol{\beta}}_k + \underline{\boldsymbol{\beta}}_k\right) - \boldsymbol{w}^\top \boldsymbol{\rho}_k, \quad \forall e_k \in [L],\ k\in[K]. \tag{A.13}$$

给定 $(\hat{\boldsymbol{\nu}}^{e_k}, \overline{\boldsymbol{\beta}}_k, \underline{\boldsymbol{\beta}}_k, \boldsymbol{\rho}_k), e_k \in [L], k \in [K]$, 通过 $\mathcal{W}_k$ 的锥结构 (4.24), 对于每个 $k\in[K]$, 我们总能找到一个可行的 $\boldsymbol{d}^\diamond \in \mathcal{D}$ 和足够大的 $\boldsymbol{w}^\diamond$ 使得 $(\boldsymbol{d}^\diamond, \boldsymbol{w}^\diamond)$ 为集合 $\mathcal{W}_k$ 的一个严格内点. 因此, 利用锥强对偶定理, 我们得到约束 (A.13) 右侧最大化问题的对偶形式:

$$\begin{aligned}
\min\quad & \boldsymbol{b}_k^\top \boldsymbol{\xi}^{e_k} + \gamma^{e_k} \\
\text{s.t.}\quad & \boldsymbol{\psi}^{e_k} - \boldsymbol{A}_k^\top \boldsymbol{\xi}^{e_k} - \hat{\boldsymbol{\nu}}^{e_k} + \boldsymbol{r} - \overline{\boldsymbol{\beta}}_k + \underline{\boldsymbol{\beta}}_k = \boldsymbol{0}, \\
& (\boldsymbol{\psi}^{e_k}, \boldsymbol{\rho}_k, \gamma^{e_k}) \in \mathcal{K}_k^*, \qquad\qquad \forall e_k \in [L],\ k \in [K], \\
& \boldsymbol{\xi}^{e_k} \in \mathbb{R}_+^{m_k}, \boldsymbol{\psi}^{e_k} \in \mathbb{R}^J, \gamma^{e_k} \in \mathbb{R},
\end{aligned} \tag{A.14}$$

其中, 对于每个 $e_k \in [L]$, $k \in [K]$, $\boldsymbol{\xi}^{e_k}, \boldsymbol{\psi}^{e_k}$ 和 γ^{e_k} 是辅助变量, 且 $\mathcal{K}_k^*$ 为集合 $\mathcal{K}_k$ 的对偶锥. 这使得我们可以将约束 (A.13) 转化为如下形式:

$$\begin{aligned}
& \alpha_k - \boldsymbol{b}_k^\top \boldsymbol{\xi}^{e_k} - \gamma^{e_k} + \sum_{i\in[I]} q_i y_i \hat{\lambda}_i^{e_k} \geqslant 0, \\
& \boldsymbol{\psi}^{e_k} - \boldsymbol{A}_k^\top \boldsymbol{\xi}^{e_k} - \overline{\boldsymbol{\beta}}_k + \underline{\boldsymbol{\beta}}_k + \boldsymbol{r} - \hat{\boldsymbol{\nu}}^{e_k} = \boldsymbol{0}, \qquad \forall e_k \in [E],\ k \in [K]. \\
& (\boldsymbol{\psi}^{e_k}, \boldsymbol{\rho}_k, \gamma^{e_k}) \in \mathcal{K}_k^*, \\
& \boldsymbol{\xi}^{e_k} \in \mathbb{R}_+^{m_k}, \boldsymbol{\psi}^{e_k} \in \mathbb{R}^J, \gamma^{e_k} \in \mathbb{R}.
\end{aligned} \tag{A.15}$$

给定集合 $\mathcal{H}$ 极值点集的子集 $[L]$ 以及上述约束 (A.15), 我们可以得出鲁棒优化问题 (4.33) 与混合整数线性锥规划问题 (4.34) 等价.

推论 4.2 的证明

利用命题 4.2 的证明逻辑, 可以证得式 (4.36) 和式 (4.37) 成立. 为了证明式 (4.38) 成立, 首先由式 (4.37) 得到 $\sup_{\mathbb{P}\in\mathcal{F}_k} \mathbb{E}_{\mathbb{P}}[\mathrm{Q}^L(\boldsymbol{y}, \tilde{\boldsymbol{d}})]$ 是关于 $\boldsymbol{y}$ 的凸函数. 因此, 为了证明式 (4.38) 成立, 只需要证明给定 $\boldsymbol{y}$, 有

$$\sup_{\mathbb{P}\in\mathcal{F}_k} \mathbb{E}_{\mathbb{P}}\left[\mathrm{Q}^L(\hat{\boldsymbol{y}}, \tilde{\boldsymbol{d}})\right]$$

$$\geqslant \sup_{\mathbb{P}\in\mathcal{F}_k} \mathbb{E}_{\mathbb{P}}\left[\mathrm{Q}^L(\boldsymbol{y},\tilde{\boldsymbol{d}})\right]+\sum_{i\in[I]}\left(-q_i\sum_{e^k\in[L]}\hat{\lambda}_i^e\zeta_\star^{e_k}\right)(\hat{y}_i-y_i),\ \forall\hat{\boldsymbol{y}}\in\mathcal{Y},\ k\in[K]. \tag{A.16}$$

给定的 $\boldsymbol{y}$, 令 $(\zeta_\star^{e_k},\boldsymbol{\eta}_\star^{e_k})$, $e_k\in[L]$, $k\in[K]$ 为问题 (4.37) 对应的最优解. 此时, 不等式 (A.16) 的右侧变为

$$\begin{aligned}&\sum_{e_k\in[L]}(\boldsymbol{r}-\hat{\boldsymbol{\nu}}^{e_k})^\top\boldsymbol{\eta}_\star^{e_k}-\sum_{e_k\in[L]}\sum_{i\in[I]}q_iy_i\hat{\lambda}_i^{e_k}\zeta_\star^{e_k}+\sum_{i\in[I]}\left(-q_i\sum_{e^k\in[L]}\hat{\lambda}_i^e\zeta_\star^{e_k}\right)(\hat{y}_i-y_i)\\ =&\sum_{e_k\in[L]}(\boldsymbol{r}-\hat{\boldsymbol{\nu}}^{e_k})^\top\boldsymbol{\eta}_\star^{e_k}-\sum_{i\in[I]}q_i\hat{y}_i\sum_{e_k\in[L]}\hat{\lambda}_i^{e_k}\zeta_\star^{e_k}.\end{aligned} \tag{A.17}$$

此外, 容易验证 $(\zeta_\star^{e_k},\boldsymbol{\eta}_\star^{e_k})$ 是给定 $\hat{\boldsymbol{y}}$ 下问题 (4.37) 的最优解. 因此, 问题

$$\sup_{\mathbb{P}\in\mathcal{F}_k}\mathbb{E}_{\mathbb{P}}\left[\mathrm{Q}^L(\hat{\boldsymbol{y}},\tilde{\boldsymbol{d}})\right]$$

的最优值大于等于给定解 $(\zeta_\star^{e_k},\boldsymbol{\eta}_\star^{e_k})$ 的目标值, 即 (A.17). 因此, 式 (A.16) 成立.

引理 4.3 的证明

对于每个 $k\in[K]$, 给定 $\boldsymbol{y}\in\mathcal{Y}$, 问题 $\sup_{\mathbb{P}\in\mathcal{F}_k}\mathbb{E}_{\mathbb{P}}[\mathrm{Q}^L(\boldsymbol{y},\tilde{\boldsymbol{d}})]$ 的上界有限, 这是因为对任意 $\boldsymbol{d}\in\mathcal{D}_k$, 有 $\mathrm{Q}^L(\boldsymbol{y},\boldsymbol{d})\leqslant\mathrm{Q}(\boldsymbol{y},\boldsymbol{d})<\infty$ 且集合 $\mathcal{D}_k$ 是一个有界多面体. 也就是说,

$$\sup_{\mathbb{P}\in\mathcal{F}_k}\mathbb{E}_{\mathbb{P}}[\mathrm{Q}^L(\boldsymbol{y},\tilde{\boldsymbol{d}})]\leqslant\max_{\boldsymbol{d}\in\mathcal{D}_k}\mathrm{Q}^L(\boldsymbol{y},\tilde{\boldsymbol{d}})<\infty.$$

此外, 问题 $\sup_{\mathbb{P}\in\mathcal{F}_k}\mathbb{E}_{\mathbb{P}}[\mathrm{Q}^L(\boldsymbol{y},\tilde{\boldsymbol{d}})]$ 的下界也是有限的. 因为对任意的 $k\in[K]$, 我们总是可以找到一个可行单点分布 $\mathbb{P}_k^\dagger[\tilde{\boldsymbol{d}}=\boldsymbol{d}_k^\dagger]=1$, 其中, $\boldsymbol{d}_k^\dagger\in[\boldsymbol{d}_k^-,\boldsymbol{d}_k^+]$, 即 $\mathbb{P}_k^\dagger\in\mathcal{F}_k$. 因此,

$$\sup_{\mathbb{P}\in\mathcal{F}_k}\mathbb{E}_{\mathbb{P}}[\mathrm{Q}^L(\boldsymbol{y},\tilde{\boldsymbol{d}})]\geqslant\mathbb{E}_{\mathbb{P}_k^\dagger}[\mathrm{Q}^L(\boldsymbol{y},\tilde{\boldsymbol{d}})]\geqslant\boldsymbol{r}^\top\boldsymbol{d}_k^\dagger+\max_{e\in L}\left\{-\left[\boldsymbol{d}_k^\dagger\right]^\top\boldsymbol{\nu}^e-\sum_{i\in I}q_iy_i\lambda_i^e\right\}>-\infty.$$

命题 4.5 的证明

利用推论 4.2 的式 (4.38) 以及 $\sup_{\mathbb{P}\in\mathcal{F}_k}\mathbb{E}_{\mathbb{P}}[\mathrm{Q}^L(\boldsymbol{y},\tilde{\boldsymbol{d}})]$ 关于 $\boldsymbol{y}$ 的凸性, 给定 $\boldsymbol{y}\in\mathcal{Y}$, 我们有

$$\sup_{\mathbb{P}\in\mathcal{F}_k}\mathbb{E}_{\mathbb{P}}[\mathrm{Q}^L(\boldsymbol{y},\tilde{\boldsymbol{d}})]$$

$$
\begin{aligned}
&\geqslant \sup_{\mathbb{P}\in\mathcal{F}_k} \mathbb{E}_{\mathbb{P}}[\mathrm{Q}^L(\boldsymbol{y}^\diamond, \tilde{\boldsymbol{d}})] \\
&\quad + \sum_{i\in[I]} \left(-q_i \sum_{e^k\in L} \hat{\lambda}_i^e \zeta_\star^{e_k}(\boldsymbol{y}^\diamond) \right) (y_i - y_i^\diamond), \quad \forall \boldsymbol{y}^\diamond \in \mathcal{Y},\ k\in[K],
\end{aligned}
$$

且当 $\boldsymbol{y}^\diamond = \boldsymbol{y}$ 时, 上述不等式变为等式. 上述不等式等价于如下不等式:

$$
\begin{aligned}
\sup_{\mathbb{P}\in\mathcal{F}_k} \mathbb{E}_{\mathbb{P}}[\mathrm{Q}^L(\boldsymbol{y}, \tilde{\boldsymbol{d}})] \geqslant & \sum_{e_k\in[L]} (\boldsymbol{r} - \hat{\boldsymbol{\nu}}^{e_k})^\top \boldsymbol{\eta}_\star^{e_k}(\boldsymbol{y}^\diamond) - \sum_{e_k\in[L]} \sum_{i\in[I]} q_i y_i^\diamond \hat{\lambda}_i^{e_k} \zeta_\star^{e_k}(\boldsymbol{y}^\diamond) \\
& + \sum_{i\in[I]} \left(-q_i \sum_{e^k\in[L]} \hat{\lambda}_i^e \zeta_\star^{e_k}(\boldsymbol{y}^\diamond) \right) (y_i - y_i^\diamond), \quad \forall \boldsymbol{y}^\diamond \in \mathcal{Y},\ k\in[K],
\end{aligned}
$$

因此,

$$
\begin{aligned}
\sup_{\mathbb{P}\in\mathcal{F}_k} \mathbb{E}_{\mathbb{P}}[\mathrm{Q}^L(\boldsymbol{y}, \tilde{\boldsymbol{d}})] \geqslant & \sum_{e_k\in[L]} (\boldsymbol{r} - \hat{\boldsymbol{\nu}}^{e_k})^\top \boldsymbol{\eta}_\star^{e_k}(\boldsymbol{y}^\diamond) \\
& - \sum_{i\in[I]} q_i y_i \sum_{e_k\in[L]} \hat{\lambda}_i^{e_k} \zeta_\star^{e_k}(\boldsymbol{y}^\diamond), \quad \forall \boldsymbol{y}^\diamond \in \mathcal{Y},\ k\in[K].
\end{aligned}
$$

利用 θ_k 来表示 $\sup_{\mathbb{P}\in\mathcal{F}_k} \mathbb{E}_{\mathbb{P}}[\mathrm{Q}^L(\boldsymbol{y}, \tilde{\boldsymbol{d}})]$, 可以将松弛主问题 (4.33) 转化为问题 (4.40).

命题 4.6 的证明

首先证明下述引理.

引理 A.1 最大化问题 (4.19) 的最优解 $(\boldsymbol{\nu}^\star, \boldsymbol{\lambda}^\star)$ 满足下述不等式:

$$
\nu_j^\star \leqslant \bar{r}, \quad \lambda_i^\star \leqslant \bar{r}, \quad \forall i\in[I], j\in[J],
$$

其中 $\bar{r} = \max_{i\in[I], j\in[J]} \{0,\ r_j - c_{ij}\}$.

引理 A.1 的证明

式 (4.19) 可以等价表示为

$$
\mathrm{Q}(\boldsymbol{y}, \boldsymbol{d}) = \boldsymbol{r}^\top \boldsymbol{d} - \min_{(\boldsymbol{\nu}, \boldsymbol{\lambda})\in\mathcal{H}} \left\{ \boldsymbol{\nu}^\top \boldsymbol{d} + \sum_{i\in[I]} q_i y_i \lambda_i \right\}, \tag{A.18}
$$

因此, 最大化问题 (4.19) 的最优解可以通过求解如下最小化问题得到

$$\begin{aligned} \min \quad & \boldsymbol{\nu}^\top \boldsymbol{d} + \sum_{i\in I} q_i y_i \lambda_i \\ \text{s.t.} \quad & \nu_j + \lambda_i \geqslant r_j - c_{ij},\ \forall i \in [I],\ j \in [J], \\ & \boldsymbol{\nu} \in \mathbb{R}_+^J,\ \boldsymbol{\lambda} \in \mathbb{R}_+^I. \end{aligned} \tag{A.19}$$

假设问题 (A.19) 存在一个最优解 $(\boldsymbol{\nu}^\star, \boldsymbol{\lambda}^\star)$, 且对某些 $j' \in [J]$, 该解满足 $\nu_{j'}^\star > \bar{r}$. 我们构造一个解 $(\boldsymbol{\nu}', \boldsymbol{\lambda}')$: $\nu_j' = \nu_j^\star, j \neq j', j \in [J]$, $\nu_{j'}' = \bar{r}$ 和 $\boldsymbol{\lambda}' = \boldsymbol{\lambda}^\star$.

由于 $\bar{r} = \max\{0,\ r_j - c_{ij}, \forall i \in [I], j \in [J]\}$, 且 $\boldsymbol{\lambda}' \geqslant \boldsymbol{0}$, 我们可知 $\nu_{j'}'$ 和 $\boldsymbol{\lambda}'$ 满足如下不等式:

$$\nu_{j'}' + \lambda_i' \geqslant r_{j'} - c_{ij'}, \quad \forall i \in [I]. \tag{A.20}$$

此外, 因为 $(\boldsymbol{\nu}^\star, \boldsymbol{\lambda}^\star)$ 是问题 (A.19) 的最优解, 所以 $(\boldsymbol{\nu}', \boldsymbol{\lambda}')$ 也满足如下不等式:

$$\nu_j' + \lambda_i' \geqslant r_j - c_{ij}, \quad \forall i \in [I], j \in [J], j \neq j'. \tag{A.21}$$

根据不等式 (A.20) 和 (A.21), 我们可知 $(\boldsymbol{\nu}', \boldsymbol{\lambda}')$ 是问题 (A.19) 的一个可行解. 因为 $\nu' < \nu^\star$, $d_{j'} > 0$, $\boldsymbol{d}$ 和 $q_i y_i, \forall i \in [I]$ 都是非负的, 我们可以得出如下不等式:

$$\boldsymbol{d}^\top \boldsymbol{\nu}^\star + \sum_{i\in[I]} q_i y_i \lambda_i^\star \geqslant \boldsymbol{d}^\top \boldsymbol{\nu}' + \sum_{i\in[I]} q_i y_i \lambda_i',$$

因此, 给定问题 (A.19) 的一个最优解 $(\boldsymbol{\nu}^\star, \boldsymbol{\lambda}^\star)$, 其中, 对于某些 $j' \in J$ 满足 $\nu_{j'}^\star > \bar{r}$, 我们可以构造另一个解满足 $\nu_{j'}^\star \leqslant \bar{r}$, $\forall j' \in [J]$ 且不会影响目标函数值. 因此, 问题 (A.19) 总存在一个最优解满足上述条件. 可以类似地证明问题 (A.19) 总存在一个最优解满足 $\lambda_i^\star \leqslant \bar{r}, i \in [I]$. 因此, 引理 A.1 的结果成立.

现在继续证明命题 4.6.

给定 $(\boldsymbol{d}, \boldsymbol{w}) \in \mathcal{W}_k$, 问题 (4.44) 中的最小化问题, 即

$$\min_{\boldsymbol{x} \in \mathcal{X}(\boldsymbol{y}^\star, \boldsymbol{d})} \left\{ \sum_{i\in[I]} \sum_{j\in[J]} (c_{ij} - r_j) x_{ij} \right\} \tag{A.22}$$

是一个具有有限最优值 (和非空可行集) 的线性规划问题. 所以, 原问题和对偶问题的最优解满足 KKT 条件且其是充要条件. 换句话说, 可以用 KKT 条件替换问题 (4.44) 中的最大化问题. 对于线性规划问题 (A.22), KKT 条件由原始和对偶可行性约束以及互补松弛条件组成:

$$\left\{\boldsymbol{\nu}\in\mathbb{R}_+^J,\boldsymbol{\lambda}\in\mathbb{R}_+^I,\boldsymbol{x}\in\mathbb{R}_+^{I\times J}\ \middle|\ \begin{array}{l}\sum\limits_{i\in[I]}x_{ij}\leqslant d_j,\ \forall j\in[J],\\ \sum\limits_{j\in[J]}x_{ij}\leqslant q_iy_i^\star,\ \forall i\in[I],\\ \nu_j\Big(d_j-\sum\limits_{i\in[I]}x_{ij}\Big)=0,\ \forall j\in[J],\\ \lambda_i\Big(q_iy_i^\star-\sum\limits_{j\in[J]}x_{ij}\Big)=0,\ \forall i\in[I],\\ x_{ij}(c_{ij}-r_j+\nu_j+\lambda_i)=0,\ \forall i\in[I],\ j\in[J],\\ c_{ij}-r_j+\nu_j+\lambda_i\geqslant 0,\ \forall i\in[I],\ j\in[J],\\ x_{ij}\geqslant 0,\nu_j\geqslant 0,\lambda_i\geqslant 0,\ \forall i\in[I],\ j\in[J]\end{array}\right\},\tag{A.23}$$

其中, $x_{ij}, i\in[I],\ j\in[J]$ 为运输决策, ν_j ($j\in[J]$) 和 λ_i ($i\in[I]$) 分别为需求和产能约束对应的对偶变量. 接下来线性化上述约束(A.23) 中的非线性项:

$$\begin{cases}0\leqslant\nu_j\leqslant M_1\pi_j^1,\ \forall j\in[J],\\ 0\leqslant\lambda_i\leqslant M_2\pi_i^2,\ \forall i\in[I],\\ 0\leqslant x_{ij}\leqslant M_3(1-\pi_{ij}^3),\ \forall i\in[I],\ j\in[J],\\ 0\leqslant d_j-\sum\limits_{i\in I}x_{ij}\leqslant M_4(1-\pi_j^1),\ \forall j\in[J],\\ 0\leqslant q_iy_i^\star-\sum\limits_{j\in[J]}x_{ij}\leqslant M_5(1-\pi_i^2),\ \forall i\in[I],\\ 0\leqslant c_{ij}-r_j+\nu_j+\lambda_i\leqslant M_6\pi_{ij}^3,\ \forall i\in[I],\ j\in[J]\\ \pi_j^1,\pi_i^2,\pi_{ij}^3\in\{0,1\},\ \forall i\in[I],\ j\in[J],\end{cases}\tag{A.24}$$

其中, 对于每个 $t\in\{1,2,\cdots,6\}$, M_t 为一个足够大的正常数. 利用 KKT 条件和上述线性化形式 (A.24), 问题 (4.44) 可以转化为 (4.45).

最后, 我们给出 $M_t, t\in\{1,2,\cdots,6\}$ 的有效界限. 根据引理 A.1, 有 $\nu_j^\star\leqslant\overline{r}$, $\lambda_i^\star\leqslant\overline{r}$, $\forall i\in[I], j\in[J]$ 和 $\overline{r}=\max\{0,\ r_j-c_{ij},\forall i\in[I],j\in[J]\}$. 如果 $M_1=M_2=\overline{r}$, $M_6=2\overline{r}-\underline{r}$ 且 $\underline{r}=\min\{0,\ r_j-c_{ij},\forall i\in[I],j\in[J]\}$, 则 (A.24) 的第一、第二和第六个约束成立. 在证明 (A.24) 的第三、第四和第五个约束之前, 定义如下符号:

$$\overline{q}=\max\{\ q_iy_i^\star,\ \forall i\in[I]\}\quad 和\quad \overline{d}=\max\{\ \overline{d}_{jk},\ \forall j\in[J],k\in[K]\},$$

其中,

$$\overline{d}_{jk} = \max\left\{\boldsymbol{e}_j^\top \boldsymbol{d} \;\middle|\; \boldsymbol{A}_k\boldsymbol{d} \leqslant \boldsymbol{b}_k,\ \boldsymbol{d} \in \mathbb{R}_+^J\right\}.$$

因为 $x_{ij}, \forall i \in [I], j \in [J]$ 是非负的, 如果 $M_4 = \overline{d}$ 且 $M_5 = \overline{q}$, 则第四和第五个约束成立. 根据条件 (A.23) 的第一和第二个约束, 则当 $M_3 = \overline{q}$ 时, (A.24) 的第三个约束成立. 因此, 如果 $M_1 = M_2 = \overline{r}$, $M_3 = M_5 = \overline{q}$, $M_4 = \overline{d}$ 且 $M_6 = 2\overline{r} - \underline{r}$, 则约束 (A.24) 成立. 因此, 我们验证了约束 (4.45) 在上述所有 M 取值下都成立.

命题 4.7 的证明

对于结论 (1), 考虑某个 $k \in [K]$. 根据 Ω_k 的定义, 可以得到如下结果:

$$\begin{aligned}
\Omega_k &= \max_{(\boldsymbol{d},\boldsymbol{w})\in\mathcal{W}_k}\left\{\min_{\boldsymbol{x}\in\mathcal{X}(\boldsymbol{y}^\star,\boldsymbol{d})}\left\{\sum_{i\in[I]}\sum_{j\in[J]}(c_{ij}-r_j)x_{ij}\right\} + (\boldsymbol{r} - \overline{\boldsymbol{\beta}}_k^\star + \underline{\boldsymbol{\beta}}_k^\star)^\top\boldsymbol{d} - \boldsymbol{w}^\top\boldsymbol{\rho}_k^\star\right\} \\
&= \max_{(\boldsymbol{d},\boldsymbol{w})\in\mathcal{W}_k}\left\{\mathrm{Q}(\boldsymbol{y}^\star,\boldsymbol{d}) + (\boldsymbol{r} - \overline{\boldsymbol{\beta}}_k^\star + \underline{\boldsymbol{\beta}}_k^\star)^\top\boldsymbol{d} - \boldsymbol{w}^\top\boldsymbol{\rho}_k^\star\right\} \\
&\geqslant \max_{(\boldsymbol{d},\boldsymbol{w})\in\mathcal{W}_k}\left\{\mathrm{Q}^{\mathrm{L}}(\boldsymbol{y}^\star,\boldsymbol{d}) + (\boldsymbol{r} - \overline{\boldsymbol{\beta}}_k^\star + \underline{\boldsymbol{\beta}}_k^\star)^\top\boldsymbol{d} - \boldsymbol{w}^\top\boldsymbol{\rho}_k^\star\right\} \\
&= \alpha_k^\star.
\end{aligned}$$

对于结论 (2), 利用上述关于 $\Omega_k, k\in[K]$ 的不等式, 可知 $((\Omega_k, \overline{\boldsymbol{\beta}}_k^\star, \underline{\boldsymbol{\beta}}_k^\star, \boldsymbol{\rho}_k^\star)_{k\in K}, \boldsymbol{y}^\star)$ 是问题 (4.28) 的一个可行解. 因此, 整体问题的上界是

$$\boldsymbol{f}^\top\boldsymbol{y}^\star + \sum_{k\in[K]} p_k\left(\Omega_k + (\overline{\boldsymbol{\beta}}_k^\star)^\top\boldsymbol{d}_k^+ - (\underline{\boldsymbol{\beta}}_k^\star)^\top\boldsymbol{d}_k^- + (\boldsymbol{\rho}_k^\star)^\top\boldsymbol{\sigma}_k\right).$$

此外, 因为 $((\alpha_k^\star, \overline{\boldsymbol{\beta}}_k^\star, \underline{\boldsymbol{\beta}}_k^\star, \boldsymbol{\rho}_k^\star)_{k\in K}, \boldsymbol{y}^\star)$ 是松弛主问题的最优解, 所以, 可以利用它为整体问题构造一个下界:

$$\boldsymbol{f}^\top\boldsymbol{y}^\star + \sum_{k\in K} p_k\left(\alpha_k^\star + (\overline{\boldsymbol{\beta}}_k^\star)^\top\boldsymbol{d}_k^+ - (\underline{\boldsymbol{\beta}}_k^\star)^\top\boldsymbol{d}_k^- + (\boldsymbol{\rho}_k^\star)^\top\boldsymbol{\sigma}_k\right).$$

因此, 整体问题的最优性间隙由上述上界和下界之差限制, 即

$$\sum_{k\in K} p_k(\Omega_k - \alpha_k^\star).$$

命题 4.8 的证明

给定 $\boldsymbol{y}^\star$ 和 $\boldsymbol{d}^\star$, 由式 (4.8) 和式 (4.9) 可知第二阶段问题 $\mathrm{Q}(\boldsymbol{y}^\star, \boldsymbol{d}^\star)$ 是一个具有非空可行集的线性规划问题. 因此, 该问题的强对偶性成立.

根据问题 (4.45) 的第一和第二个约束, 容易验证 $\boldsymbol{x}^\star$ 是问题 $\mathrm{Q}(\boldsymbol{y}^\star, \boldsymbol{d}^\star)$ 的一个可行解. 根据第三个约束, 我们可以知道 $(\boldsymbol{\nu}^\star, \boldsymbol{\lambda}^\star)$ 是一个对偶可行解. 根据和 $\boldsymbol{x}^\star$、$\boldsymbol{\nu}^\star$ 和 $\boldsymbol{\lambda}^\star$ 相关的其他约束, 我们可以得到如下方程组:

$$\begin{cases} \nu_j^\star\Big(d_j^\star - \sum\limits_{i\in[I]} x_{ij}^\star\Big) = 0, & \forall j \in [J], \\ \lambda_i^\star\Big(q_i y_i^\star - \sum\limits_{j\in[J]} x_{ij}^\star\Big) = 0, & \forall i \in [I], \\ x_{ij}^\star(c_{ij} - r_j + \nu_j^\star + \lambda_i^\star) = 0, & \forall i \in [I],\ j \in [J]. \end{cases} \tag{A.25}$$

利用方程组 (A.25), 可以得到如下等式:

$$\sum_{i\in[I]}\sum_{j\in[J]}(c_{ij} - r_j)x_{ij}^\star = -\sum_{j\in[J]}\nu_j^\star d_j^\star - \sum_{i\in[I]} q_i y_i^\star \lambda^\star,$$

这表明问题 $\mathrm{Q}(\boldsymbol{y}^\star, \boldsymbol{d}^\star)$ 在点 $\boldsymbol{x}^\star$ 处的目标值等于其对偶问题在 $(\boldsymbol{\nu}^\star, \boldsymbol{\lambda}^\star)$ 点的目标函数值. 因此, $(\boldsymbol{\nu}^\star, \boldsymbol{\lambda}^\star)$ 是问题 $\mathrm{Q}(\boldsymbol{y}^\star, \boldsymbol{d}^\star)$ 对偶问题的一个最优解且是集合 $\mathcal{H}$ 的一个极值点.

接下来证明线性规划 (4.47) 的任何基本可行解都是多面体集合 $\mathcal{H}$ 的一个极值点. 证明过程借鉴了 (Bertsimas and Tsitsiklis, 1997) 中的定理 2.7. 首先, 用 $\mathcal{L}$ 表示线性规划 (4.47) 的可行集:

$$\mathcal{L} = \left\{(\boldsymbol{\nu}, \boldsymbol{\lambda}) \in \mathbb{R}_+^J \times \mathbb{R}_+^I \,\middle|\, (\boldsymbol{\nu}, \boldsymbol{\lambda}) \in \mathcal{H}, \boldsymbol{\nu}^\top \boldsymbol{d}^\star + \sum_{i\in[I]} q_i y_i^\star \lambda_i = \boldsymbol{\nu}^{\star\top}\boldsymbol{d}^\star + \sum_{i\in[I]} q_i y_i^\star \lambda_i^\star\right\}.$$

因为 $(\boldsymbol{\nu}^\star, \boldsymbol{\lambda}^\star) \in \mathcal{H}$ 且集合 $\mathcal{H}$ 具有多面体结构, 所以可以得出集合 $\mathcal{L}$ 是一个非空多面体. 因此, 我们只需要证明多面体集 $\mathcal{L}$ 的任意一个极值点都是多面体集 $\mathcal{H}$ 的极值点. 令 $(\boldsymbol{\nu}', \boldsymbol{\lambda}')$ 为集合 $\mathcal{L}$ 的某个极值点, 下面证明 $(\boldsymbol{\nu}', \boldsymbol{\lambda}')$ 也是 $\mathcal{H}$ 的极值点. 使用反证法来证明. 假设 $(\boldsymbol{\nu}', \boldsymbol{\lambda}')$ 不是 $\mathcal{H}$ 的极值点. 那么存在 $(\boldsymbol{\nu}_1, \boldsymbol{\lambda}_1) \in \mathcal{H}$, $(\boldsymbol{\nu}_2, \boldsymbol{\lambda}_2) \in \mathcal{H}$ 和某个 $\theta \in [0, 1]$ 使得 $(\boldsymbol{\nu}', \boldsymbol{\lambda}') = \theta(\boldsymbol{\nu}_1, \boldsymbol{\lambda}_1) + (1-\theta)(\boldsymbol{\nu}_2, \boldsymbol{\lambda}_2)$ 成立. 接下来, 我们有

$$\boldsymbol{\nu}^{\star\top}\boldsymbol{d}^\star + \sum_{i\in[I]} q_i y_i^\star \lambda_i^\star = \boldsymbol{\nu}'^{\top}\boldsymbol{d}^\star + \sum_{i\in[I]} q_i y_i^\star \lambda_i'$$

$$=\theta\left(\boldsymbol{\nu}_1^\top \boldsymbol{d}^\star+\sum_{i\in[I]} q_i y_i^\star \lambda_{i1}\right)+(1-\theta)\left(\boldsymbol{\nu}_2^\top \boldsymbol{d}^\star+\sum_{i\in[I]} q_i y_i^\star \lambda_{i2}\right). \tag{A.26}$$

因为 $\boldsymbol{\nu}^{\star\top}\boldsymbol{d}^\star+\sum_{i\in[I]} q_i y_i^\star \lambda_i^\star$ 是给定 $(\boldsymbol{y}^\star,\boldsymbol{d}^\star)$ 下子问题 (4.19) 的最优值, 所以有

$$\boldsymbol{\nu}_1^\top \boldsymbol{d}^\star+\sum_{i\in[I]} q_i y_i^\star \lambda_{i1} \geqslant \boldsymbol{\nu}^{\star\top}\boldsymbol{d}^\star+\sum_{i\in[I]} q_i y_i^\star \lambda_i^\star$$

和

$$\boldsymbol{\nu}_2^\top \boldsymbol{d}^\star+\sum_{i\in[I]} q_i y_i^\star \lambda_{i2} \geqslant \boldsymbol{\nu}^{\star\top}\boldsymbol{d}^\star+\sum_{i\in[I]} q_i y_i^\star \lambda_i^\star.$$

成立. 利用上述等式 (A.26), 可以推出

$$\boldsymbol{\nu}^{\star\top}\boldsymbol{d}^\star+\sum_{i\in[I]} q_i y_i^\star \lambda_i^\star=\boldsymbol{\nu}_1^\top \boldsymbol{d}^\star+\sum_{i\in[I]} q_i y_i^\star \lambda_{i1}=\boldsymbol{\nu}_2^\top \boldsymbol{d}^\star+\sum_{i\in[I]} q_i y_i^\star \lambda_{i2}.$$

因此, 有 $(\boldsymbol{\nu}_1,\boldsymbol{\lambda}_1)\in\mathcal{L}$ 且 $(\boldsymbol{\nu}_2,\boldsymbol{\lambda}_2)\in\mathcal{L}$ 成立, 而这与 $(\boldsymbol{\nu}',\boldsymbol{\lambda}')$ 是 $\mathcal{L}$ 的极值点这一事实相矛盾. 故 $(\boldsymbol{\nu}',\boldsymbol{\lambda}')$ 是 $\mathcal{H}$ 的极值点.

定理 4.1 的证明

给定集合 $\mathcal{H}$ 的任何极值点子集 $[L]$, 因为可行集

$$\mathcal{Y}=\left\{\boldsymbol{y}\in\{0,1\}^I \mid \mathbf{G}\boldsymbol{y}\geqslant \boldsymbol{h}\right\}$$

的元素是有限的, 故问题 (4.40) 有有限个次梯度割平面, 并且算法 2 只需要有限次迭代. 进一步, 因为集合 $\mathcal{H}$ 的极值点个数是有限的, 这保证了算法 3 至多执行有限次算法 1 来求得最优解.

A.2.2 状态依赖分布不确定集的构造

在本小节中, 我们讨论如何利用历史需求数据来估计状态依赖分布不确定集的参数.

(1) 状态已知的状态依赖分布不确定集. 许多实际情况下的状态都是已知的. 例如, 当状态指的是季节、工作条件 (例如温度) 或天气时. 在这些情况下, 我们很自然地按照每个状态 (例如, 四个季节之一) 对样本进行分组, 并估计每个状态下需求的 (条件) 均值、散度和支撑集等样本统计量. 此外, 散度统计量有多种选择. 例如, 我们可以考虑例 9 中的绝对离差和例 10 中的 Mahalanobis 距离. 进一

步, 对于每个状态, 可以分别通过对相应样本均值和样本散度按比例增加或减少, 来估计需求均值的上下界以及散度的上界.

(2) 状态未知的状态依赖分布不确定集. 在某些情况下, 我们无法观察到状态变量. 在这种情况下, 可以首先采用一些无监督机器学习技术来学习不同的状态. 比如, 我们可以使用 K-means 聚类方法来将样本分类为不同状态 (Chen et al., 2020; Gambella et al., 2021; Hao et al., 2020; Perakis et al., 2022). 一旦确定了样本所处的状态, 就可以按照上述状态已知情况中提到的方法来构造状态依赖分布不确定集. 一个实际问题是样本数量和状态数量 K 的选择. 如果聚类后的状态太多, 则某些状态下的样本量可能不足以估计状态依赖分布不确定参数. 因此, 如果聚类后的状态数量相对较大, 可以采用一些统计方法如自举法 (Bootstrapping) (Samanta and Welsh, 2013) 或一些机器学习方法来处理数据量不足的状态集 (Haixiang et al., 2017).

(3) 状态为预测值的状态依赖分布不确定集. 正如在例 11 中所提到的, 当决策者直接使用每个预测值作为状态时, 他可以将状态依赖分布不确定集构造为 ∞ 型 Wasserstein 球. 在这种情况下, 分布不确定集参数, 即 Wasserstein 球半径 θ 可以通过交叉验证方法来确定, 该方法常用于机器学习中的超参数选择 (Mohajerin Esfahani and Kuhn, 2018).

A.3 第 5 章的证明及其他相关内容

本节将提供第 5 章中引理、命题和定理的证明以及其他相关内容.

A.3.1 第 5 章的证明

引理 5.1 的证明

给定 $\beta_2 \leqslant \beta_2'$, 根据 $U(\beta_2)$ 的定义 (5.6), 可得 $U(\beta_2) \subseteq U(\beta_2')$. 因此, 根据 $\beta_1^*[\boldsymbol{x}|\beta_2]$ 的定义 (5.7), 对任意的 $\beta_1 \in [0,1]$, 若其满足对任意的 $(\boldsymbol{v},\boldsymbol{u}) \in V(\beta_1) \times U(\beta_2')$ 都有 $\phi(\boldsymbol{x};\boldsymbol{v},\boldsymbol{u}) \geqslant \tau_0$, 则一定有对任意 $(\boldsymbol{v},\boldsymbol{u}) \in V(\beta_1) \times U(\beta_2)$ 都有 $\phi(\boldsymbol{x};\boldsymbol{v},\boldsymbol{u}) \geqslant \tau_0$. 由此可得 $\beta_1^*[\boldsymbol{x}|\beta_2'] \leqslant \beta_1^*[\boldsymbol{x}|\beta_2]$. 类似地, 我们可以证明当 $\beta_1 \leqslant \beta_1'$ 时, $\beta_2^*[\boldsymbol{x}|\beta_1'] \leqslant \beta_2^*[\boldsymbol{x}|\beta_1]$.

命题 5.1 的证明

为了证明简洁, 考虑如下不包含固定成本 $\boldsymbol{c}^\top \boldsymbol{x}$ 的最小最大化利润问题:

$$\min_{(\boldsymbol{v},\boldsymbol{u}) \in V(\beta_1) \times U(\beta_2)} \max_{\boldsymbol{y},\boldsymbol{z}} \quad \sum_{i\in[I]} \sum_{j\in[J]} y_{ij}\kappa_{ij}(\boldsymbol{u}) - \sum_{i\in[I]} h_i z_i \tag{A.27}$$

$$\text{s.t.} \quad \sum_{j\in[J]} y_{ij} + z_i = v_i, \ \forall\, i \in [I], \tag{A.28}$$

$$\sum_{i\in[I]} y_{ij} \leqslant s_j x_j, \qquad \forall\, j \in [J], \tag{A.29}$$

$$\boldsymbol{y} \in \mathbb{R}_+^{I\times J}, \boldsymbol{z} \in \mathbb{R}_+^{I} \tag{A.30}$$

记以 β_1 为参数的初始 (primitive) 原料质量因子 $\boldsymbol{\xi}$ 的超立方体 (hypercube) 为

$$V_0(\beta_1) := \left\{\boldsymbol{\xi} \in \mathbb{R}^K \;\middle|\; \xi_k \in \left[\xi_k^-(\beta_1), \xi_k^+(\beta_1)\right],\ \forall\, k \in [K]\right\}, \tag{A.31}$$

并对问题 (A.27)-(A.30) 中内层最大化问题做对偶, 则可以将问题 (A.27)-(A.30) 转化为如下形式:

$$\begin{aligned}
&\min_{\boldsymbol{u}\in U(\beta_2)} \min_{\boldsymbol{v}\in V(\beta_1)} \min_{(\boldsymbol{d},\boldsymbol{\pi})\in Z(\boldsymbol{u})} \sum_{i\in[I]} v_i d_i + \sum_{j\in[J]} s_j x_j \pi_j \\
&= \min_{\boldsymbol{u}\in U(\beta_2)} \min_{(\boldsymbol{d},\boldsymbol{\pi})\in Z(\boldsymbol{u})} \min_{\boldsymbol{\xi}\in V_0(\beta_1)} \sum_{k\in[K]} \xi_k \left(\sum_{i\in[I]} \alpha_{ik} d_i\right) + \sum_{j\in[J]} s_j x_j \pi_j \\
&= \min_{\boldsymbol{u}\in U(\beta_2)} \min_{(\boldsymbol{d},\boldsymbol{\pi})\in Z(\boldsymbol{u})} \sum_{k\in[K]} \overline{\xi}_k(\beta_1) \left(\sum_{i\in[I]} \alpha_{ik} d_i\right) \\
&\quad + \sum_{j\in[J]} s_j x_j \pi_j - \sum_{k\in[K]} \left|\underline{\xi}_k(\beta_1) \left(\sum_{i\in[I]} \alpha_{ik} d_i\right)\right|,
\end{aligned}$$

其中, $\boldsymbol{d}, \boldsymbol{\pi}$ 是对偶变量, 对偶可行域如下:

$$Z(\boldsymbol{u}) := \Big\{(\boldsymbol{d}, \boldsymbol{\pi}) \mid d_i + \pi_j \geqslant \kappa_{ij}(\boldsymbol{u}),\ d_i + h_i \geqslant 0,\ d_i \in \mathbb{R}, \pi_j \geqslant 0,\ \forall\, i \in [I], j \in [J]\Big\},$$

且 $\overline{\xi}_k(\beta_1)$ 和 $\underline{\xi}_k(\beta_1)$ 由式 (5.14) 给出. 此外, 上述问题中的最大化绝对值项

$$\max \sum_{k\in[K]} \left|\underline{\xi}_k(\beta_1) \left(\sum_{i\in[I]} \alpha_{ik} d_i\right)\right|$$

可以利用整数规划线性化方法转化为如下形式:

$$\max \left\{ \sum_{k\in[K]} \gamma_k \;\middle|\; \begin{array}{l} \gamma_k \leqslant \underline{\xi}_k(\beta_1) \displaystyle\sum_{i\in[I]} \alpha_{ik} d_i + W\lambda_k,\ \forall\, k \in [K] \\ \gamma_k \leqslant -\underline{\xi}_k(\beta_1) \displaystyle\sum_{i\in[I]} \alpha_{ik} d_i + (1-\lambda_k)W,\ \forall\, k \in [K] \\ \lambda_k \in \{0, 1\}, \gamma_k \geqslant 0,\ \forall\, k \in [K] \end{array} \right\},$$

其中, $\lambda_k, \forall k \in [K]$ 为二元整数变量, W 为足够大的正数. 基于上述结果, 问题 (5.12) 可以转化为混合整数规划问题 (5.15), 证毕.

命题 5.2 的证明

给定 $\boldsymbol{x}$, 定义集合 $\mathcal{Y}(\boldsymbol{x}) := \{\boldsymbol{y} \in \mathbb{R}_+^{I\times J}, \boldsymbol{z} \in \mathbb{R}_+^I \mid \text{(A.28)-(A.29)}\}$. 则问题 (5.12) 可以写为

$$\begin{aligned}
&\min_{\boldsymbol{u}\in U(\beta_2)} \max_{\boldsymbol{y},\boldsymbol{z}\in\mathcal{Y}(\boldsymbol{x})} \sum_{i\in[I]}\sum_{j\in[J]} \kappa_{ij}(\boldsymbol{u})y_{ij} - \boldsymbol{h}^\top\boldsymbol{z} - \boldsymbol{c}'\boldsymbol{x} \\
=& \max_{\boldsymbol{y},\boldsymbol{z}\in\mathcal{Y}(\boldsymbol{x})} \min_{\boldsymbol{u}\in U(\beta_2)} \sum_{i\in[I]}\sum_{j\in[J]} \kappa_{ij}(\boldsymbol{u})y_{ij} - \boldsymbol{h}^\top\boldsymbol{z} - \boldsymbol{c}'\boldsymbol{x},
\end{aligned} \tag{A.32}$$

其中, 上述目标函数是关于变量 $\boldsymbol{u}$ 的仿射函数, 因此目标函数关于变量 $\boldsymbol{u}$ 是凸且连续的. 此外, 目标函数关于变量 $\boldsymbol{y}$ 是仿射的, 因此其关于变量 $\boldsymbol{y}$ 是凹的, 并且集合 $U(\beta_2)$ 和 $\mathcal{Y}(\boldsymbol{x})$ 都是有界凸集. 所以, 最小化和最大化算子的互换可以根据最小最大化定理 (Sion, 1958) 得到.

利用线性规划的强对偶定理, 问题 (A.32) 的内层最小化问题的对偶问题为

$$\begin{aligned}
\max_{d_i,d_0}\quad & \sum_{i\in[I]} d_i + I\beta_2 d_0 \\
\text{s.t.}\quad & d_i + d_0\delta_m \leqslant \sum_{j\in[J]} q_{mj}y_{ij},\ \forall\, i\in[I],\ m\in[M], \\
& d_0 \geqslant 0, d_i \in \mathbb{R},\ \forall\, i \in [I],
\end{aligned}$$

其中, $d_0, d_i, i \in [I]$ 是对偶变量. 引入辅助变量 γ, 则得到命题 5.2 中的结果.

命题 5.3 的证明

首先, 当不同收集点的原料质量相互独立时, 则由式 (5.19) 给出的原料质量下界 $v_i^-(\beta_1), \forall i \in [I]$ 都可以单独取到.

其次, 给定 $(\boldsymbol{x};\boldsymbol{v},\boldsymbol{u})$, 当问题 $\phi(\boldsymbol{x};\boldsymbol{v},\boldsymbol{u})$ 的第一个约束放松为 (5.18) 时, 则过量原料变量 $\boldsymbol{z}$ 将不再起作用 ($\boldsymbol{z} \equiv \boldsymbol{0}$), 因此运营问题 $\phi(\boldsymbol{x};\boldsymbol{v},\boldsymbol{u})$ 的可行域转化为 $\mathcal{Y}(\boldsymbol{v}) := \{\boldsymbol{y} \in \mathbb{R}_+^{I\times J} \mid \text{(A.29)}, \text{(5.18)}\}$. 由此, 我们可得

$$\mathcal{Y}(\boldsymbol{v}^-(\beta_1)) \subseteq \mathcal{Y}(\boldsymbol{v}),\ \forall\, \boldsymbol{v} \in V(\beta_1),$$

其中, $\boldsymbol{v}^-(\beta_1) = \left[v_1^-(\beta_1), v_2^-(\beta_1), \cdots, v_I^-(\beta_1)\right]^\top$. 根据上述集合关系, 有如下等式:

$$\min_{\boldsymbol{v}\in V(\beta_1)} \phi(\boldsymbol{x};\boldsymbol{v},\boldsymbol{u}) = \min_{\boldsymbol{v}\in V(\beta_1)} \max_{\boldsymbol{y}\in\mathcal{Y}(\boldsymbol{v})} \sum_{i\in[I]}\sum_{j\in[J]} \kappa_{ij}(\boldsymbol{u})y_{ij} - \boldsymbol{c}'\boldsymbol{x}$$

$$= \max_{\boldsymbol{y}\in\mathcal{Y}(\boldsymbol{v}^-(\beta_1))} \sum_{i\in[I]}\sum_{j\in[J]} \kappa_{ij}(\boldsymbol{u})y_{ij} - \boldsymbol{c}'\boldsymbol{x} = \phi(\boldsymbol{x};\boldsymbol{v}^-(\beta_1),\boldsymbol{u}).$$

结合命题 5.2, 可以得到问题 (5.12) 等价于线性规划问题 (5.16), 其中决策 $\boldsymbol{z}\equiv\boldsymbol{0}$, 且 $\boldsymbol{v}\equiv\boldsymbol{v}^-(\beta_1)$. 证明完毕.

命题 5.4 的证明

给定 $\boldsymbol{x}$ 和 β_1, 当不同收集点的原料质量不相关, 并原料组成成分 $\boldsymbol{u}\in\mathcal{O}(\boldsymbol{x},\beta_1)$, 那么我们有

$$\min_{\boldsymbol{v}\in V(\beta_1)} \phi(\boldsymbol{x};\boldsymbol{v},\boldsymbol{u}) = \phi(\boldsymbol{x};\boldsymbol{v}^-(\beta_1),\boldsymbol{u}). \tag{A.33}$$

接下来, 考虑如下放松后的问题 (此时决策 $\boldsymbol{z}$ 恒为零):

$$\max_{\boldsymbol{y}} \left\{ \sum_{i\in[I]}\sum_{j\in[J]} y_{ij}\kappa_{ij}(\boldsymbol{u}) \middle| \boldsymbol{y}\in\left\{\boldsymbol{y}\in\mathbb{R}_+^{I\times J} \mid (\text{A.29}),(5.18)\right\} \right\}.$$

根据定义 (5.20)-(5.22), 如果有 $\boldsymbol{u}\in\mathcal{O}(\boldsymbol{x},\beta_1)$, 那么对于每个 $i\in[I]$, 都有 $\mathcal{J}_{\boldsymbol{x}}(i,\boldsymbol{u})\neq\varnothing$. 因此, 给定任意 $\boldsymbol{v}\in V(\beta_1)$, 上述问题可以被等价地转化为

$$\begin{aligned}
\max_{\boldsymbol{y}\in\mathbb{R}_+^{I\times J}} \quad & \sum_{i\in[I]}\left[\sum_{j\in\mathcal{J}_{\boldsymbol{x}}(i,\boldsymbol{u})} y_{ij}\kappa_{ij}(\boldsymbol{u}) + \sum_{j\notin\mathcal{J}_{\boldsymbol{x}}(i,\boldsymbol{u})} y_{ij}\kappa_{ij}(\boldsymbol{u})\right] \\
\text{s.t.} \quad & \sum_{j\notin\mathcal{J}_{\boldsymbol{x}}(i,\boldsymbol{u})} y_{ij} + \sum_{j\in\mathcal{J}_{\boldsymbol{x}}(i,\boldsymbol{u})} y_{ij} \leqslant v_i, \ \forall\, i\in[I], \\
& \sum_{i\in[I]} y_{ij} \leqslant s_j x_j, \qquad \forall\, j\in[J],
\end{aligned}$$

其中, 最优解满足

$$\begin{cases} y_{ij}\geqslant 0, & \text{如果 } j\in\mathcal{J}_{\boldsymbol{x}}(i,\boldsymbol{u}), \\ y_{ij}=0, & \text{其他情况.} \end{cases}$$

接下来, 证明最优解仅在如下约束是紧的时取到, 即

$$\sum_{j\in\mathcal{J}_{\boldsymbol{x}}(i,\boldsymbol{u})} y_{ij}^\star = v_i, \quad \forall\, i\in[I]. \tag{A.34}$$

利用反证法来证明. 假设对于最优解 $\boldsymbol{y}^\star$, 存在某个 $i'\in[I]$ 使得

$$\sum_{j\in\mathcal{J}_{\boldsymbol{x}}(i',\boldsymbol{u})} y_{i'j}^\star < v_{i'}. \tag{A.35}$$

那么可以定义一个新的解 $\boldsymbol{y}^{\star\star}$, 使得

$$\begin{cases} y_{i'j}^{\star\star} \geqslant y_{i'j}^{\star}, & \forall\, j \in \mathcal{J}_{\boldsymbol{x}}(i', \boldsymbol{u}), \\ y_{i'j}^{\star\star} = 0 = y_{i'j}^{\star}, & \forall\, j \neq \mathcal{J}_{\boldsymbol{x}}(i', \boldsymbol{u}), \\ y_{ij}^{\star\star} = y_{ij}^{\star}, & \forall\, i \neq i', i \in [I], j \in [J], \\ \displaystyle\sum_{j \in \mathcal{J}_{\boldsymbol{x}}(i', \boldsymbol{u})} y_{i'j}^{\star\star} = v_{i'}, \end{cases} \tag{A.36}$$

这里条件 $y_{i'j}^{\star\star} = 0 = y_{i'j}^{\star}$, $\forall\, j \neq \mathcal{J}_{\boldsymbol{x}}(i', \boldsymbol{u})$ 和 $\sum_{j \in \mathcal{J}_{\boldsymbol{x}}(i', \boldsymbol{u})} y_{i'j}^{\star\star} = v_{i'}$ 成立由如下不等式保证:

$$\sum_{j \in \mathcal{J}_{\boldsymbol{x}}(i, \boldsymbol{u})} y_{ij}^{\star\star} \leqslant v_i^{+}(\beta_1) \leqslant \sum_{j \in \mathcal{J}_{\boldsymbol{x}}(i, \boldsymbol{u})} s_j, \quad \forall\, i \in [I], \tag{A.37}$$

其中, 根据 $\mathcal{O}(\boldsymbol{x}, \beta_1)$ 的假设可得第二个不等式成立. 也就是说, 条件 (A.37) 保证了正盈利的资源回收设施具有充足的容量来处理收集点 i' 的原料质量 $v_{i'}$, 且不需要从其余设施 $j \neq \mathcal{J}_{\boldsymbol{x}}(i', \boldsymbol{u})$ 处获取额外的容量.

接下来, 我们证明总可以找到一个满足约束 (A.36) 的解 $\boldsymbol{y}^{\star\star}$, 且其满足如下条件:

$$\sum_{i \in [I]} y_{ij}^{\star\star} = \sum_{i \neq i'} y_{ij}^{\star\star} + y_{i'j}^{\star\star} \leqslant s_j, \quad \forall\, j \in \mathcal{J}_{\boldsymbol{x}}(\boldsymbol{u}). \tag{A.38}$$

如果存在某些 j 使得约束条件 (A.38) 不满足, 那么可以在条件 $\sum_{j \in \mathcal{J}_{\boldsymbol{x}}(i', \boldsymbol{u})} y_{i'j}^{\star\star} = v_{i'}$ 下调整 $y_{i'j}^{\star\star}, j \in \mathcal{J}_{\boldsymbol{x}}(i', \boldsymbol{u})$ 的值使得条件 (A.38) 成立, 此外, 如果

$$\sum_{i \in [I]} y_{ij}^{\star\star} \geqslant s_j, \quad \forall\, j \in \mathcal{J}_{\boldsymbol{x}}(\boldsymbol{u}),$$

其中, 至少一个不等式严格成立. 上述情况是不可能存在的, 因为它违背了集合 $\mathcal{O}(\boldsymbol{x}, \beta_1)$ 中如下假设:

$$\sum_{i \in [I]} \sum_{j \in \mathcal{J}_{\boldsymbol{x}}(i, \boldsymbol{u})} y_{ij}^{\star\star} \leqslant \sum_{i \in [I]} v_i^{+}(\beta_1) \leqslant \sum_{j \in \mathcal{J}_{\boldsymbol{x}}(\boldsymbol{u})} s_j. \tag{A.39}$$

因为对任意 $j \in \mathcal{J}_{\boldsymbol{x}}(i, \boldsymbol{u})$ 都有 $\kappa_{ij}(\boldsymbol{u}) > 0$, 且存在至少一个 $j \in \mathcal{J}_{\boldsymbol{x}}(i', \boldsymbol{u})$ 使得 $y_{i'j}^{\star\star} > y_{ij}^{\star\star}$, 所以得到如下不等式:

$$\sum_{i \in [I]} \sum_{j \in \mathcal{J}_{\boldsymbol{x}}(i, \boldsymbol{u})} y_{ij}^{\star\star} \kappa_{ij}(\boldsymbol{u}) > \sum_{i \in [I]} \sum_{j \in \mathcal{J}_{\boldsymbol{x}}(i, \boldsymbol{u})} y_{ij}^{\star} \kappa_{ij}(\boldsymbol{u}),$$

这与 $\boldsymbol{y}^{\star}$ 是最优解相矛盾, 因此我们证明了最优解在不等式约束满足 (A.34) 时取到.

特别地, 如果有多个 i' 满足严格不等式约束 (A.35), 则可以重复上述过程 (每次处理一个 i') 来证明最优解在约束满足 (A.34) 时取到. 因此, 对优化问题中任意 $\boldsymbol{v} \in V(\beta_1)$, 可以得到如下两个约束是等价的:

$$\sum_{j\in[J]} y_{ij} \leqslant v_i \Leftrightarrow \sum_{j\in[J]} y_{ij} = v_i, \quad \forall\, i \in [I].$$

根据命题 5.3, 可得等式 (A.33) 成立. 最后, 如果 $U(\beta_2) \subseteq \mathcal{O}(\boldsymbol{x}, \beta_1)$, 那么问题 (5.12) 中的 $\mathcal{Z}(\boldsymbol{x}; \beta_1, \beta_2)$ 可以转化为一个线性规划问题 (5.16)-(5.17), 其中 $\boldsymbol{z} \equiv \mathbf{0}$, 且 $\boldsymbol{v} \equiv \boldsymbol{v}^{-}(\beta_1)$.

命题 5.5 的证明

首先证明如下概率不等式成立:

$$\mathbb{P}\Big[\boldsymbol{u} \in U(\beta_2)\Big] \geqslant \left[1 - \frac{(I-1)\Omega^2 + \Omega}{I\beta_2^2}\right]^+, \quad \forall\, \beta_2 \in (0,1), \tag{A.40}$$

其中, $\Omega := \sum_{m\in[M]}(1-\delta_m)\bar{\mu}_{mi}$, 且对于任意的 $m \in [M]$ 和 $i \in [I]$, 都有 $\mathbb{E}[u_{mi}] \leqslant \bar{\mu}_{mi}$. 对任意 $i \in [I]$, 定义如下两个随机变量 ζ_i 和 η_i: $\zeta_i := \sum_{m\in[M]} \delta_m u_{mi}$, 以及 $\eta_i := 1 - \zeta_i = \sum_{m\in[M]}(1-\delta_m)u_{mi}$. 由此可得, 对于任意的 $i \in [I]$, 都有 $\mathbb{E}[\zeta_i] = \sum_{m\in[M]} \delta_m \mathbb{E}[u_{mi}]$, 且 $\mathbb{E}[\eta_i] = \sum_{m\in[M]}(1-\delta_m)\mathbb{E}[u_{mi}]$.

根据上述定义, 概率 $\mathbb{P}[\boldsymbol{u} \in U(\beta_2)]$ 可以转化为如下形式:

$$\mathbb{P}\Big[\boldsymbol{u} \in U(\beta_2)\Big] = \mathbb{P}\left[\frac{1}{I}\sum_{i\in[I]} \zeta_i \geqslant 1 - \beta_2\right] = 1 - \mathbb{P}\left[\frac{1}{I}\sum_{i\in[I]} \eta_i \geqslant \beta_2\right]. \tag{A.41}$$

利用马尔可夫不等式, 可以得到如下不等式

$$\begin{aligned}
\mathbb{P}\left[\frac{1}{I}\sum_{i\in[I]} \eta_i \geqslant \beta_2\right] &\leqslant \frac{\mathbb{E}\left[\left(\frac{1}{I}\sum_{i\in[I]} \eta_i\right)^2\right]}{\beta_2^2} \\
&= \frac{\mathbb{E}^2\left[\frac{1}{I}\sum_{i\in[I]} \eta_i\right] + \mathbb{V}\mathrm{ar}\left[\frac{1}{I}\sum_{i\in[I]} \eta_i\right]}{\beta_2^2} \\
&= \frac{\mathbb{E}^2\left[\frac{1}{I}\sum_{i\in[I]} \eta_i\right] + \frac{1}{I}\left(\mathbb{E}\left[\eta_i^2\right] - \mathbb{E}^2\left[\eta_i\right]\right)}{\beta_2^2}
\end{aligned} \tag{A.42}$$

$$\leqslant \frac{\mathbb{E}^2\left[\frac{1}{I}\sum_{i\in[I]}\eta_i\right]+\frac{1}{I}\left(\mathbb{E}\left[\eta_i\right]-\mathbb{E}^2\left[\eta_i\right]\right)}{\beta_2^2} \tag{A.43}$$

$$= \frac{(I-1)\mathbb{E}^2[\eta_i]+\mathbb{E}[\eta_i]}{I\beta_2^2},$$

其中, 根据随机变量 $\eta_i, i \in [I]$ 的相互独立性可得等式(A.42), 不等式 (A.43) 由 $0 \leqslant \eta_i \leqslant 1$ 几乎处处成立可推出. 最后, 因为 $\mathbb{E}[u_{mi}]$ 的上界为 $\bar{\mu}_{mi}$, 所以 $\dfrac{(I-1)\mathbb{E}^2[\eta_i]+\mathbb{E}[\eta_i]}{I\beta_2^2} \leqslant \dfrac{(I-1)\Omega^2+\Omega}{I\beta_2^2}$, 此外, 利用等式 (A.41), 我们可以推出不等式 (A.40).

接下来, 给定任意 $\beta_1 \in (0,1)$, 根据 $\beta_2^*\left[\boldsymbol{x}|\beta_1\right]$ 的定义, 我们有

$$\boldsymbol{u} \in U\Big(\beta_2^*\left[\boldsymbol{x}|\beta_1\right]\Big) \Rightarrow \phi(\boldsymbol{x};\boldsymbol{v},\boldsymbol{u}) \geqslant \tau_0, \quad \forall\ \boldsymbol{v} \in V(\beta_1),$$

进而推出概率不等式 (5.24).

命题 5.6 的证明

首先, 对于任意 $\beta_1 \in (0,1)$, 如果对于任意 $k \in [K]$ 有 $\mathbb{P}\big[\xi_k^-(\beta_1) \leqslant \xi_k \leqslant \xi_k^+(\beta_1)\big] \geqslant 1-(1-\beta_1)/K$, 则由 Bonferroni 不等式, 我们有

$$\mathbb{P}\Big[\boldsymbol{v} \in V(\beta_1)\Big] = \mathbb{P}\Big[\xi_k \in \left[\xi_k^-(\beta_1), \xi_k^+(\beta_1)\right], k \in [K]\Big] \geqslant \beta_1. \tag{A.44}$$

其次, 给定任意 $\beta_2 \in (0,1)$, 根据 $\beta_1^*\left[\boldsymbol{x}|\beta_2\right]$ 的定义 (5.7), 可以得到

$$\boldsymbol{v} \in V\Big(\beta_1^*\left[\boldsymbol{x}|\beta_2\right]\Big) \Rightarrow \phi(\boldsymbol{x};\boldsymbol{v},\boldsymbol{u}) \geqslant \tau_0, \quad \forall\ \boldsymbol{u} \in U(\beta_2).$$

最后, 利用上式和不等式 (A.44), 可得

$$\beta_1^*\left[\boldsymbol{x}|\beta_2\right] \leqslant \mathbb{P}\left[\begin{array}{l}\phi(\boldsymbol{x};\boldsymbol{v},\boldsymbol{u}) \geqslant \tau_0, \\ \forall\ \boldsymbol{u} \in U(\beta_2)\end{array}\right] \leqslant \min_{\boldsymbol{u}\in U(\beta_2)} \mathbb{P}\Big[\phi(\boldsymbol{x};\boldsymbol{v},\boldsymbol{u}) \geqslant \tau_0\Big],\ \forall\ \beta_2 \in (0,1).$$

命题 5.7 的证明

根据复合鲁棒性指标的定义 5.2, 有

$$\pi(\boldsymbol{x}) = \max_{\alpha,\beta_2\in[0,1]}\Big\{\alpha\ \Big|\ \theta_1\beta_1^*\left[\boldsymbol{x}|\beta_2\right] \geqslant \alpha,\ \theta_2\beta_2 \geqslant \alpha\Big\}.$$

此外, 根据鲁棒值函数 $\beta_1^*[\boldsymbol{x}|\beta_2]$ 的定义, 上述问题可转化为如下形式:

$$\max_{\alpha,\beta_1,\beta_2\in[0,1]} \quad \alpha \tag{A.45}$$

$$\text{s.t.} \quad \theta_1\beta_1 \geqslant \alpha, \tag{A.46}$$

$$\theta\beta_2 \geqslant \alpha, \tag{A.47}$$

$$\min_{(\boldsymbol{v},\boldsymbol{u})\in V(\beta_1)\times U(\beta_2)} \tau(\boldsymbol{x};\boldsymbol{v},\boldsymbol{u}) \geqslant \tau_0. \tag{A.48}$$

上述模型 (A.45)-(A.48) 的前两个约束可以分别写为 $\beta_1 \geqslant \dfrac{\alpha}{\theta_1}$ 和 $\beta_2 \geqslant \dfrac{\alpha}{\theta_2}$. 已知不确定集 $V(\beta_1)$ 和 $U(\beta_2)$ 的大小分别关于 β_1 和 β_2 是单调不减的, 因此, 上述约束 (A.46)-(A.48) 等价于

$$\min_{(\boldsymbol{v},\boldsymbol{u})\in V\left(\frac{\alpha}{\theta_1}\right)\times U\left(\frac{\alpha}{\theta_2}\right)} \tau(\boldsymbol{x};\boldsymbol{v},\boldsymbol{u}) \geqslant \tau_0.$$

因此, 模型 (A.45)-(A.48) 可以转化为问题 (5.27).

命题 5.8 的证明

首先, 给定 $\boldsymbol{x}$, 根据命题 5.2 的证明过程可知, 不含固定成本项 $\boldsymbol{c}^\top\boldsymbol{x}$ 的最小最大化利润问题 (5.12) 可转化为如下问题:

$$\min_{\boldsymbol{v}\in V(\beta_1)} \max_{\boldsymbol{y},\boldsymbol{z}\in\mathcal{Y}(\boldsymbol{x},\boldsymbol{v})} \min_{\boldsymbol{u}\in U(\beta_2)} \sum_{i\in[I]}\sum_{j\in[J]} \kappa_{ij}(\boldsymbol{u})y_{ij} - \sum_{i\in[I]} h_i z_i, \tag{A.49}$$

其中, 对任意给定的 $\boldsymbol{x}$ 和 $\boldsymbol{v}\in V(\beta_1)$, $\mathcal{Y}(\boldsymbol{x},\boldsymbol{v}) := \{\boldsymbol{y}\in\mathbb{R}_+^{I\times J}, \boldsymbol{z}\in\mathbb{R}^I \mid \text{(A.28)-(A.29)}\}$. 对上述问题的内层最小化问题做对偶, 并代入 $v_i = \boldsymbol{\alpha}_i^\top\boldsymbol{\xi}, i\in[I]$, 则问题 (A.49) 可以被转化为

$$\begin{aligned}
\min_{\boldsymbol{\xi}\in V_0(\beta_1)} \max_{\boldsymbol{y},\boldsymbol{z},\boldsymbol{d}} \quad & \sum_{i\in[I]} d_i + I\beta_2 d_0 - \sum_{i\in[I]} h_i z_i \\
\text{s.t.} \quad & d_i + d_0\delta_m \leqslant \sum_{j\in[J]} q_{mj}y_{ij},\ \forall\, i\in[I], m\in[M], \\
& \sum_{j\in[J]} y_{ij} + z_i = \boldsymbol{\alpha}_i^\top\boldsymbol{\xi},\ \forall\, i\in[I], \\
& \boldsymbol{y}\in\left\{\boldsymbol{y}\in\mathbb{R}_+^{I\times J} \mid \text{(A.29)}\right\}, \boldsymbol{z}\in\mathbb{R}_+^I, d_0\geqslant 0, d_i\in\mathbb{R},\ \forall\, i\in[I],
\end{aligned}$$

其中, $V_0(\beta_1)$ 由式 (A.31) 给出.

接下来, 利用对偶 (dualization)、极值点 (extreme point) 和线性化 (linearization) 等方法, 我们可以将问题 (A.49) 转化为问题 (5.31). 主要步骤如下:

(1) 对内层最大化问题做对偶, 且所有在原问题右端的项 $\left(\boldsymbol{\alpha}_i^\top\boldsymbol{\xi}\right)$ 都被转移到对偶向量为 $\boldsymbol{p}$ 的对偶问题的 (最小化) 目标函数中;

(2) 交换关于变量 $\boldsymbol{\xi}$ 的最小化算子和关于变量 $\boldsymbol{p}$ 的最小化算子;

(3) 因为目标函数关于变量 $\boldsymbol{\xi}$ 是仿射函数, 所以最小值在集合 $V_0(\beta_1)$ 的端点 $\hat{\boldsymbol{\xi}}^e=(\hat{\xi}_1^e,\hat{\xi}_2^e,\cdots,\hat{\xi}_K^e), e\in\{1,2,\cdots,E\}, E=2^K$ 处取到, 其中 $\hat{\xi}_k^e=\xi_k^-(\beta_1)$ 或 $\xi_k^+(\beta_1)$. 因此, 可以用离散变量 $\hat{\boldsymbol{\xi}}^e, e\in\{1,2,\cdots,E\}$ 的最小化代替变量 $\boldsymbol{\xi}\in V_0(\beta_1)$ 的最小化;

(4) 将关于变量 $\boldsymbol{\xi}$ 的最小化算子和关于变量 $\boldsymbol{p}$ 的最小化算子再次交换, 并且对内层中关于变量 $\boldsymbol{d}$ 的最小化问题做对偶, 得到关于对每个 e 的 $(\boldsymbol{y}^e,\boldsymbol{z}^e,\boldsymbol{d}^e)$ 的最大化问题;

(5) 对于每个 $(\boldsymbol{y}^e,\gamma^e)$, 利用命题 5.2 中的 $\Gamma^e(\beta_2)$ 问题, 通过以下方式将目标函数线性化得到

$$\min_{1\leqslant i\leqslant n} g_i=\max\{r\mid r\leqslant g_i, i=1,2,\cdots,n\}.$$

最后, 通过在问题 (5.31) 中引入固定成本项 $\boldsymbol{c}^\top\boldsymbol{x}$ 可求解最小最大化利润问题 (5.12).

A.3.2　原料不确定集的构建

将不确定集分别构建为 V 和 U 的一个好处是可以将模型转化为高效计算形式. 假设原料质量和原料组成成分都使用同一个变量 $\boldsymbol{w}=(w_{mi})_{M\times I}$ 来建模, 其中, w_{mi} 是原料收集点 i 收集的第 m 种原料组成成分的量. 记 y_{ij} 为原料从收集点 i 运输到资源回收系统 j 的比例决策, 则给定决策 $\boldsymbol{x}$ 下资源回收系统的运营问题可以转化为如下线性规划问题:

$$\begin{aligned}
\phi(\boldsymbol{x};\boldsymbol{w}):=\max_{\boldsymbol{y},\boldsymbol{z}}\quad & \sum_{i\in[I]}\sum_{j\in[J]}\left(\sum_{m\in[M]}w_{mi}q_{mj}\right)y_{ij}-\sum_{i\in[I]}\sum_{m\in[M]}w_{mi}h_iz_i-\boldsymbol{c}^\top\boldsymbol{x}\\
\text{s.t.}\quad & \sum_{j\in[J]}y_{ij}+z_i=1,\ \forall\, i\in[I],\\
& \sum_{i\in[I]}\sum_{m\in[M]}w_{mi}y_{ij}\leqslant s_jx_j,\ \forall\, j\in[J],\\
& \boldsymbol{y},\boldsymbol{z}\geqslant\boldsymbol{0}.
\end{aligned}$$

当原料状况 $\boldsymbol{w}$ 是不确定变量, 且决策变量 $(\boldsymbol{y},\boldsymbol{z})$ 关于 $\boldsymbol{w}$ 具有自适应性时, 上述模型一般很难被高效求解. 假设集合 $\mathcal{W}$ 是关于 $\boldsymbol{w}$ 的一个不确定集, 则极值情

况下第二阶段运营问题具有如下形式:

$$\begin{aligned}
\min_{\boldsymbol{w}\in\mathcal{W}}\max_{\boldsymbol{y},\boldsymbol{z}}\quad & \sum_{i\in[I]}\sum_{j\in[J]}\left(\sum_{m\in[M]}w_{mi}q_{mj}\right)y_{ij}-\sum_{i\in[I]}\sum_{m\in[M]}w_{mi}h_iz_i-\boldsymbol{c}^\top\boldsymbol{x}\\
\text{s.t.}\quad & \sum_{j\in[J]}y_{ij}+z_i=1,\ \forall\, i\in[I],\\
& \sum_{i\in[I]}\sum_{m\in[M]}w_{mi}y_{ij}\leqslant s_jx_j,\ \forall\, j\in[J],\\
& \boldsymbol{y},\boldsymbol{z}\geqslant\boldsymbol{0},
\end{aligned}$$

其中, 第二组约束中决策变量 $\boldsymbol{y}$ 的系数是不确定变量 $\boldsymbol{w}$, 这能说明上述问题具有不确定补偿矩阵 (uncertain recourse matrix) 结构. 一般来说, 具有不确定补偿矩阵的两阶段鲁棒优化模型是难以求解的 (Ben-Tal et al., 2004). 为了处理上述问题, 我们重新定义决策变量 $\boldsymbol{y}$ 和 $\boldsymbol{z}$ 分别为原料运输量和未处理原料量, 进而构建如下等价形式:

$$\min_{\boldsymbol{w}\in\mathcal{W}}\max_{\boldsymbol{y},\boldsymbol{z}}\quad \sum_{i\in[I]}\sum_{j\in[J]}\left(\frac{w_{mi}}{\sum_{m\in[M]}w_{mi}}\right)q_{mj}y_{ij}-\sum_{i\in[I]}h_iz_i-\boldsymbol{c}^\top\boldsymbol{x} \tag{A.50}$$

$$\text{s.t.}\quad \sum_{j\in[J]}y_{ij}+z_i=\sum_{m\in[M]}w_{mi},\ \forall\, i\in[I], \tag{A.51}$$

$$\sum_{i\in[I]}y_{ij}\leqslant s_jx_j,\ \forall\, j\in[J], \tag{A.52}$$

$$\boldsymbol{y},\boldsymbol{z}\geqslant\boldsymbol{0}, \tag{A.53}$$

其中, $\dfrac{w_{mi}}{\sum_{m\in[M]}w_{mi}}$ 是收集点 i 处的原料组成成分 m 的占其原料总质量的比例, $\sum_{m\in[M]}w_{mi}$ 是收集点 i 处原料的总质量. 利用上述变量, 问题不再具有不确定补偿矩阵, 但是有非线性项 $\dfrac{w_{mi}}{\sum_{m\in[M]}w_{mi}}$. 进一步, 对于所有的 $m\in[M]$ 和 $i\in[I]$, 我们用 $\boldsymbol{v}$ 和 $\boldsymbol{u}$ 分别表示原料质量 (即, $v_i=\sum_{m\in[M]}w_{mi}$) 以及原料组成成分 (即 $u_{mi}=\dfrac{w_{mi}}{\sum_{m\in[M]}w_{mi}}$), 则上述的最小最大化问题 (A.50)-(A.53)可以转化为 5.3.2 小节中的问题 (5.12):

$$\min_{\boldsymbol{v}\in V(\beta_1)}\min_{\boldsymbol{u}\in U(\beta_2)}\max_{\boldsymbol{y},\boldsymbol{z}}\quad \sum_{i\in[I]}\sum_{j\in[J]}q_{mj}u_{mi}y_{ij}-\sum_{i\in[I]}h_iz_i-\boldsymbol{c}^\top\boldsymbol{x}$$

$$\begin{aligned} \text{s.t.} \quad & \sum_{j\in[J]} y_{ij} + z_i = v_i, \ \forall\, i \in [I], \\ & \sum_{i\in[I]} y_{ij} \leqslant s_j x_j, \ \forall\, j \in [J], \\ & \boldsymbol{y}, \boldsymbol{z} \geqslant \boldsymbol{0}. \end{aligned}$$

在上述问题中, 不确定变量 $\boldsymbol{u}$ 出现在目标函数中, 且不确定变量 $\boldsymbol{v}$ 仅以线性方式出现在约束右端项中. 给定 $\boldsymbol{v}$, 利用最小最大化定理, 可以交换关于 $\boldsymbol{u}$ 的最小化算子和关于 $(\boldsymbol{y}, \boldsymbol{z})$ 的最大化算子顺序. 此外, 将 V 定义为式 (5.4), 可以将问题转化为一个混合整数线性规划问题, 该问题可以使用商业求解器高效求解. 此结果已在命题 5.1 给出.

将原料质量和组成成分不确定性分别建模的另一个好处是可以更直观地描述能源回收系统在这两个原料不确定性维度上的表现. 第 5 章提出的复合鲁棒性指标考虑了系统在原料质量和组成成分不确定性上的权衡, 这有很强的实际应用价值. 因为能源回收方案受原料分类的群众参与率与分离纯度的影响, 这分别对应于原料质量 $\boldsymbol{v}$ 和原料组成成分 $\boldsymbol{u}$. 在第 5 章, 我们通过参数 β_1 和 β_2 控制上述两种不确定性.

A.3.3 问题 $\max_{\boldsymbol{x}} \mathcal{Z}(\boldsymbol{x}, \beta_1, \beta_2)$ 的 Benders 分解算法

根据命题 5.8, 我们可以定义问题 (5.30) 的主问题 (master problem) 为如下混合整数线性规划问题:

$$\max_{\boldsymbol{x}, r} \ r - \boldsymbol{c}^\top \boldsymbol{x} \tag{A.54}$$

$$\text{s.t.} \ \ \boldsymbol{x} \in \mathcal{X}, \ (\boldsymbol{x}, r) \in \boldsymbol{\Pi}\left(\hat{\boldsymbol{\xi}}^e\right), \ e = 1, 2, \cdots, E', \tag{A.55}$$

其中, 每个 $\boldsymbol{\Pi}(\hat{\boldsymbol{\xi}}^e)$ 是割平面集合, 其定义如下:

$$\boldsymbol{\Pi}(\hat{\boldsymbol{\xi}}^e) := \left\{ \boldsymbol{x}, r \ \middle| \ \begin{array}{l} r \leqslant \gamma^e - \displaystyle\sum_{i\in[I]} h_i z_i^e, \\ \boldsymbol{y}^e, \boldsymbol{z}^e, \gamma^e, \boldsymbol{x} \in \{(\boldsymbol{y}^e, \boldsymbol{z}^e, \gamma^e, \boldsymbol{x}) \mid (5.31)\} \end{array} \right\}, \tag{A.56}$$

且 $E' \leqslant E$. 当 $E' = E$ 时, 根据命题 5.8, 可知上述问题等价于自适应鲁棒优化问题 (5.30). 当 $E' < E$ 时, 上述问题为原问题 (5.30) 松弛主问题 (relaxed master problem).

记问题 (5.30) 的最优目标函数值为 $\mathcal{Z}^\star = \max_{\boldsymbol{x}} \mathcal{Z}(\boldsymbol{x}, \beta_1, \beta_2)$, 则对于松弛主问题 (A.54)-(A.55) 的任意最优解 $(\bar{\boldsymbol{x}}, \bar{r})$, 有

$$\bar{r} - \boldsymbol{c}^\top \bar{\boldsymbol{x}} \geqslant \mathcal{Z}^\star. \tag{A.57}$$

此外, 从命题 5.1 的证明中可以得到, 问题 (5.12) 不含固定成本项 $\boldsymbol{c}^\top\boldsymbol{x}$ 的最小最大化利润子问题等价于如下混合整数线性规划问题:

$$\mathcal{R}_{\text{MIP}}(\boldsymbol{x}) := \min \quad \sum_{k\in[K]} \bar{\xi}_k(\beta_1)\left(\sum_{i\in[I]} \alpha_{ik} d_i\right) + \sum_{j\in[J]} s_j x_j \pi_j - \sum_{k\in[K]} \gamma_k \tag{A.58}$$

$$\text{s.t.} \quad \boldsymbol{d}, \boldsymbol{\pi}, \boldsymbol{u}, \boldsymbol{\gamma}, \boldsymbol{\lambda} \in \{\boldsymbol{d}, \boldsymbol{\pi}, \boldsymbol{u}, \boldsymbol{\gamma}, \boldsymbol{\lambda} \mid \text{ 问题 (5.15) 的约束}\}. \tag{A.59}$$

因此, 问题 (5.30) 可以转化为如下形式:

$$\max_{\boldsymbol{x}, r} \ r - \boldsymbol{c}^\top \boldsymbol{x} \tag{A.60}$$

$$\text{s.t.} \ r \leqslant \mathcal{R}_{\text{MIP}}(\boldsymbol{x}), \ \boldsymbol{x} \in \mathcal{X}. \tag{A.61}$$

对于松弛主问题 (A.54)-(A.55) 的任意解 $(\bar{\boldsymbol{x}}, \bar{r})$, 如果这个解满足 $\bar{r} \leqslant \mathcal{R}_{\text{MIP}}(\bar{\boldsymbol{x}})$, 那么它是问题 (A.60)-(A.61) 的一个可行解, 因此可以得到 $\bar{r} - \boldsymbol{c}^\top\bar{\boldsymbol{x}} \leqslant \mathcal{Z}^\star$. 结合式 (A.57), 我们有 $\bar{r} - \boldsymbol{c}^\top\bar{\boldsymbol{x}} = \mathcal{Z}^\star$. 因此, 不等式 $\bar{r} \leqslant \mathcal{R}_{\text{MIP}}(\bar{\boldsymbol{x}})$ 可以作为终止条件. 进一步, 在整个迭代过程中, 利用松弛主问题 (A.54)-(A.55) 的解 $(\bar{\boldsymbol{x}}, \bar{r})$, 可以得到主问题的上下界 (分别记为 UB 和 LB) 为

$$\text{UB} = \bar{r} - \boldsymbol{c}^\top\bar{\boldsymbol{x}} \quad \text{与} \quad \text{LB} = \mathcal{R}_{\text{MIP}}(\bar{\boldsymbol{x}}) - \boldsymbol{c}^\top\bar{\boldsymbol{x}}.$$

利用上下界的相对差 $r_G := (\text{UB} - \text{LB})/\text{LB} \times 100\%$, 并结合下界 LB, 可以得到一个可接受的 (acceptable) 近似解 $\bar{\boldsymbol{x}}$. 上述求解算法总结为如下 Benders 分解算法, 该算法可以在有限步内求得最优解.

算法 A.1 $\max_{\boldsymbol{x}} \mathcal{Z}(\boldsymbol{x}, \boldsymbol{\beta}_1, \boldsymbol{\beta}_2)$ 的 Benders 分解

输入: 模型参数, 最优性容差 $\epsilon > 0$

初始化: $\boldsymbol{\Pi} \leftarrow \varnothing$, $e \leftarrow 0$.

1. $e \leftarrow e + 1$, $\boldsymbol{\Pi} \leftarrow \boldsymbol{\Pi} \cup \boldsymbol{\Pi}(\hat{\boldsymbol{\xi}}^e)$.
2. 基于 $\boldsymbol{\Pi}$ 求解松弛主问题 (A.54)-(A.55), 得到 $(\bar{r}, \bar{\boldsymbol{x}})$.
3. 计算上界 UB $\leftarrow \bar{r} - c^\top\bar{\boldsymbol{x}}$.
4. 通过求解混合整数线性规划问题 (A.58)-(A.59), 得到 $\bar{\boldsymbol{x}}$, 确定 $\mathcal{R}_{\text{MIP}}(\bar{\boldsymbol{x}})$ 的值.
5. 计算下界 LB $\leftarrow \mathcal{R}_{\text{MIP}}(\bar{\boldsymbol{x}}) - c^\top\bar{\boldsymbol{x}}$.
6. 如果 $\bar{r} \leqslant \mathcal{R}_{\text{MIP}}(\bar{\boldsymbol{x}})$ 或 $e = E$(或 $r_G < \epsilon$), 则算法停止; 否则, 返回第 1 步.

输出: 最优解 $\boldsymbol{x}^\star \leftarrow \bar{\boldsymbol{x}}$.

A.3.4 不确定集 $\boldsymbol{V}(\boldsymbol{\beta}_1)$ 和 $\boldsymbol{U}(\boldsymbol{\beta}_2)$ 的例子

第 5 章基于可调整不确定集 $V(\beta_1)$ 和 $U(\beta_2)$ 构建了鲁棒资源回收设施选址模型, 其中不确定集 $V(\beta_1)$ 和 $U(\beta_2)$ 分别由式 (5.4) 和 (5.6) 给出. 通过考虑不同的原料数据信息, 介绍如下关于不确定集 $V(\beta_1)$ 和 $U(\beta_2)$ 的例子.

例 22　下面给出原料质量不确定集 $V(\beta_1)$ 的一些实用例子:

(1) 基于 (置信) 区间的原料质量不确定集:

$$V_{\mathrm{I}}(\beta_1) := \left\{ \boldsymbol{v} \in \mathbb{R}_+^I \,\middle|\, \frac{|v_i - \hat{v}_i|}{\Delta_i^v} \leqslant \beta_1, \ \forall\, i \in [I] \right\}, \tag{A.62}$$

其中, $\hat{v}_i$ 是收集点 i 处原料质量的估计名义值 (estimated nominal value), $\Delta_i^v > 0$ 是原料质量 v_i 关于名义值 $\hat{v}_i$ 的偏离程度统计量. 参数 $\beta_1 \in [0,1]$ 控制所有收集点 i 处原料质量 v_i 的偏离程度. 上述不确定集是由式 (5.4) 定义的 $V(\beta_1)$ 的一个特例.

(2) 基于预算的 (budgeted) 原料质量不确定集:

$$V_{\mathrm{B}}(\beta_1) := \left\{ \boldsymbol{v} \in \mathbb{R}_+^I \,\middle|\, |v_i - \hat{v}_i| \leqslant \Delta_i^v, \forall\, i \in [I], \sum_{i \in [I]} \frac{|v_i - \hat{v}_i|}{\Delta_i^v} \leqslant \beta_1 I \right\}, \tag{A.63}$$

其中, 参数 $\hat{v}_i$ 和 $\Delta_i^v > 0$ 与不确定集 (A.62) 的参数定义相同, 参数 $\beta_1 \in [0,1]$ 控制 $\boldsymbol{v}$ 关于 $\hat{\boldsymbol{v}}$ 的平均偏离程度, 也称为不确定预算 (uncertainty budget, Bertsimas and Sim, 2004).

(3) 基于 Mahalanobis 距离的原料质量不确定集: 如果决策者能够估计未来原料质量 $\boldsymbol{v}$ 概率分布 $\mathbb{P}_{\boldsymbol{v}}$ 的均值 $\boldsymbol{\mu}_{\boldsymbol{v}}$ 和协方差 $\boldsymbol{\Sigma}_{\boldsymbol{v}}$, 则不确定集 $V(\beta_1)$ 可以定义为如下形式:

$$V_{\mathrm{M}}(\beta_1) := \left\{ \boldsymbol{v} \in \mathbb{R}_+^I \,\middle|\, \underbrace{\sqrt{(\boldsymbol{v} - \boldsymbol{\mu}_{\boldsymbol{v}})^\top \boldsymbol{\Sigma}_{\boldsymbol{v}}^{-1} (\boldsymbol{v} - \boldsymbol{\mu}_{\boldsymbol{v}})}}_{d_{\mathrm{M}}(\boldsymbol{v}, \mathbb{P}_{\boldsymbol{v}})} \leqslant \beta_1 D_{\mathrm{M}}^{\boldsymbol{v}} \right\}, \tag{A.64}$$

其中, $d_{\mathrm{M}}(\boldsymbol{v}, \mathbb{P}_{\boldsymbol{v}}) := \sqrt{(\boldsymbol{v} - \boldsymbol{\mu}_{\boldsymbol{v}})^\top \boldsymbol{\Sigma}_{\boldsymbol{v}}^{-1} (\boldsymbol{v} - \boldsymbol{\mu}_{\boldsymbol{v}})}$ 是关于观测值 $\boldsymbol{v}$ 和概率分布 $\mathbb{P}_{\boldsymbol{v}}$ 之间的 Mahalanobis 距离, 它考虑了不同原料收集点之间原料质量的协方差信息. 这里假设 $\boldsymbol{\Sigma}_{\boldsymbol{v}} \succ \boldsymbol{0}$, 且 $(\boldsymbol{\mu}_{\boldsymbol{v}}, \boldsymbol{\Sigma}_{\boldsymbol{v}})$ 都可以通过对未来原料质量 $\boldsymbol{v}$ 的预测来估计. 此外, $D_{\mathrm{M}}^{\boldsymbol{v}}$ 是 Mahalanobis 距离 $d_{\mathrm{M}}(\boldsymbol{v}, \mathbb{P}_{\boldsymbol{v}})$ 的上界.

例 23　下面给出原料组成成分不确定集 $U(\beta_2)$ 的一些实用例子:

(1) 基于预算的原料组成成分不确定集:

$$U_{\mathrm{B}}(\beta_2) := \left\{ \boldsymbol{u} \in \mathbb{R}_+^{M \times I} \,\middle|\, \begin{array}{l} \displaystyle\sum_{m \in [M]} u_{mi} = 1, |u_{mi} - \hat{u}_{mi}| \leqslant \Delta_{mi}^c, \forall\, i \in [I], \\ \displaystyle\sum_{i \in [I]} \sum_{m \in [M]} \frac{|u_{mi} - \hat{u}_{mi}|}{\Delta_{mi}^c} \leqslant \beta_2 IM \end{array} \right\}, \tag{A.65}$$

其中, $\hat{u}_{mi}$ 为收集点 i 处第 m 种原料组分成分的估计名义值, $\Delta_i^c > 0$ 为偏离程度统计量. 参数 $\beta_2 \in [0,1]$ 反映了原料组成 $\boldsymbol{u}$ 相对于 $\hat{\boldsymbol{u}}$ 的平均偏离程度, 这与不确定集 $U(\beta_2)$ (5.6) 的原料纯度水平参数不同.

(2) 基于 Mahalanobis 距离的原料组成成分不确定集: 与原料质量不确定集 (A.64)类似, 我们可以定义基于 Mahalanobis 距离 $d_{\mathrm{M}}(\boldsymbol{u}, \mathbb{P}_{\boldsymbol{u}})$ 的原料组成成分不确定集 $U(\beta_2)$ 如下:

$$U_{\mathrm{M}}(\beta_2) := \left\{ \boldsymbol{u} \in \mathbb{R}_+^{M \times I} \,\middle|\, \sqrt{(\boldsymbol{u} - \boldsymbol{\mu}_{\boldsymbol{u}})^\top \boldsymbol{\Sigma}_{\boldsymbol{u}}^{-1} (\boldsymbol{u} - \boldsymbol{\mu}_{\boldsymbol{u}})} \leqslant \beta_1 D_{\mathrm{M}}^{\boldsymbol{u}} \right\}, \tag{A.66}$$

其中, 矩阵 $\boldsymbol{u}$ 可以按列顺序转化为一个向量.

(3) 基于 φ-散度的原料组成成分不确定集:

$$U_\varphi(\beta_2) := \left\{ \boldsymbol{u} \in \mathbb{R}_+^{M \times I} \,\middle|\, \sum_{m \in [M]} u_{mi} = 1, d_\varphi(\boldsymbol{u}_i, \hat{\boldsymbol{u}}_i) \leqslant D_i^\varphi, \forall i \in [I], \sum_{i \in [I]} \frac{d_\varphi(\boldsymbol{u}_i, \hat{\boldsymbol{u}}_i)}{D_i^\varphi} \leqslant \beta_2 I \right\}, \tag{A.67}$$

其中, $\hat{\boldsymbol{u}}_i = (\hat{u}_{1i}, \hat{u}_{2i}, \cdots, \hat{u}_{Mi})^\top$ 为在收集点 i 处原料组成成分的估计名义值. 此外我们定义 $d_\varphi(\boldsymbol{u}_i, \hat{\boldsymbol{u}}_i) := \sum_{m \in [M]} \hat{u}_{mi} \varphi(u_{mi}/\hat{u}_{mi})$ 为 $\boldsymbol{u}_i$ 和 $\hat{\boldsymbol{u}}_i$ 关于散度函数 (divergence function) φ 的 φ-散度度量, 散度函数 φ 可以有不同的形式 (Ben-Tal et al., 2013). 这里我们考虑常用的 Kullback-Leibler 散度 (Kuhlback-Leibler-divergence):

$$d_{\varphi_{\mathrm{KL}}}(\boldsymbol{u}_i, \hat{\boldsymbol{u}}_i) := \sum_{m \in [M]} \hat{u}_{mi} \log\left(\frac{\hat{u}_{mi}}{u_{mi}}\right), \quad \forall i \in [I].$$

它的上境图 (epigraph) 具有指数锥 (exponential conic) 表示形式. 我们也可以使用 χ^2-散度:

$$d_{\varphi_{\chi^2}}(\boldsymbol{u}_i, \hat{\boldsymbol{u}}_i) := \sum_{m \in [M]} \frac{(u_{mi} - \hat{u}_{mi})^2}{u_{mi}}, \forall i \in [I],$$

它的上境图具有二阶锥 (second-order conic) 表示形式. 参数 D_i^φ 是 $d_\varphi(\boldsymbol{u}_i, \hat{\boldsymbol{u}}_i)$ 的上界. 在不确定集 $U(\beta_2)$ 中, 参数 $\beta_2 \in [0,1]$ 控制不同收集点处原料组成成分与估计名义值之间的平均 φ-偏离水平.

基于上述不确定集 $V(\beta_1)$ 和 $U(\beta_2)$ 构建的鲁棒资源回收系统选址模型的高效计算形式不同. 接下来, 通过讨论一些代表性的不确定集选择来简要介绍模型的高效计算形式, 并在表 A.1 中给出了所有可能不确定集下模型的高效计算形式. 为了表述简洁, 用 $V_{\mathrm{F}}(\beta_1)$ 和 $V_{\mathrm{Q}}(\beta_2)$ 分别表示由式 (5.4) 和式 (5.6) 定义的不确定集.

表 A.1　原料不确定集 $V(\beta_1)$ 和 $U(\beta_2)$ 的不同选择以及对应的模型计算形式

$V(\beta_1)$	$U(\beta_2)$	$\mathcal{Z}(\boldsymbol{x};\beta_1,\beta_2)$	$\beta_1^*[\boldsymbol{x}\|\beta_2]$ 和 $\pi(\boldsymbol{x})$	$\max_{\boldsymbol{x}\in\mathcal{X}}\pi(\boldsymbol{x})$
$V_{\mathrm{F}}(\beta_1)$ 或 $V_{\mathrm{I}}(\beta_1)$	$U_{\mathrm{Q}}(\beta_2)$ 或 $U_{\mathrm{B}}(\beta_2)$	MILP	$\mathcal{BS}$[1] + MILPs	$\mathcal{BS}+\mathcal{BD}$ (MILP, MILP)[2]
$V_{\mathrm{F}}(\beta_1)$ 或 $V_{\mathrm{I}}(\beta_1)$	$U_{\mathrm{M}}(\beta_2)$ 或 $U_{\varphi_{\chi^2}}(\beta_2)$	MISOCP[3]	$\mathcal{BS}$ + MISOCPs	$\mathcal{BS}+\mathcal{BD}$ (MISOCP, MISOCP)
$V_{\mathrm{F}}(\beta_1)$ 或 $V_{\mathrm{I}}(\beta_1)$	$U_{\varphi_{\mathrm{KL}}}(\beta_2)$	MIECP[4]	$\mathcal{BS}$ + MIECPs	$\mathcal{BS}+\mathcal{BD}$ (MIECP, MIECP)
$V_{\mathrm{B}}(\beta_1)$	$U_{\mathrm{Q}}(\beta_2)$ 或 $U_{\mathrm{B}}(\beta_2)$	BLP[5]	$\mathcal{BS}$ + BLP	$\mathcal{BS}+\mathcal{BD}$ (BLP, MILP)
$V_{\mathrm{B}}(\beta_1)$	$U_{\mathrm{M}}(\beta_2)$	BSOCP[6]	$\mathcal{BS}$ + BSOCP	$\mathcal{BS}+\mathcal{BD}$ (BSOCP, MISOCP)
$V_{\mathrm{B}}(\beta_1)$	$U_{\varphi_{\mathrm{KL}}}(\beta_2)$	BECP[7]	$\mathcal{BS}$ + BECP	$\mathcal{BS}+\mathcal{BD}$ (BECP, MIECP)
$V_{\mathrm{M}}(\beta_1)$	$U_{\mathrm{Q}}(\beta_2)$, $U_{\mathrm{B}}(\beta_2)$, $U_{\mathrm{M}}(\beta_2)$ 或 $U_{\varphi_{\chi^2}}(\beta_2)$	BSOCP	$\mathcal{BS}$ + BSOCP	—[9]
$V_{\mathrm{M}}(\beta_1)$	$U_{\varphi_{\mathrm{KL}}}(\beta_2)$	BSOC-ECP[8]	$\mathcal{BS}$ + BSOC-ECP	—

1 $\mathcal{BS}$: 二元搜索 (二分搜索) 算法, 当最优性间隙为 ϵ 时, 至多需要 $\log(1/\epsilon)$ 步迭代.
2 $\mathcal{BD}$: 有限次迭代的 Benders 分解算法, 其中每次迭代需求解下界 (LB) 问题和上界 (UB) 问题.
3 MISOCP: 混合整数二阶锥规划.
4 MIECP: 混合整数指数锥规划 (mixed integer exponential conic program), 其中指数锥规划在理论上是可求解的, 可以用多项式时间算法求解 (Nesterov and Nemirovskii, 1994).
5 BLP: 双线性规划 (bilinear program).
6 BSOCP: 双二阶锥规划 (bi-second order conic program).
7 BECP: 双指数锥规划 (bi-exponential conic program).
8 BSOC-ECP: 包含二阶锥约束和指数锥约束的双凸规划 (bi-convex program) 问题.
9 '—': 此问题的求解比其他类别问题相比较困难, BD 算法无法直接求解.

(1) 当 $V(\beta_1) = V_{\mathrm{F}}(\beta_1)$ 或 $V_{\mathrm{I}}(\beta_1)$, 且 $U(\beta_2) = U_{\mathrm{Q}}(\beta_2)$ 或 $U_{\mathrm{B}}(\beta_2)$ 时, 由式 (5.12) 给出的最小最大化利润问题 $\mathcal{Z}(\boldsymbol{x};\beta_1,\beta_2)$ 等价于一个混合整数线性规划问题, 其中整数变量的个数等于不确定集 $V(\beta_1) = V_{\mathrm{F}}(\beta_1)$ 的因子个数, 或者等于不确定集 $V(\beta_1) = V_{\mathrm{I}}(\beta_1)$ 的原料收集点的个数. 此外, 给定决策 $\boldsymbol{x}$, 计算鲁棒性函数 $\beta_1^*[\boldsymbol{x}|\beta_2]$ 和复合鲁棒性指标 $\pi(\boldsymbol{x})$ 等价于在最优间隙 ϵ 下, 利用二分搜索算法求解 $\log(1/\epsilon)$ 个混合整数线性规划问题. 最后, 利用二分搜索算法和 Benders 分解算法 (见 5.4.2 节) 可以求解问题 $\max_{\boldsymbol{x}\in\{0,1\}^I}\pi(\boldsymbol{x})$.

(2) 当 $V(\beta_1) = V_{\mathrm{F}}(\beta_1)$ 或 $V_{\mathrm{I}}(\beta_1)$, 且 $U(\beta_2) = U_{\mathrm{M}}(\beta_2)$ 或 $U_{\varphi_{\chi^2}}(\beta_2)$ 时, 问题 $\mathcal{Z}(\boldsymbol{x};\beta_1,\beta_2)$ 等价于一个混合整数二阶锥规划 (mixed integer second-order cone program, MISOCP) 问题, 它可以被优化求解器直接求解, 其中整数变量的个数等于不确定集 $V(\beta_1) = V_{\mathrm{F}}(\beta_1)$ 的因子个数或 $V(\beta_1) = V_{\mathrm{I}}(\beta_1)$ 的原料收集点数. 此外, 给定决策 $\boldsymbol{x}$ 下, $\beta_1^*[\boldsymbol{x}|\beta_2]$ 和 $\pi(\boldsymbol{x})$ 的求解等价于在最优间隙为 ϵ 下使用二分搜索算法求解 $\log(1/\epsilon)$ 个混合整数二阶锥规划问题. 最后, 利用二分搜索算法和 Benders 分解算法来求解问题 $\max_{\boldsymbol{x}\in\{0,1\}^I}\pi(\boldsymbol{x})$, 其中 Benders 分解算法需要求解有限个混合整数二阶锥规划问题.

(3) 当 $V(\beta_1) = V_{\mathrm{B}}(\beta_1)$, 且 $U(\beta_2) = U_{\mathrm{Q}}(\beta_2)$ 或 $U_{\mathrm{B}}(\beta_2)$ 时, 问题 $\mathcal{Z}(\boldsymbol{x};\beta_1,\beta_2)$ 的目标函数是双线性的且可行域是一个多面体, 该问题是一个双线性规划问题, 因此难以求解. 此外, 给定决策 $\boldsymbol{x}$ 下, 求解 $\beta_1^*[\boldsymbol{x}|\beta_2]$ 和 $\pi(\boldsymbol{x})$ 等价于在最优性间隙 ϵ 下进行 $\log(1/\epsilon)$ 次二分搜索算法迭代, 其中每次迭代需要求解一个类似问题 $\mathcal{Z}(\boldsymbol{x};\beta_1,\beta_2)$ 的问题. 最后, 虽然利用二分搜索算法和 Benders 分解算法可以求解 $\max_{\boldsymbol{x}\in\{0,1\}^I}\pi(\boldsymbol{x})$, 但是在每一次迭代中, 需要求解一个混合整数线性规划问题来获得上界 (UB), 并且求解一个双线性规划问题 (或 $\mathcal{Z}(\boldsymbol{x};\beta_1,\beta_2)$) 来获得下界 (LB).

表 A.1 的结果表明: 模型的计算形式受不确定集 $V(\beta_1)$ 结构的影响比 $U(\beta_2)$ 的影响更大, 体现在以下三个方面:

(1) 当 $V(\beta_1) = V_{\mathrm{F}}(\beta_1)$ 或 $V_{\mathrm{I}}(\beta_1)$ 时, 复合鲁棒性指标和鲁棒性函数可以转化为具有多项式时间可解的混合整数凸问题 (MILP、MISOCP 或 MIECP), 并且可通过二分搜索算法迭代求解. 此外, 鲁棒资源回收系统选址模型可由精确迭代求解方法求解, 其中外层迭代采用二分搜索算法, 内层迭代采用 Benders 分解算法, 且 Benders 分解中的 LB 问题和 UB 问题均为多项式时间可解的混合整数凸问题. 虽然上述问题在理论上都是可高效求解的, 但是选择不同的不确定集 $U(\beta_2)$ 会导致问题的求解困难程度不同, 例如, 当 $U(\beta_2) = U_{\varphi_{\mathrm{KL}}}(\beta_2)$ 时, 模型等价于一个 MIECP 问题.

(2) 当 $V(\beta_1) = V_{\mathrm{B}}(\beta_1)$ 时, 鲁棒性函数和复合鲁棒性指标的估计等价于求解一个双凸规划问题 (BLP、BSOCP 和 BECP). 虽然可以利用 BS 与 Benders 分解算法来求解模型, 但是 LB 问题都是双凸规划问题, 这会极大降低算法迭代的效率.

(3) 当 $V(\beta_1) = V_{\mathrm{M}}(\beta_1)$ 时, 鲁棒性函数和复合鲁棒性指标的估计等价于一个更加复杂的双凸问题 (例如, 当 $U(\beta_2) = U_{\varphi_{\mathrm{KL}}}(\beta_2)$ 时, 是 BSOC-ECP 问题), Benders 分解算法无法直接求解此类问题.

A.4　第 6 章的证明及其他相关内容

本节将提供第 6 章中引理、命题和定理的证明以及其他相关内容.

A.4.1　第 6 章的证明

命题 6.1 的证明

因为 $0 \leqslant \lambda_1 < \lambda_2 < 1$, 所以我们有如下不等式:

$$\inf_{\mathbb{P}\in\mathbb{F}(\boldsymbol{s}_2^\star)} \mathbb{E}_{\mathbb{P}}\left[\phi\left(\boldsymbol{x}_2^\star, \tilde{\boldsymbol{v}}, \tilde{\boldsymbol{u}}\right)\right] - \lambda_1 \boldsymbol{c}^\top \boldsymbol{x}_2^\star > \inf_{\mathbb{P}\in\mathbb{F}(\boldsymbol{s}_2^\star)} \mathbb{E}_{\mathbb{P}}\left[\phi\left(\boldsymbol{x}_2^\star, \tilde{\boldsymbol{v}}, \tilde{\boldsymbol{u}}\right)\right] - \lambda_2 \boldsymbol{c}^\top \boldsymbol{x}_2^\star \geqslant \tau.$$

根据上述不等式, 我们可知当 $\lambda = \lambda_1$ 时, $((\boldsymbol{s}_2^\star, (1-\lambda_1)\boldsymbol{c}^\top \boldsymbol{x}_2^\star), \boldsymbol{x}_2^\star)$ 是问题 (6.6)-(6.7) 的一个可行解. 根据定义, 当 $\lambda = \lambda_1$ 时, $(\boldsymbol{s}_1^\star, \xi_1^\star = (1-\lambda_1)\boldsymbol{c}^\top \boldsymbol{x}_1^\star)$ 是地方政府 (LA) 问题 (6.7) 的最优解, 则有如下不等式:

$$(1-\lambda_1)\boldsymbol{c}^\top \boldsymbol{x}_2^\star + \sup_{\mathbb{P}\in\mathbb{F}(\boldsymbol{s}_2^\star)} \mathbb{E}_{\mathbb{P}}\left[\ell(\boldsymbol{s}_2^\star, \tilde{\boldsymbol{v}})\right] \geqslant \xi_1^\star + \sup_{\mathbb{P}\in\mathbb{F}(\boldsymbol{s}_1^\star)} \mathbb{E}_{\mathbb{P}}\left[\ell(\boldsymbol{s}_1^\star, \tilde{\boldsymbol{v}})\right].$$

因为 $\xi_2^\star = (1-\lambda_2)\boldsymbol{c}^\top \boldsymbol{x}_2^\star$, 所以, 可以推出如下不等式:

$$\sup_{\mathbb{P}\in\mathbb{F}(\boldsymbol{s}_2^\star)} \mathbb{E}_{\mathbb{P}}\left[\ell(\boldsymbol{s}_2^\star, \tilde{\boldsymbol{v}})\right] \geqslant \sup_{\mathbb{P}\in\mathbb{F}(\boldsymbol{s}_1^\star)} \mathbb{E}_{\mathbb{P}}\left[\ell(\boldsymbol{s}_1^\star, \tilde{\boldsymbol{v}})\right] + \left[\xi_1^\star - \xi_2^\star \left(\frac{1-\lambda_1}{1-\lambda_2}\right)\right],$$

成立. 因此不等式 (6.9) 成立.

定理 6.1 的证明

假设 $\boldsymbol{s}^\star$ 和 $\boldsymbol{x}^\star$ 为模型 (6.10) 的最优解. 首先, 我们用反证法证明结论 (1). 分别用 $\mathcal{Z}^\star$ 和 $\mathcal{Z}^{\star\star}$ 表示问题 (6.10) 和博弈问题 (6.6)-(6.7) 的最优值. 因为问题 (6.10) 关于变量 $(\boldsymbol{x}, \boldsymbol{s})$ 的可行集包含领导者-跟随者问题 (6.6)-(6.7) 关于变量 $(\boldsymbol{x}, \boldsymbol{s})$ 的可行集, 所以如下不等式成立:

$$\mathcal{Z}^\star \leqslant \mathcal{Z}^{\star\star}. \tag{A.68}$$

此外, 根据问题结构, 如果 $\boldsymbol{s}^\star$ 是问题 (6.10) 的最优解, 则它必是博弈模型 (6.6)-(6.7) 的最优解. 然后, 我们只需要证明 $\boldsymbol{x}^\star$ 同样是博弈模型 (6.6)-(6.7) 的最优解.

我们用反证法证明. 假设 $(\boldsymbol{s}^\star, \boldsymbol{x}^\star)$ 满足 $\xi^\star = (1-\lambda)\boldsymbol{c}^\top \boldsymbol{x}^\star$ 且不是问题 (6.6)-(6.7) 的最优解. 则必存在另一个最优选址, 记为 $\boldsymbol{x}^\sharp \in \{0,1\}^J$, 其中 $\boldsymbol{x}^\sharp \neq \boldsymbol{x}^\star$, 对于问题 (6.6)-(6.7), 其满足如下不等式:

$$
\begin{aligned}
& -\lambda \boldsymbol{c}^\top \boldsymbol{x}^\sharp + \inf_{\mathbb{P}\in\mathbb{F}(\boldsymbol{s}^\star)} \mathbb{E}_{\mathbb{P}}\left[\phi\left(\boldsymbol{x}^\sharp, \tilde{\boldsymbol{v}}, \tilde{\boldsymbol{u}}\right)\right] \\
= & \max_{\boldsymbol{x}\in\{0,1\}^J} \left\{ -\lambda \boldsymbol{c}^\top \boldsymbol{x} + \inf_{\mathbb{P}\in\mathbb{F}(\boldsymbol{s}^\star)} \mathbb{E}_{\mathbb{P}}\left[\phi\left(\boldsymbol{x}, \tilde{\boldsymbol{v}}, \tilde{\boldsymbol{u}}\right)\right] \middle| (1-\lambda)\boldsymbol{c}^\top \boldsymbol{x} \leqslant \xi^\star \right\} \\
> & -\lambda \boldsymbol{c}^\top \boldsymbol{x}^\star + \inf_{\mathbb{P}\in\mathbb{F}(\boldsymbol{s}^\star)} \mathbb{E}_{\mathbb{P}}\left[\phi\left(\boldsymbol{x}^\star, \tilde{\boldsymbol{v}}, \tilde{\boldsymbol{u}}\right)\right] \geqslant \tau.
\end{aligned} \tag{A.69}
$$

不等式 (A.69) 成立的原因是: 如果有

$$
\begin{aligned}
& -\lambda \boldsymbol{c}^\top \boldsymbol{x}^\star + \inf_{\mathbb{P}\in\mathbb{F}(\boldsymbol{s}^\star)} \mathbb{E}_{\mathbb{P}}\left[\phi\left(\boldsymbol{x}^\star, \tilde{\boldsymbol{v}}, \tilde{\boldsymbol{u}}\right)\right] \\
\geqslant & \max_{\boldsymbol{x}\in\{0,1\}^J} \left\{ -\lambda \boldsymbol{c}^\top \boldsymbol{x} + \inf_{\mathbb{P}\in\mathbb{F}(\boldsymbol{s}^\star)} \mathbb{E}_{\mathbb{P}}\left[\phi\left(\boldsymbol{x}, \tilde{\boldsymbol{v}}, \tilde{\boldsymbol{u}}\right)\right] \middle| (1-\lambda)\boldsymbol{c}^\top \boldsymbol{x} \leqslant \xi^\star \right\},
\end{aligned}
$$

再结合不等式 (A.68), 可得 $(\boldsymbol{s}^\star, \xi^\star = (1-\lambda)\boldsymbol{c}^\top \boldsymbol{x}^\star)$ 和 $\boldsymbol{x}^\star$ 构成了博弈问题 (6.6)-(6.7) 的最优解. 这与我们之前关于 $(\boldsymbol{s}^\star, \xi^\star = (1-\lambda)\boldsymbol{c}^\top \boldsymbol{x}^\star, \boldsymbol{x}^\star)$ 不是问题 (6.6)-(6.7) 最优解的假设相矛盾. 因此, 在 (6.10) 的最优解 $(\boldsymbol{s}^\star, \boldsymbol{x}^\star)$ 所构建的解 $(\boldsymbol{s}^\star,\ \xi^\star = (1-\lambda)\boldsymbol{c}^\top \boldsymbol{x}^\star, \boldsymbol{x}^\star)$ 不是领导者-跟随者模型 (6.6)-(6.7) 最优解的假设下, 最优解 $\boldsymbol{x}^\sharp \in \{0,1\}^J$ 对于领导者-跟随者模型 (6.6)-(6.7) 存在有效约束 (A.69).

接下来, 根据解 $(\boldsymbol{s}^\star, \boldsymbol{x}^\sharp)$ 的最优性, 以及针对领导者-跟随者模型 (6.6)-(6.7) 满足条件 $\xi^\sharp = (1-\lambda)\boldsymbol{c}^\top \boldsymbol{x}^\sharp$ 和结论 (1) 的条件, 即 $\boldsymbol{c}^\top \boldsymbol{x} \neq \boldsymbol{c}^\top \boldsymbol{x}'$, $\forall\, \boldsymbol{x}, \boldsymbol{x}' \in \{0,1\}^J, \boldsymbol{x} \neq \boldsymbol{x}'$, 那么我们有

$$
(1-\lambda)\boldsymbol{c}^\top \boldsymbol{x}^\sharp < \xi^\star = (1-\lambda)\boldsymbol{c}^\top \boldsymbol{x}^\star,
$$

这表明 $(\boldsymbol{s}^\star, \boldsymbol{x}^\sharp)$ 对应问题 (6.10) 的目标函数值比 $(\boldsymbol{s}^\star, \boldsymbol{x}^\star)$ 对应问题 (6.10) 的目标函数值更小. 这与 $(\boldsymbol{s}^\star, \boldsymbol{x}^\star)$ 是问题 (6.10) 的最优解相矛盾. 因此, 假设 $(\boldsymbol{s}^\star, \boldsymbol{x}^\sharp)$ 与 $\xi^\sharp = (1-\lambda)\boldsymbol{c}^\top \boldsymbol{x}^\sharp$, 其中 $\boldsymbol{x}^\sharp \neq \boldsymbol{x}^\star$, 是领导者-跟随者模型 (6.6)-(6.7) 的最优解不成立. 进而证明了结论 (1).

对于结论 (2), 选址决策 $\boldsymbol{x}^{\star\star}$ 仅是寻找如下跟随者问题

$$
\max_{\boldsymbol{x}\in\{0,1\}^J} -\lambda \boldsymbol{c}^\top \boldsymbol{x} + \inf_{\mathbb{P}\in\mathbb{F}(\boldsymbol{s}^\star)} \mathbb{E}_{\mathbb{P}}\left[\phi\left(\boldsymbol{x}, \tilde{\boldsymbol{v}}, \tilde{\boldsymbol{u}}\right)\right]
$$

的最优值, 同时保持领导者问题的最优值

$$
(1-\lambda)\boldsymbol{c}^\top \boldsymbol{x}^\star + \sup_{\mathbb{P}\in\mathbb{F}(\boldsymbol{s}^\star)} \mathbb{E}_{\mathbb{P}}\left[\ell(\boldsymbol{s}^\star, \tilde{\boldsymbol{v}})\right]
$$

不变. 因此, 它是领导者-追随者问题 (6.6)-(6.7) 的解.

命题 6.2 的证明

根据定理 6.1, 可以将参数 $\lambda \in [0,1]$ 看作问题 (6.10) 的一个决策变量. 我们可以通过考虑如下约束的上境图:

$$(1-\lambda)\boldsymbol{c}^\top \boldsymbol{x} \leqslant \xi_{\mathrm{LA}}, \tag{A.70}$$

以及

$$\lambda \boldsymbol{c}^\top \boldsymbol{x} \leqslant \xi_{\mathrm{PO}} \tag{A.71}$$

来提升 (lift) 问题. 然后分别在目标函数和约束 (6.10) 中替换 $(1-\lambda)\boldsymbol{c}^\top \boldsymbol{x}$ 为 ξ_{LA}, 且替换 $\lambda \boldsymbol{c}^\top \boldsymbol{x}$ 为 ξ_{PO}. 此外, 约束 (A.70) 和 (A.71) 与 $\lambda \in [0,1]$ 等价于约束

$$\boldsymbol{c}^\top \boldsymbol{x} \leqslant \xi_{\mathrm{LA}} + \xi_{\mathrm{PO}}, \quad \xi_{\mathrm{LA}} \in \mathbb{R}_+,\ \xi_{\mathrm{PO}} \in \mathbb{R}_+,$$

进而消去了变量 λ. 因此, 我们建立了以 $\lambda \in [0,1]$ 为决策变量的问题 (6.10) 与问题 (6.15) 之间的等价关系, 以及最优解之间的关系:

$$\lambda^\star = \frac{\xi_{\mathrm{PO}}^\star}{\xi_{\mathrm{LA}}^\star + \xi_{\mathrm{PO}}^\star}.$$

给定 $\lambda^\star$ 下, 私人运营商 (PO) 问题的最佳选址决策 $\boldsymbol{x}^\star$ 可以使用定理 6.1 的结论 (2) 获得. 证明完毕.

命题 6.3 的证明

极值概率分布下期望分类成本 (6.21) 可以表示为如下最大化问题:

$$\sup_{\mathbb{P}} \left\{ \mathbb{E}_{\mathbb{P}}\left[\sum_{i\in[I]} \sum_{h\in[H]} (\varrho_{ih}\tilde{v}_i + \iota_{ih}) s_{ih} \right] \middle| \mathbb{E}_{\mathbb{P}}\left[\tilde{v}_i\right] \in \left[\boldsymbol{\alpha}_i^\top \boldsymbol{s}_i, \boldsymbol{\beta}_i^\top \boldsymbol{s}_i\right], \forall i \in [I]; \mathbb{P}\left[\tilde{\boldsymbol{v}} \in \mathcal{V}\right] = 1 \right\}. \tag{A.72}$$

接下来, 我们考虑通过求解如下优化问题来获得极值情况下的 $\boldsymbol{v}^*$:

$$\max_{\boldsymbol{v}} \left\{ \sum_{i\in[I]} \sum_{h\in[H]} (\varrho_{ih} v_i + \iota_{ih}) s_{ih} \middle| v_i \in \left[\boldsymbol{\alpha}_i^\top \boldsymbol{s}_i, \boldsymbol{\beta}_i^\top \boldsymbol{s}_i\right], \forall i \in [I]; \boldsymbol{v} \in \mathcal{V} \right\}. \tag{A.73}$$

首先, 我们有

$$\sum_{i\in[I]} \sum_{h\in[H]} (\varrho_{ih} v_i^* + \iota_{ih}) s_{ih} \geqslant \sum_{i\in[I]} \sum_{h\in[H]} (\varrho_{ih} \mathbb{E}_{\mathbb{P}}[\tilde{v}_i] + \iota_{ih}) s_{ih}, \quad \forall \mathbb{P} \in \mathbb{F}_{\tilde{\boldsymbol{v}}}(\boldsymbol{s}).$$

其次, 构造分布 $\mathbb{P}^*$ 满足条件

$$\mathbb{P}^*\{\tilde{\boldsymbol{v}} = \boldsymbol{v}^*\} = 1,$$

则有 $\mathbb{P}^* \in \mathbb{F}_{\tilde{\boldsymbol{v}}}(\boldsymbol{s})$, 以及

$$\sum_{i\in[I]}\sum_{h\in[H]}(\varrho_{ih}\mathbb{E}_{\mathbb{P}^*}[\tilde{v}_i] + \iota_{ih})s_{ih} = \sum_{i\in[I]}\sum_{h\in[H]}(\varrho_{ih}v_i^* + \iota_{ih})s_{ih}.$$

结合第一个不等式和上述等式, 可得 $\mathbb{P}^*$ 是极值概率分布下期望分类成本问题 (A.72) 的极值概率分布, 且该问题最优值与问题 (A.73) 的最优值相等. 最后, 根据定义 (6.17), 收集点 i 原料产量的凸包可以转化为如下多面体形式:

$$\mathcal{V} := \left\{\boldsymbol{v}^0 + \sum_{k\in[K]}\gamma^k\boldsymbol{\Delta}^k \,\middle|\, \sum_{k\in[K]}\gamma^k = 1, \boldsymbol{\gamma}\in\mathbb{R}_+^K\right\}.$$

因此, 命题 6.3 的结果成立.

命题 6.4 的证明

极值概率分布下私人运营商的期望运营收益 (6.22) 可以转化为如下矩问题:

$$\begin{aligned}
\inf_{\mathbb{P}} \quad & \int_{(\tilde{\boldsymbol{v}},\tilde{\boldsymbol{u}})\in\mathcal{V}\times\mathcal{U}(\boldsymbol{s})} \phi(\boldsymbol{x},\tilde{\boldsymbol{v}},\tilde{\boldsymbol{u}})\,\mathrm{d}\mathbb{P}(\tilde{\boldsymbol{v}},\tilde{\boldsymbol{u}}) \\
\text{s.t.} \quad & \int_{(\tilde{\boldsymbol{v}},\tilde{\boldsymbol{u}})\in\mathcal{V}\times\mathcal{U}(\boldsymbol{s})} \tilde{v}_i\mathrm{d}\mathbb{P}(\tilde{\boldsymbol{v}},\tilde{\boldsymbol{u}}) \geqslant \boldsymbol{\alpha}_i^\top\boldsymbol{s}_i, \ \forall i\in[I], \\
& \int_{(\tilde{\boldsymbol{v}},\tilde{\boldsymbol{u}})\in\mathcal{V}\times\mathcal{U}(\boldsymbol{s})} \tilde{v}_i\mathrm{d}\mathbb{P}(\tilde{\boldsymbol{v}},\tilde{\boldsymbol{u}}) \leqslant \boldsymbol{\beta}_i^\top\boldsymbol{s}_i, \ \forall i\in[I], \\
& \int_{(\tilde{\boldsymbol{v}},\tilde{\boldsymbol{u}})\in\mathcal{V}\times\mathcal{U}(\boldsymbol{s})} \boldsymbol{G}\tilde{\boldsymbol{u}}_i\mathrm{d}\mathbb{P}(\tilde{\boldsymbol{v}},\tilde{\boldsymbol{u}}) \geqslant \boldsymbol{F}\boldsymbol{s}_i, \ \forall i\in[I], \\
& \int_{(\tilde{\boldsymbol{v}},\tilde{\boldsymbol{u}})\in\mathcal{V}\times\mathcal{U}(\boldsymbol{s})} \mathrm{d}\mathbb{P}(\tilde{\boldsymbol{v}},\tilde{\boldsymbol{u}}) = 1.
\end{aligned} \tag{A.74}$$

利用矩问题的强对偶定理 (Wiesemann et al., 2014), 问题 (A.74) 等价于如下优化问题:

$$\begin{aligned}
\max_{\boldsymbol{q},\underline{\boldsymbol{\delta}},\overline{\boldsymbol{\delta}},\pi} \quad & \pi + \sum_{i\in[I]}\boldsymbol{q}_i^\top\boldsymbol{F}\boldsymbol{s}_i + \sum_{i\in[I]}\overline{\delta}_i(\boldsymbol{\alpha}_i^\top\boldsymbol{s}_i) + \sum_{i\in[I]}\underline{\delta}_i(\boldsymbol{\beta}_i^\top\boldsymbol{s}_i) \\
\text{s.t.} \quad & \min_{\boldsymbol{v}\,\in\,\mathcal{V},\,\boldsymbol{u}\,\in\,\mathcal{U}(\boldsymbol{s})}\left[\phi(\boldsymbol{x},\boldsymbol{v},\boldsymbol{u}) - \sum_{i\in[I]}\boldsymbol{q}_i^\top\boldsymbol{G}\boldsymbol{u}_i - \sum_{i\in[I]}(\overline{\delta}_i + \underline{\delta}_i)v_i\right] \geqslant \pi,
\end{aligned}$$

$$\boldsymbol{q}\in\mathbb{R}_+^{L_1\times I},\underline{\boldsymbol{\delta}}\in\mathbb{R}_-^I,\overline{\boldsymbol{\delta}}\in\mathbb{R}_+^I,\pi\in\mathbb{R}.$$

注意到

$$\begin{aligned}
&\min_{\boldsymbol{v}\in\mathcal{V},\boldsymbol{u}\in\mathcal{U}(\boldsymbol{s})}\left[\phi(\boldsymbol{x},\boldsymbol{v},\boldsymbol{u})-\sum_{i\in[I]}\boldsymbol{q}_i^\top\boldsymbol{G}\boldsymbol{u}_i-\sum_{i\in[I]}(\overline{\delta}_i+\underline{\delta}_i)v_i\right]\geqslant\pi\\
\Leftrightarrow&\min_{\boldsymbol{v}\in\mathcal{V}}\left[\min_{\boldsymbol{u}\in\mathcal{U}(\boldsymbol{s})}\left[\phi(\boldsymbol{x},\boldsymbol{v},\boldsymbol{u})-\sum_{i\in[I]}\boldsymbol{q}_i^\top\boldsymbol{G}\boldsymbol{u}_i-\sum_{i\in[I]}(\overline{\delta}_i+\underline{\delta}_i)v_i\right]\right]\geqslant\pi\\
\Leftrightarrow&\min_{\boldsymbol{v}\in\mathcal{V}}\left[\max_{\boldsymbol{y}\in\{(6.5)\text{ 的可行域}\}}\left[\min_{\boldsymbol{u}\in\mathcal{U}(\boldsymbol{s})}\sum_{i\in[I]}\sum_{j\in[J]}\sum_{n\in[N]}r_{ijn}u_{in}y_{ij}-\sum_{i\in[I]}\boldsymbol{q}_i^\top\boldsymbol{G}\boldsymbol{u}_i\right]\right.\\
&\left.+c_D\sum_{i\in[I]}\sum_{j\in[J]}y_{ij}-c_D\boldsymbol{e}_I^\top\boldsymbol{v}-\sum_{i\in[I]}(\overline{\delta}_i+\underline{\delta}_i)v_i\right]\geqslant\pi,
\end{aligned}\tag{A.75}$$

其中, $\boldsymbol{e}_I\in\mathbb{R}^I$ 是元素全为 1 的向量. 进一步, 我们将上述问题的内层最小化问题转化为

$$\begin{aligned}
&\min_{\boldsymbol{u}\in\mathcal{U}(\boldsymbol{s})}\sum_{i\in[I]}\sum_{j\in[J]}\sum_{n\in[N]}r_{ijn}u_{in}y_{ij}-\sum_{i\in[I]}\boldsymbol{q}_i^\top\boldsymbol{G}\boldsymbol{u}_i\\
=&\min_{\boldsymbol{u}\in\mathcal{U}(\boldsymbol{s})}\sum_{i\in[I]}\boldsymbol{u}_i^\top\mathbf{R}_i\boldsymbol{y}_i-\sum_{i\in[I]}\boldsymbol{u}_i^\top\boldsymbol{G}^\top\boldsymbol{q}_i.
\end{aligned}\tag{A.76}$$

根据由式 (6.19) 定义的 $\mathcal{U}(\boldsymbol{s}):=\{\boldsymbol{u}\in\mathbb{R}_+^{I\times N}\mid\boldsymbol{A}\boldsymbol{u}_i\geqslant\boldsymbol{B}\boldsymbol{s}_i,\forall i\in[I];\ \sum_{n\in[N]}u_{in}=1,\forall i\in[I]\}$, 问题 (A.76) 的对偶问题具有如下形式:

$$\max_{\substack{\boldsymbol{A}^\top\boldsymbol{h}_i+t_i\boldsymbol{e}_N+\boldsymbol{G}^\top\boldsymbol{q}_i-\mathbf{R}_i\boldsymbol{y}_i\leqslant 0,\forall i\in[I]\\ \boldsymbol{h}\in\mathbb{R}_+^{L_2\times I},\boldsymbol{t}\in\mathbb{R}^I}}\sum_{i\in[I]}\boldsymbol{h}_i^\top\boldsymbol{B}\boldsymbol{s}_i+\sum_{i\in[I]}t_i,\tag{A.77}$$

其中 $\boldsymbol{e}_N\in\mathbb{R}^N$ 是元素全为 1 的向量.

现在, 我们可以将内层最大最小化问题转化为如下最大化问题:

$$\max_{\boldsymbol{y},\boldsymbol{h},\boldsymbol{t}}\quad\sum_{i\in[I]}\boldsymbol{h}_i^\top\boldsymbol{B}\boldsymbol{s}_i+\sum_{i\in[I]}t_i+c_D\sum_{i\in[I]}\sum_{j\in[J]}y_{ij}\tag{A.78}$$

$$\text{s.t.}\quad\mathbf{R}_i\boldsymbol{y}_i-\boldsymbol{A}^\top\boldsymbol{h}_i-t_i\boldsymbol{e}_N\geqslant\boldsymbol{G}^\top\boldsymbol{q}_i,\ \forall i\in[I],\tag{A.79}$$

$$\sum_{j\in[J]}y_{ij}\leqslant v_i,\ \forall i\in[I],\tag{A.80}$$

$$\sum_{i\in[I]} y_{ij} \leqslant x_j \varpi_j,\ \forall j \in [J], \tag{A.81}$$

$$\boldsymbol{y} \in \mathbb{R}_+^{I\times J}, \boldsymbol{h} \in \mathbb{R}_+^{L_2\times I}, \boldsymbol{t} \in \mathbb{R}^I \tag{A.82}$$

上述问题的对偶问题为

$$\min_{\boldsymbol{z},\boldsymbol{\kappa},\boldsymbol{d}} \quad \sum_{i\in[I]} \boldsymbol{q}_i^\top \boldsymbol{G}\boldsymbol{z}_i + \boldsymbol{v}^\top \boldsymbol{\kappa} + \boldsymbol{x}^\top \mathrm{Diag}(\boldsymbol{\varpi})\boldsymbol{d} \tag{A.83}$$

$$\text{s.t.} \quad \boldsymbol{d} + \kappa_i \boldsymbol{e}_J + \mathbf{R}_i^\top \boldsymbol{z}_i \geqslant c_D \boldsymbol{e}_J,\ \forall i \in [I], \tag{A.84}$$

$$\boldsymbol{B}\boldsymbol{s}_i + \boldsymbol{A}\boldsymbol{z}_i \leqslant \boldsymbol{0},\ \forall i \in [I], \tag{A.85}$$

$$1 + \boldsymbol{e}_N^\top \boldsymbol{z}_i = 0,\ \forall i \in [I], \tag{A.86}$$

$$\boldsymbol{z} \in \mathbb{R}_-^{N\times I}, \boldsymbol{\kappa} \in \mathbb{R}_+^I, \boldsymbol{d} \in \mathbb{R}_+^J, \tag{A.87}$$

其中, $\boldsymbol{e}_J \in \mathbb{R}^J$ 是元素全为 1 的向量. 所以, 约束 (A.75) 可以等价地写为

$$\min_{\boldsymbol{v}\in\mathcal{V}} \min_{\substack{\boldsymbol{z},\,\boldsymbol{\kappa},\,\boldsymbol{d}\,\in \\ \text{(A.84)-(A.87)}}} \left[\sum_{i\in[I]} \boldsymbol{q}_i^\top \boldsymbol{G}\boldsymbol{z}_i + \boldsymbol{v}^\top \boldsymbol{\kappa} + \boldsymbol{x}^\top \mathrm{Diag}(\boldsymbol{\varpi})\boldsymbol{d}\right] - c_D \boldsymbol{e}_I^\top \boldsymbol{v} - (\overline{\boldsymbol{\delta}} + \underline{\boldsymbol{\delta}})^\top \boldsymbol{v} \geqslant \pi, \tag{A.88}$$

上式又等价于

$$\min_{\substack{\boldsymbol{z},\,\boldsymbol{\kappa},\,\boldsymbol{d}\,\in \\ \text{(A.84)-(A.87)}}} \sum_{i\in[I]} \boldsymbol{q}_i^\top \boldsymbol{G}\boldsymbol{z}_i + \boldsymbol{x}^\top \mathrm{Diag}(\boldsymbol{\varpi})\boldsymbol{d} + \min_{\boldsymbol{v}\in\mathcal{V}} \left[\boldsymbol{v}^\top \boldsymbol{\kappa} - c_D \boldsymbol{e}_I^\top \boldsymbol{v} - (\overline{\boldsymbol{\delta}} + \underline{\boldsymbol{\delta}})^\top \boldsymbol{v}\right] \geqslant \pi. \tag{A.89}$$

最后, 因为支撑集 $\mathcal{V}$ 是基于不同收集点和时间段的有限原料产量观测样本的凸包, 即 $\mathcal{V} := \mathrm{conv}\{\boldsymbol{v}^k, k \in [K]\}$, 其中 $\boldsymbol{v}^k := \boldsymbol{v}^0 + \boldsymbol{\Delta}^k, \forall k \in [K]$, 所以问题 (A.89) 等价于如下最小化问题:

$$\begin{aligned} \min_{\boldsymbol{z},\boldsymbol{\kappa},\boldsymbol{d}} \quad & \sum_{i\in[I]} \boldsymbol{q}_i^\top \boldsymbol{G}\boldsymbol{z}_i^k + \boldsymbol{x}^\top \mathrm{Diag}(\boldsymbol{\varpi})\boldsymbol{d}^k \\ & + \left[(\boldsymbol{v}^k)^\top \boldsymbol{\kappa}^k - c_D \boldsymbol{e}_I^\top \boldsymbol{v}^k - (\overline{\boldsymbol{\delta}} + \underline{\boldsymbol{\delta}})^\top \boldsymbol{v}^k\right] \geqslant \pi, \forall k \in [K] \end{aligned}$$

$$\text{s.t.} \quad \boldsymbol{d}^k + \kappa_i^k \boldsymbol{e}_J + \mathbf{R}_i^\top \boldsymbol{z}_i^k \geqslant c_D \boldsymbol{e}_J,\ \forall i \in [I], k \in [K], \tag{A.90}$$

$$\boldsymbol{B}\boldsymbol{s}_i + \boldsymbol{A}\boldsymbol{z}_i^k \leqslant \boldsymbol{0},\ \forall i \in [I], k \in [K], \tag{A.91}$$

$$1 + \boldsymbol{e}_N^\top \boldsymbol{z}_i^k = 0,\ \forall i \in [I], k \in [K], \tag{A.92}$$

$$\boldsymbol{z} \in \mathbb{R}_-^{I\times N\times K}, \boldsymbol{\kappa} \in \mathbb{R}_+^{I\times K}, \boldsymbol{d} \in \mathbb{R}_+^{J\times K}, \tag{A.93}$$

上述问题可以等价地写为如下问题:

$$\min_{\substack{\boldsymbol{z},\boldsymbol{\kappa},\boldsymbol{d}\in \\ (\text{A.90})\text{-}(\text{A.93})}} \left[\sum_{i\in[I]} \boldsymbol{q}_i^\top \boldsymbol{G}\boldsymbol{z}_i^k + (\boldsymbol{v}^k)^\top \boldsymbol{\kappa}^k + \boldsymbol{x}^\top \mathrm{Diag}(\boldsymbol{\varpi})\boldsymbol{d}^k\right]$$
$$- c_D \boldsymbol{e}_I^\top \boldsymbol{v}^k - (\overline{\boldsymbol{\delta}} + \underline{\boldsymbol{\delta}})^\top \boldsymbol{v}^k \geqslant \pi,\ \forall k \in [K].$$

对上述问题求对偶, 可以得到如下最大化问题:

$$\max_{\substack{\boldsymbol{\nu},\boldsymbol{p},\boldsymbol{\chi}\in \\ (6.28)\text{-}(6.30)}} \left[\sum_{i\in[I]} c_D \boldsymbol{e}_J^\top \boldsymbol{\nu}_i^k - \sum_{i\in[I]} (\boldsymbol{B}\boldsymbol{s}_i)^\top \boldsymbol{p}_i^k - \boldsymbol{e}_I^\top \boldsymbol{\chi}^k\right]$$
$$- c_D \boldsymbol{e}_I^\top \boldsymbol{v}^k - (\overline{\boldsymbol{\delta}} + \underline{\boldsymbol{\delta}})^\top \boldsymbol{v}^k \geqslant \pi,\ \forall k \in [K],$$
$$\boldsymbol{\nu} \in \mathbb{R}_+^{I\times J\times K},\ \boldsymbol{p} \in \mathbb{R}_-^{I\times L_2\times K},\ \boldsymbol{\chi} \in \mathbb{R}^{I\times K}.$$

因此, 命题 6.4 的结果成立.

推论 6.1 的证明

给定问题 (6.10) 的最优解 $(\boldsymbol{s}^\star, \boldsymbol{x}^\star)$, 利用命题 6.1 和命题 6.4 结果, 最优选址 $\boldsymbol{x}^{\star\star}$ 可以通过求解如下问题得到:

$$\max_{\boldsymbol{x}\in\{0,1\}^J} \left\{ \max_{\boldsymbol{q},\underline{\boldsymbol{\delta}},\overline{\boldsymbol{\delta}},\pi} \pi + \sum_{i\in[I]} \boldsymbol{q}_i^\top \boldsymbol{F}\boldsymbol{s}_i^\star + \sum_{i\in[I]} \overline{\delta}_i(\boldsymbol{\alpha}_i^\top \boldsymbol{s}_i^\star) \right.$$
$$\left. + \sum_{i\in[I]} \underline{\delta}_i(\boldsymbol{\beta}_i^\top \boldsymbol{s}_i^\star) - \lambda \boldsymbol{c}^\top \boldsymbol{x} \ \middle|\ (6.27)\text{-}(6.31), \boldsymbol{c}^\top \boldsymbol{x} \leqslant \boldsymbol{c}^\top \boldsymbol{x}^\star \right\}.$$

定理 6.2 的证明

利用命题 6.3 的结果, 给定垃圾分类方案 $\boldsymbol{s}$, 问题 (6.21) 等价于如下最大化问题:

$$\begin{aligned} \max_{\boldsymbol{\gamma}} \quad & \sum_{i\in[I]} \sum_{h\in[H]} \left[\varrho_{ih}\left(v_i^0 + \sum_{k\in[K]} \gamma^k \Delta_i^k\right) + \iota_{ih}\right] s_{ih} \\ \text{s.t.} \quad & v_i^0 + \sum_{k\in[K]} \gamma^k \Delta_i^k \in \left[\boldsymbol{\alpha}_i^\top \boldsymbol{s}_i, \boldsymbol{\beta}_i^\top \boldsymbol{s}_i\right], \forall i \in [I], && \text{(A.94)} \\ & \sum_{k\in[K]} \gamma^k = 1, && \text{(A.95)} \end{aligned}$$

$$\boldsymbol{\gamma} \in \mathbb{R}_+^K. \tag{A.96}$$

因此, 问题 (6.10), 即

$$\begin{aligned} \min_{\boldsymbol{s},\boldsymbol{x}} \quad & (1-\lambda)\boldsymbol{c}^\top \boldsymbol{x} + \sup_{\mathbb{P}\in\mathbb{F}(\boldsymbol{s})} \mathbb{E}_{\mathbb{P}}\left[\ell(\boldsymbol{s},\tilde{\boldsymbol{v}})\right] \\ \text{s.t.} \quad & \mathbb{E}_{\mathbb{P}}\left[\phi\left(\boldsymbol{x},\tilde{\boldsymbol{v}},\tilde{\boldsymbol{u}}\right)\right] - \lambda \boldsymbol{c}^\top \boldsymbol{x} \geqslant \tau, \forall \mathbb{P} \in \mathbb{F}(\boldsymbol{s}), \qquad (\text{A.97}) \\ & \boldsymbol{s} \in \mathcal{S}, \boldsymbol{x} \in \{0,1\}^J \qquad (\text{A.98}) \end{aligned}$$

等价于如下最小化问题:

$$\min_{\boldsymbol{s},\boldsymbol{x}\in(\text{A.97})\text{-}(\text{A.98})} \quad (1-\lambda)\boldsymbol{c}^\top \boldsymbol{x} + \max_{\boldsymbol{\gamma}\in(\text{A.94})\text{-}(\text{A.96})} \sum_{i\in[I]}\sum_{h\in[H]} \left[\varrho_{ih}\left(v_i^0 + \boldsymbol{\Delta}_i^\top \boldsymbol{\gamma}\right) + \iota_{ih}\right] s_{ih},$$

其中, $\boldsymbol{\Delta}_i = (\Delta_i^1, \cdots, \Delta_i^K)^\top$, 上述问题可以重新写为

$$\min_{\boldsymbol{s},\boldsymbol{x}\in(\text{A.97})\text{-}(\text{A.98})} (1-\lambda)\boldsymbol{c}^\top \boldsymbol{x} + \sum_{i\in[I]} \left(v_i^0 \boldsymbol{\varrho}_i^\top + \boldsymbol{\iota}_i^\top\right) \boldsymbol{s}_i + \max_{\boldsymbol{\gamma}\in(\text{A.94})\text{-}(\text{A.96})} \sum_{i\in[I]} \boldsymbol{\varrho}_i^\top \boldsymbol{s}_i \boldsymbol{\Delta}_i^\top \boldsymbol{\gamma}. \tag{A.99}$$

对内层最大化问题求对偶, 我们得到如下对偶问题:

$$\begin{aligned} \min_{\underline{\boldsymbol{\rho}},\overline{\boldsymbol{\rho}},\zeta} \quad & \sum_{i\in[I]} \underline{\rho}_i(\boldsymbol{\alpha}_i^\top \boldsymbol{s}_i) + \sum_{i\in[I]} \overline{\rho}_i(\boldsymbol{\beta}_i^\top \boldsymbol{s}_i) + \zeta - \left(\underline{\boldsymbol{\rho}} + \overline{\boldsymbol{\rho}}\right)^\top \boldsymbol{v}^0 \\ \text{s.t.} \quad & (\underline{\boldsymbol{\rho}}_i + \overline{\boldsymbol{\rho}}_i)^\top \boldsymbol{\Delta}^k + \zeta \geqslant \sum_{i\in[I]} \boldsymbol{s}_i^\top \boldsymbol{\varrho}_i \Delta_i^k, \ \forall k \in [K]. \qquad (\text{A.100}) \\ & \underline{\boldsymbol{\rho}} \in \mathbb{R}_-^I, \overline{\boldsymbol{\rho}} \in \mathbb{R}_+^I, \zeta \in \mathbb{R}. \qquad (\text{A.101}) \end{aligned}$$

所以, 问题 (6.10) 等价于

$$\begin{aligned} \min_{\boldsymbol{s},\boldsymbol{x},\underline{\boldsymbol{\rho}},\overline{\boldsymbol{\rho}},\zeta} \quad & (1-\lambda)\boldsymbol{c}^\top \boldsymbol{x} + \sum_{i\in[I]} \left[\left(v_i^0 \boldsymbol{\varrho}_i^\top + \boldsymbol{\iota}_i^\top\right) \boldsymbol{s}_i + \underline{\rho}_i(\boldsymbol{\alpha}_i^\top \boldsymbol{s}_i) + \overline{\rho}_i(\boldsymbol{\beta}_i^\top \boldsymbol{s}_i)\right] \\ & + \zeta - \left(\underline{\boldsymbol{\rho}} + \overline{\boldsymbol{\rho}}\right)^\top \boldsymbol{v}^0 \qquad (\text{A.102}) \\ \text{s.t.} \quad & (\text{A.97})\text{-}(\text{A.98}), (\text{A.100}), (\text{A.101}). \end{aligned}$$

利用命题 6.4 的结果, 问题 (A.102) 进一步等价于如下最小化问题:

$$\begin{aligned} \min_{\boldsymbol{s},\boldsymbol{x},\underline{\boldsymbol{\rho}},\overline{\boldsymbol{\rho}},\zeta} \quad & (1-\lambda)\boldsymbol{c}^\top \boldsymbol{x} + \sum_{i\in[I]} \left[\left(v_i^0 \boldsymbol{\varrho}_i^\top + \boldsymbol{\iota}_i^\top\right) \boldsymbol{s}_i + \underline{\rho}_i(\boldsymbol{\alpha}_i^\top \boldsymbol{s}_i) + \overline{\rho}_i(\boldsymbol{\beta}_i^\top \boldsymbol{s}_i)\right] \\ & + \zeta - \left(\underline{\boldsymbol{\rho}} + \overline{\boldsymbol{\rho}}\right)^\top \boldsymbol{v}^0 \qquad (\text{A.103}) \end{aligned}$$

$$
\text{s.t.} \quad (\text{A.98}), (6.27)\text{-}(6.31), (\text{A.100}), (\text{A.101}),
$$

$$
\pi + \sum_{i\in[I]} \boldsymbol{q}_i^\top \boldsymbol{F}\boldsymbol{s}_i + \sum_{i\in[I]} \overline{\delta}_i(\boldsymbol{\alpha}_i^\top \boldsymbol{s}_i) + \sum_{i\in[I]} \underline{\delta}_i(\boldsymbol{\beta}_i^\top \boldsymbol{s}_i) - \lambda \boldsymbol{c}^\top \boldsymbol{x} \geqslant \tau. \quad (\text{A.104})
$$

现在, 我们考虑线性化约束 (6.27), 目标函数 (A.103) 和约束 (A.104) 中的非线性项. 我们定义式 (6.19) 中矩阵 $\boldsymbol{B} \in \mathbb{R}_+^{L_2\times H}$ 的元素为 $b_{\ell h}$. 我们可以将约束 (6.27) 重写为

$$
\sum_{i\in[I]}\sum_{h\in[H]}\left(\sum_{\ell\in[L_2]} b_{\ell h} p_{i\ell}^k\right) s_{ih}
$$

$$
\leqslant \sum_{i\in[I]} c_D \boldsymbol{e}_J^\top \boldsymbol{\nu}_i^k - \boldsymbol{e}_I^\top \boldsymbol{\chi}^k - c_D \boldsymbol{e}_I^\top \boldsymbol{v}^k - (\overline{\boldsymbol{\delta}} + \underline{\boldsymbol{\delta}})^\top \boldsymbol{v}^k - \pi, \quad \forall k \in [K], \quad (\text{A.105})
$$

其中, $\boldsymbol{e}_I \in \mathbb{R}^I$ 和 $\boldsymbol{e}_J \in \mathbb{R}^J$ 是元素全为 1 的向量. 注意到 $\sum_{h\in[H]} s_{ih} = 1, \forall i \in [I]$, 通过引入变量 $\boldsymbol{\psi} \in \mathbb{R}^{I\times K}$ 和一个充分大的正数 M, 上述约束等价于如下约束:

$$
\begin{cases}
\displaystyle\sum_{\ell\in[L_2]} b_{\ell h} p_{i\ell}^k + M(s_{ih} - 1) \leqslant \psi_i^k, \ \forall i \in [I], h \in [H], k \in [K], \\
\displaystyle\sum_{i\in[I]} \psi_i^k \leqslant \sum_{i\in[I]} c_D \boldsymbol{e}_J^\top \boldsymbol{\nu}_i^k - \boldsymbol{e}_I^\top \boldsymbol{\chi}^k - c_D \boldsymbol{e}_I^\top \boldsymbol{v}^k - (\overline{\boldsymbol{\delta}} + \underline{\boldsymbol{\delta}})^\top \boldsymbol{v}^k - \pi, \ \forall k \in [K],
\end{cases} \quad (\text{A.106})
$$

也可以写为如下形式:

$$
\begin{cases}
\boldsymbol{b}_h^\top \boldsymbol{p}_i^k + M(s_{ih} - 1) \leqslant \psi_i^k, \ \forall i \in [I], h \in [H], k \in [K], \\
\displaystyle\sum_{i\in[I]} c_D \boldsymbol{e}_J^\top \boldsymbol{\nu}_i^k - \boldsymbol{e}_I^\top \boldsymbol{\chi}^k - c_D \boldsymbol{e}_I^\top \boldsymbol{v}^k - (\overline{\boldsymbol{\delta}} + \underline{\boldsymbol{\delta}})^\top \boldsymbol{v}^k - \pi \geqslant \boldsymbol{e}_I^\top \boldsymbol{\psi}^k, \ \forall k \in [K].
\end{cases} \quad (\text{A.107})
$$

类似地, 我们定义式 (6.18) 中矩阵 $\boldsymbol{F} \in \mathbb{R}_+^{L_1\times H}$ 的元素为 $f_{\ell h}$, 则可以将约束 (A.104) 重写为

$$
\sum_{i\in[I]}\sum_{h\in[H]}\left(\sum_{\ell\in[L_1]} f_{\ell h} q_{i\ell} + \overline{\delta}_i \alpha_{ih} + \underline{\delta}_i \beta_{ih}\right) s_{ih} \geqslant \tau + \lambda \boldsymbol{c}^\top \boldsymbol{x} - \pi. \quad (\text{A.108})
$$

通过引入变量 $\boldsymbol{\varphi} \in \mathbb{R}^I$ 和足够大的正数 M, 约束 (A.108) 等价于如下形式:

$$
\begin{cases}
\boldsymbol{f}_h^\top \boldsymbol{q}_i + \overline{\delta}_i \alpha_{ih} + \underline{\delta}_i \beta_{ih} + M(1 - s_{ih}) \geqslant \varphi_i, \quad \forall i \in [I], h \in [H], \\
\boldsymbol{e}_I^\top \boldsymbol{\varphi} \geqslant \tau + \lambda \boldsymbol{c}^\top \boldsymbol{x} - \pi.
\end{cases} \quad (\text{A.109})
$$

最后, 为了线性化目标函数 (A.103), 我们首先将其重新表述为

$$\min_{\boldsymbol{s},\boldsymbol{x},\underline{\boldsymbol{\rho}},\overline{\boldsymbol{\rho}},\zeta}(1-\lambda)\boldsymbol{c}^\top\boldsymbol{x}+\sum_{i\in[I]}\left(v_i^0\boldsymbol{\varrho}_i^\top+\boldsymbol{\iota}_i^\top\right)\boldsymbol{s}_i$$
$$+\sum_{i\in[I]}\sum_{h\in[H]}\left(\alpha_{ih}\underline{\rho}_i+\beta_{ih}\overline{\rho}_i\right)s_{ih}+\zeta-\left(\underline{\boldsymbol{\rho}}+\overline{\boldsymbol{\rho}}\right)^\top\boldsymbol{v}^0.$$

然后, 对于任意的 $i\in[I]$, 我们用 t_i 表示双线性项 $\sum_{h\in[H]}(\alpha_{ih}\underline{\rho}_i+\beta_{ih}\overline{\rho}_i)s_{ih}$. 其中, 则目标函数中的双线性项

$$\sum_{i\in[I]}\sum_{h\in[H]}\left(\alpha_{ih}\underline{\rho}_i+\beta_{ih}\overline{\rho}_i\right)s_{ih}$$

可以被线性化为如下形式:

$$\begin{cases}\displaystyle\sum_{i\in[I]}t_i,\\ \quad\left(\alpha_{ih}\underline{\rho}_i+\beta_{ih}\overline{\rho}_i\right)-M(1-s_{ih})\leqslant t_i\\ \leqslant\left(\alpha_{ih}\underline{\rho}_i+\beta_{ih}\overline{\rho}_i\right)+M(1-s_{ih}),\ \forall i\in[I],h\in[H],\end{cases}\tag{A.110}$$

其中, M 是充分大的正数.

A.4.2 BRRP 模型中 PPP 合同类型

本节对 6.2 节提出的 BRRP 模型 (6.6)-(6.7) 中三种不同类型的 PPP 合同进行具体分析.

(1) 建设-经营-转让 (BOT) 合同 ($\lambda=1$). 该合同要求私人运营商在合同期内承担建设和运营资源回收系统的所有成本. 有一个关于该类合同的实际案例, 当地政府与一家私人环保股份有限公司签订 BOT 合同, 来建设和运营垃圾焚烧发电厂 (Song et al., 2013). 根据合同到期后, 资源回收系统的所有权是否会从私人运营商转移到地方政府, BOT 合同可分为三种形式, 包括建设-经营-自有 (build-operate-own, BOO) 合同、建设-自有-经营-转让 (build-own-operate-transfer, BOOT) 合同和设计-建设-经营 (design-build-operate, DBO) 合同. 私人运营商负责资源回收系统设施建设和运营的情况对应于 BRRP 模型 (6.6)-(6.7) 的参数 $\lambda=1$, 则模型退化为如下形式:

$$\min_{\boldsymbol{s}}\left\{\sup_{\mathbb{P}\in\mathbb{F}(\boldsymbol{s})}\mathbb{E}_{\mathbb{P}}\left[\ell(\boldsymbol{s},\tilde{\boldsymbol{v}})\right]\ \middle|\ \mathcal{Q}^1_{\mathbb{F}(\boldsymbol{s})}\geqslant\tau,\boldsymbol{s}\in\mathcal{S}\right\},$$

这里

$$\mathcal{Q}^1_{\mathbb{F}(\boldsymbol{s})} = \max_{\boldsymbol{x}\in\{0,1\}^J} \left\{ \inf_{\mathbb{P}\in\mathbb{F}(\boldsymbol{s})} \mathbb{E}_{\mathbb{P}}\left[\phi\left(\boldsymbol{x}, \tilde{\boldsymbol{v}}, \tilde{\boldsymbol{u}}\right)\right] - \boldsymbol{c}^\top \boldsymbol{x} \right\},$$

其中, 地方当局仅通过确定 $\boldsymbol{s}$ 来最小化预期分类成本, 而私人运营商决定设施选址 $\boldsymbol{x}$ 以最大化他的预期收入, 该收入在给定受分类方案影响的原料条件 $\mathbb{F}(\boldsymbol{s})$ 下, 计算总固定成本净额来得到.

(2) 集中管理与控制 (CMC) 合同 ($\lambda = 0$). 该合同要求地方政府全权负责资源回收系统设施的建设. 通过与地方政府签订管理合同, 私人运营商仅在合同期内管理已建设的资源回收系统的运营和维护. 这对应于 BRRP 模型 (6.6)-(6.7) 参数 $\lambda = 0$ 的情况. 在这种情况下, 原料分类和选址的决策变量 $(\boldsymbol{s}, \boldsymbol{x})$ 都由地方政府确定, BRRP 模型实际上变成了地方政府问题:

$$\min_{\boldsymbol{s},\boldsymbol{x}} \left\{ \sup_{\mathbb{P}\in\mathbb{F}(\boldsymbol{s})} \mathbb{E}_{\mathbb{P}}\left[\ell(\boldsymbol{s}, \tilde{\boldsymbol{v}})\right] - \boldsymbol{c}^\top \boldsymbol{x} \;\middle|\; \inf_{\mathbb{P}\in\mathbb{F}(\boldsymbol{s})} \mathbb{E}_{\mathbb{P}}\left[\phi\left(\boldsymbol{x}, \tilde{\boldsymbol{v}}, \tilde{\boldsymbol{u}}\right)\right] \geqslant \tau, \boldsymbol{s} \in \mathcal{S} \right\}.$$

在实际应用中, 这种合同只有在系统建设成本非常低, 或者地方政府预算充足的情况下才可实现.

(3) 特许权协议 (CA) 合同 ($\lambda \in (0,1)$). 特许权协议合同是一种谈判合同, 允许私人运营商根据地方政府的独家许可, 自行承担商业风险投资、管理和维修资源回收设施. 这对应 BRRP 模型 (6.6)-(6.7) 参数 $\lambda \in (0,1)$ 的情况, 双方根据合同中确定的分摊成本支付系统建设的固定成本 $\boldsymbol{c}^\top \boldsymbol{x}$. 该合同允许地方政府在高质量的垃圾分类规划和为私人运营商提供更多的投资之间进行资源配置, 且上述两种方式都能有效提升能源回收效率.

A.4.3 BRRP 模型的等价形式

鲁棒 BRRP 模型 (6.7) 的目标是在私人运营商经济可行性限制 (求解一个两阶段分布鲁棒优化问题) 下, 最小化极值概率分布下的期望成本. 在本节中, 我们将鲁棒 BRRP 模型等价地转化为单个分布鲁棒优化问题. 为了表述简洁, 我们定义如下集合:

$$\mathcal{A}(\xi) := \{\boldsymbol{x} \in \{0,1\}^J : (1-\lambda)\boldsymbol{c}^\top \boldsymbol{x} \leqslant \xi\}. \tag{A.111}$$

给定预算和原料分类决策 $(\xi, \boldsymbol{s})$, 原料条件概率分布 $\mathbb{P} \in \mathbb{F}(\boldsymbol{s})$ 受分类决策 $\boldsymbol{s}$ 影响, 且选址决策 $\boldsymbol{x} \in \mathcal{A}(\xi)$ 受分摊成本预算决策 ξ 影响. 如果 $\mathbb{E}_{\mathbb{P}}[\phi(\boldsymbol{x}, \tilde{\boldsymbol{v}}, \tilde{\boldsymbol{u}})] - \lambda \boldsymbol{c}^\top \boldsymbol{x} < \tau$, 即私人运营商经济不可行, 则我们认为在分布 $\mathbb{P}$ 下, 决策 $(\xi, \boldsymbol{s})$ 对应选址决策 $\boldsymbol{x}$ 是不可行的. 因此, 我们定义如下地方政府的拓展损失函数 (extended loss function), 它包含私人运营商的经济不可行信息:

$$\Pi_{\mathbb{P}}(\xi, \boldsymbol{s}; \boldsymbol{x}) := \begin{cases} \xi + \mathbb{E}_{\mathbb{P}}\left[\ell(\boldsymbol{s}, \tilde{\boldsymbol{v}})\right], & \text{如果 } \mathbb{E}_{\mathbb{P}}\left[\phi(\boldsymbol{x}, \tilde{\boldsymbol{v}}, \tilde{\boldsymbol{u}})\right] - \lambda \boldsymbol{c}^{\top}\boldsymbol{x} \geqslant \tau, \\ \infty, & \text{否则}, \end{cases} \tag{A.112}$$

其中, $\phi(\boldsymbol{x}, \tilde{\boldsymbol{v}}, \tilde{\boldsymbol{u}})$ 是给定选址 $\boldsymbol{x}$ 下, 关于不确定原料条件 $(\tilde{\boldsymbol{v}}, \tilde{\boldsymbol{u}})$ 的资源回收系统选择问题 (6.5) 的最优值. 利用上述符号, 可以将鲁棒 BRRP 模型 (6.7) 转化为如下等价形式.

命题 A.1 鲁棒 BRRP 模型 (6.7) 等价于如下分布鲁棒优化问题:

$$\min_{\xi \geqslant 0, \boldsymbol{s} \in \mathcal{S}} \min_{\boldsymbol{x} \in \mathcal{A}(\xi)} \sup_{\mathbb{P} \in \mathbb{F}(\boldsymbol{s})} \Pi_{\mathbb{P}}(\xi, \boldsymbol{s}; \boldsymbol{x}), \tag{A.113}$$

其中, $\mathcal{A}(\xi)$ 和 $\Pi_{\mathbb{P}}(\xi, \boldsymbol{s}; \boldsymbol{x})$ 分别由式 (A.111) 和 (A.112) 给出.

证明 对于任意的 $(\xi, \boldsymbol{s}) \in \mathbb{R}_+ \times \mathcal{S}$, $\boldsymbol{x} \in \mathcal{A}(\xi)$, 根据 $\Pi_{\mathbb{P}}(\xi, \boldsymbol{s}; \boldsymbol{x})$ 的定义, 目标函数

$$\begin{aligned} &\sup_{\mathbb{P} \in \mathbb{F}(\boldsymbol{s})} \Pi_{\mathbb{P}}(\xi, \boldsymbol{s}; \boldsymbol{x}) \\ =& \begin{cases} \xi + \sup\limits_{\mathbb{P} \in \mathbb{F}(\boldsymbol{s})} \mathbb{E}_{\mathbb{P}}\left[\ell(\boldsymbol{s}, \tilde{\boldsymbol{v}})\right], & \text{如果 } \inf\limits_{\mathbb{P} \in \mathbb{F}(\boldsymbol{s})} \mathbb{E}_{\mathbb{P}}\left[\phi(\boldsymbol{x}, \tilde{\boldsymbol{v}}, \tilde{\boldsymbol{u}})\right] - \lambda \boldsymbol{c}^{\top}\boldsymbol{x} \geqslant \tau, \\ +\infty, & \text{否则}. \end{cases} \end{aligned}$$

接下来, 我们将问题

$$\min_{\boldsymbol{x} \in \mathcal{A}(\xi)} \sup_{\mathbb{P} \in \mathbb{F}(\boldsymbol{s})} \Pi_{\mathbb{P}}(\xi, \boldsymbol{s}; \boldsymbol{x})$$

转化为如下形式:

$$\begin{cases} \xi + \sup\limits_{\mathbb{P} \in \mathbb{F}(\boldsymbol{s})} \mathbb{E}_{\mathbb{P}}\left[\ell(\boldsymbol{s}, \tilde{\boldsymbol{v}})\right], & \text{如果 } \max\limits_{\boldsymbol{x} \in \mathcal{A}(\xi)} \left\{ \inf\limits_{\mathbb{P} \in \mathbb{F}(\boldsymbol{s})} \mathbb{E}_{\mathbb{P}}\left[\phi(\boldsymbol{x}, \tilde{\boldsymbol{v}}, \tilde{\boldsymbol{u}})\right] - \lambda \boldsymbol{c}^{\top}\boldsymbol{x} \right\} \geqslant \tau, \\ +\infty, & \text{否则}. \end{cases}$$

因为鲁棒 BRRP 模型 (6.7) 中问题

$$\mathcal{Q}^{\lambda}_{\mathbb{F}(\boldsymbol{s})}(\xi) = \max_{\boldsymbol{x} \in \mathcal{A}(\xi)} \left\{ \inf_{\mathbb{P} \in \mathbb{F}(\boldsymbol{s})} \mathbb{E}_{\mathbb{P}}\left[\phi(\boldsymbol{x}, \tilde{\boldsymbol{v}}, \tilde{\boldsymbol{u}})\right] - \lambda \boldsymbol{c}^{\top}\boldsymbol{x} \right\}.$$

所以我们证明了模型 (A.113) 与模型 (6.7) 的等价性.

如 6.2 节所述, 可以从上述等价形式 (A.113) 中看出, 地方政府和私人运营商共同优化与一个极值概率分布相关的极值情况成本 $\Pi_{\mathbb{P}}(\xi, \boldsymbol{s}; \boldsymbol{x})$.

A.4.4　分类一致性条件下的分类成本分析

本小节在分类一致性条件下, 对分类成本进行分析. 我们考虑如下分类一致性条件: 对于任意的 $\boldsymbol{s}_a, \boldsymbol{s}_b \in \mathcal{S}$,

$$\begin{aligned}&\exists\, \boldsymbol{x}^\star,\ \text{s.t.}\ \inf_{\mathbb{P}\in\mathbb{F}(\boldsymbol{s}_a)} \mathbb{E}_{\mathbb{P}}\left[\phi\left(\boldsymbol{x}^\star, \tilde{\boldsymbol{v}}, \tilde{\boldsymbol{u}}\right)\right] > \inf_{\mathbb{P}\in\mathbb{F}(\boldsymbol{s}_b)} \mathbb{E}_{\mathbb{P}}\left[\phi\left(\boldsymbol{x}^\star, \tilde{\boldsymbol{v}}, \tilde{\boldsymbol{u}}\right)\right]\\ \Rightarrow\ &\sup_{\mathbb{P}\in\mathbb{F}(\boldsymbol{s}_a)} \mathbb{E}_{\mathbb{P}}\left[\ell(\boldsymbol{s}_a, \tilde{\boldsymbol{v}})\right] > \sup_{\mathbb{P}\in\mathbb{F}(\boldsymbol{s}_b)} \mathbb{E}_{\mathbb{P}}\left[\ell(\boldsymbol{s}_b, \tilde{\boldsymbol{v}})\right],\end{aligned} \tag{A.114}$$

其中, $\boldsymbol{x}^\star$ 是在某些 λ 下最优选址. 上述一致性条件 (A.114) 要求给定一些合理的选址决策, 如果使用分类方案 $\boldsymbol{s}_a$ 可以获得比使用分类方案 $\boldsymbol{s}_b$ 更高的资源回收收益, 那么方案 $\boldsymbol{s}_a$ 的投资应该比方案 $\boldsymbol{s}_b$ 的投资更高. 否则, 将不利于对分类方案的投资, 且不符合激励相容 (incentive compatible) 性. 在实践中, 分类一致性条件是合理的.

命题 A.2　令 $0 \leqslant \lambda_1 < \lambda_2 < 1$, $((\boldsymbol{s}_1^\star, \xi_1^\star), \boldsymbol{x}_1^\star)$ 与 $((\boldsymbol{s}_2^\star, \xi_2^\star), \boldsymbol{x}_2^\star)$ 是问题 (6.6)-(6.7) 分别关于 λ_1 和 λ_2 的最优解, 假设分类一致性条件 (A.114) 成立. 那么, 如果私人运营商的经济可行性约束在 λ_1 的问题最优时是紧的, 且

$$\inf_{\mathbb{P}\in\mathbb{F}(\boldsymbol{s}_1^\star)} \mathbb{E}_{\mathbb{P}}\left[\phi\left(\boldsymbol{x}_2^\star, \tilde{\boldsymbol{v}}, \tilde{\boldsymbol{u}}\right)\right] \leqslant \inf_{\mathbb{P}\in\mathbb{F}(\boldsymbol{s}_1^\star)} \mathbb{E}_{\mathbb{P}}\left[\phi\left(\boldsymbol{x}_1^\star, \tilde{\boldsymbol{v}}, \tilde{\boldsymbol{u}}\right)\right], \tag{A.115}$$

我们有

$$\sup_{\mathbb{P}\in\mathbb{F}(\boldsymbol{s}_2^\star)} \mathbb{E}_{\mathbb{P}}\left[\ell(\boldsymbol{s}_2^\star, \tilde{\boldsymbol{v}})\right] > \sup_{\mathbb{P}\in\mathbb{F}(\boldsymbol{s}_1^\star)} \mathbb{E}_{\mathbb{P}}\left[\ell(\boldsymbol{s}_1^\star, \tilde{\boldsymbol{v}})\right]. \tag{A.116}$$

证明　在 $\boldsymbol{c}^\top \boldsymbol{x}_1^\star > \boldsymbol{c}^\top \boldsymbol{x}_2^\star$ 的情况下, 根据命题 6.1, 可得分类成本不等式 (A.116) 成立, 因此有 $\xi_1^\star - \xi_2^\star\,(1-\lambda_1/1-\lambda_2) > 0$. 则我们只需在条件

$$\boldsymbol{c}^\top \boldsymbol{x}_1^\star \leqslant \boldsymbol{c}^\top \boldsymbol{x}_2^\star \tag{A.117}$$

下, 证明结果 (A.116) 成立. 因为私人运营商的经济可行性约束在 λ_1 的问题最优时是紧的, 即

$$\inf_{\mathbb{P}\in\mathbb{F}(\boldsymbol{s}_1^\star)} \mathbb{E}_{\mathbb{P}}\left[\phi\left(\boldsymbol{x}_1^\star, \tilde{\boldsymbol{v}}, \tilde{\boldsymbol{u}}\right)\right] - \lambda_1 \boldsymbol{c}^\top \boldsymbol{x}_1^\star = \tau,$$

且在条件 (A.117) 以及 $\lambda_2 > \lambda_1$ 的假设下, $\lambda_2 \boldsymbol{c}^\top \boldsymbol{x}_2^\star > \lambda_1 \boldsymbol{c}^\top \boldsymbol{x}_1^\star$. 为了达到利润目标 τ, 我们必须在 λ_2 的问题中有如下不等式:

$$\inf_{\mathbb{P}\in\mathbb{F}(\boldsymbol{s}_2^\star)} \mathbb{E}_{\mathbb{P}}\left[\phi\left(\boldsymbol{x}_2^\star, \tilde{\boldsymbol{v}}, \tilde{\boldsymbol{u}}\right)\right] > \inf_{\mathbb{P}\in\mathbb{F}(\boldsymbol{s}_1^\star)} \mathbb{E}_{\mathbb{P}}\left[\phi\left(\boldsymbol{x}_1^\star, \tilde{\boldsymbol{v}}, \tilde{\boldsymbol{u}}\right)\right]. \tag{A.118}$$

结合假设 (A.115), 我们得到不等式:

$$\inf_{\mathbb{P}\in\mathbb{F}(\boldsymbol{s}_2^\star)} \mathbb{E}_{\mathbb{P}}\Big[\phi\left(\boldsymbol{x}_2^\star, \tilde{\boldsymbol{v}}, \tilde{\boldsymbol{u}}\right)\Big] > \inf_{\mathbb{P}\in\mathbb{F}(\boldsymbol{s}_1^\star)} \mathbb{E}_{\mathbb{P}}\Big[\phi\left(\boldsymbol{x}_2^\star, \tilde{\boldsymbol{v}}, \tilde{\boldsymbol{u}}\right)\Big]. \tag{A.119}$$

根据分类一致性条件 (A.114), 可得分类成本不等式 (A.116) 成立.

命题 A.2 直观地展现了资源回收系统应用中一种常见情况. 当更多比例的固定成本投资被强加给私人运营商时, 一旦超过某个"阈值水平"(例如, $\lambda = \lambda_1$ 的情况, 在命题 A.2 假设中私人运营商的经济可行性约束是紧的), 如果条件 (A.115) 成立, 即新的选址决策不会在以前的原料条件 (或以前的分类方案) 下提高私人运营商的收益, 那么提高私人运营商的成本分摊比率将迫使地方政府改进分类方案. 这可能与之前的原料条件极差有关, 或在提高私人运营商的成本分摊比例后, 选址决策没有改变 (如 6.5.1 小节的表 6.5).

此外, 我们还指出, 当地方政府将过多的固定成本分摊比例转移到运营商时, 其总支出会 (意外地) 变高. 例如, 从命题 A.2 的条件和结果来看, 当分摊比率的"阈值水平"(紧度), 即 λ_1 已经很高时, 将私人运营商的分摊比率从 λ_1 提高到 λ_2 时, 需要极高质量的原料输入才能满足私人运营商的经济可行性. 这使得地方政府不得不采用一些昂贵的分类方案, 且在这种情况下对系统的处理能力或容量进行更多投资 (例如, 开设更多资源回收设施) 可能会进一步增加私人运营商成本负担, 并导致私人运营商经济上的不可行. 因此, 在这种情况下, 设施固定成本保持不变 (即, $\boldsymbol{c}^\top\boldsymbol{x}_2^\star = \boldsymbol{c}^\top\boldsymbol{x}_1^\star$), 且在任何时候有

$$\sup_{\mathbb{P}\in\mathbb{F}(\boldsymbol{s}_2^\star)} \mathbb{E}_{\mathbb{P}}\left[\ell(\boldsymbol{s}_2^\star, \tilde{\boldsymbol{v}})\right] > \sup_{\mathbb{P}\in\mathbb{F}(\boldsymbol{s}_1^\star)} \mathbb{E}_{\mathbb{P}}\left[\ell(\boldsymbol{s}_1^\star, \tilde{\boldsymbol{v}})\right] + \Delta^\star,$$

其中, $\Delta^\star = (\lambda_2 - \lambda_1)\boldsymbol{c}^\top\boldsymbol{x}_1^\star$, 则有如下不等式:

$$\sup_{\mathbb{P}\in\mathbb{F}(\boldsymbol{s}_2^\star)} \mathbb{E}_{\mathbb{P}}\left[\ell(\boldsymbol{s}_2^\star, \tilde{\boldsymbol{v}})\right] + (1-\lambda_2)\boldsymbol{c}^\top\boldsymbol{x}_1^\star > \sup_{\mathbb{P}\in\mathbb{F}(\boldsymbol{s}_1^\star)} \mathbb{E}_{\mathbb{P}}\left[\ell(\boldsymbol{s}_1^\star, \tilde{\boldsymbol{v}})\right] + (1-\lambda_1)\boldsymbol{c}^\top\boldsymbol{x}_1^\star.$$

上述不等式表明地方政府的总成本增加了. 数值实验结果也验证了这一点 (在 6.5.1 小节的表 6.5 中 λ 从 40% 增加到 50% 时的结果). 因此, 将过多的设施建设成本负担转移给私人运营商实际上可能最终导致地方政府需要支付更多的费用.

A.5 第 7 章的证明及其他相关内容

本节将提供第 7 章中引理、命题和定理的证明以及其他相关内容.

A.5.1　第 7 章的证明

命题 7.1 的证明

首先, 给定选址决策 $\boldsymbol{x}$ 和原料产量预测值 $\boldsymbol{\xi} \in \mathcal{U}^{\Gamma}$, 根据资源回收设施选址模型的结构 (7.3) 和式 (7.2), 有如下等式成立:

$$\max_{(\boldsymbol{y},\boldsymbol{z})\in\mathcal{Y}(\boldsymbol{x},\boldsymbol{\xi})} \phi(\boldsymbol{x},\boldsymbol{y},\boldsymbol{z}) = \sum_{t\in[T]} \Upsilon_t(\boldsymbol{x}_{[t]},\boldsymbol{\xi}_t).$$

其次, 我们证明

$$\begin{aligned}\mathcal{Z}_{\mathrm{NPV}}(\boldsymbol{x}) = & \min_{\boldsymbol{\xi}_1\in\mathcal{U}_1^{\Gamma}(\widehat{\boldsymbol{\xi}}_0)} \min_{\boldsymbol{\xi}_2\in\mathcal{U}_2^{\Gamma}(\boldsymbol{\xi}_{[1]})} \cdots \min_{\boldsymbol{\xi}_T\in\mathcal{U}_T^{\Gamma}(\boldsymbol{\xi}_{[T-1]})} \\ & \cdot \Big[\Upsilon_1(\boldsymbol{x}_{[1]},\boldsymbol{\xi}_1) + \Upsilon_2(\boldsymbol{x}_{[2]},\boldsymbol{\xi}_2) + \cdots + \Upsilon_T(\boldsymbol{x}_{[T]},\boldsymbol{\xi}_T)\Big]. \end{aligned} \tag{A.120}$$

给定 $\boldsymbol{x}$ 以及任意的 $\boldsymbol{\xi} = (\boldsymbol{\xi}_1,\boldsymbol{\xi}_2,\cdots,\boldsymbol{\xi}_T)$, 问题

$$\sum_{t\in[T]} \Upsilon_t(\boldsymbol{x}_{[t]},\boldsymbol{\xi}_t),$$

等价于求解 T 个独立的子问题. 因此, 对于给定的原料产量预测路径 $\boldsymbol{\xi}_{[T-1]}$,

$$\begin{aligned}&\min_{\boldsymbol{\xi}_T\in\mathcal{U}_T^{\Gamma}(\boldsymbol{\xi}_{[T-1]})} \Big[\Upsilon_1(\boldsymbol{x}_{[1]},\boldsymbol{\xi}_1) + \Upsilon_2(\boldsymbol{x}_{[2]},\boldsymbol{\xi}_2) + \cdots + \Upsilon_T(\boldsymbol{x}_{[T]},\boldsymbol{\xi}_T)\Big] \\ &= \Upsilon_1(\boldsymbol{x}_{[1]},\boldsymbol{\xi}_1) + \Upsilon_2(\boldsymbol{x}_{[2]},\boldsymbol{\xi}_2) + \cdots + \min_{\boldsymbol{\xi}_T\in\mathcal{U}_T^{\Gamma}(\boldsymbol{\xi}_{[T-1]})} \Upsilon_T(\boldsymbol{x}_{[T]},\boldsymbol{\xi}_T).\end{aligned}$$

此外, 给定原料产量预测路径 $\boldsymbol{\xi}_{[T-2]}$, 我们有

$$\begin{aligned}&\min_{\boldsymbol{\xi}_{T-1}\in\mathcal{U}_{T-1}^{\Gamma}(\boldsymbol{\xi}_{[T-2]})} \min_{\boldsymbol{\xi}_T\in\mathcal{U}_T^{\Gamma}(\boldsymbol{\xi}_{[T-1]})} \Big[\Upsilon_1(\boldsymbol{x}_{[1]},\boldsymbol{\xi}_1) + \Upsilon_2(\boldsymbol{x}_{[2]},\boldsymbol{\xi}_2) + \cdots + \Upsilon_T(\boldsymbol{x}_{[T]},\boldsymbol{\xi}_T)\Big] \\ &= \min_{\boldsymbol{\xi}_{T-1}\in\mathcal{U}_{T-1}^{\Gamma}(\boldsymbol{\xi}_{[T-2]})} \Big[\Upsilon_1(\boldsymbol{x}_{[1]},\boldsymbol{\xi}_1) + \Upsilon_2(\boldsymbol{x}_{[2]},\boldsymbol{\xi}_2) + \cdots + \min_{\boldsymbol{\xi}_T\in\mathcal{U}_T^{\Gamma}(\boldsymbol{\xi}_{[T-1]})} \Upsilon_T(\boldsymbol{x}_{[T]},\boldsymbol{\xi}_T)\Big] \\ &= \Upsilon_1(\boldsymbol{x}_{[1]},\boldsymbol{\xi}_1) + \cdots \\ &\quad + \min_{\boldsymbol{\xi}_{T-1}\in\mathcal{U}_{T-1}^{\Gamma}(\boldsymbol{\xi}_{[T-2]})} \Big[\Upsilon_{T-1}(\boldsymbol{x}_{[T-1]},\boldsymbol{\xi}_{T-1}) + \Big[\min_{\boldsymbol{\xi}_T\in\mathcal{U}_T^{\Gamma}(\boldsymbol{\xi}_{[T-1]})} \Upsilon_T(\boldsymbol{x}_{[T]},\boldsymbol{\xi}_T)\Big]\Big].\end{aligned}$$

递归地重复上述过程, 可得式 (A.120) 成立.

最后, 证明预测不确定集 $\mathcal{U}^{\Gamma}$ 等价于如下预测路径依赖嵌套不确定集:

$$\left\{\boldsymbol{\xi}=(\boldsymbol{\xi}_1,\boldsymbol{\xi}_2,\cdots,\boldsymbol{\xi}_T)\,\middle|\,\boldsymbol{\xi}_1\in\mathcal{U}_1^{\Gamma}(\widehat{\boldsymbol{\xi}}_0),\ \boldsymbol{\xi}_2\in\mathcal{U}_2^{\Gamma}(\boldsymbol{\xi}_{[1]}),\ \cdots,\ \boldsymbol{\xi}_T\in\mathcal{U}_T^{\Gamma}(\boldsymbol{\xi}_{[T-1]})\right\}.$$

对于每个 $\boldsymbol{\xi}=(\boldsymbol{\xi}_1,\boldsymbol{\xi}_2,\cdots,\boldsymbol{\xi}_T)$, 且 $\boldsymbol{\xi}_t\in\mathcal{U}_t^{\Gamma}(\boldsymbol{\xi}_{[t-1]}),t\in[T]$, 我们有

$$\xi_{i1}=\widehat{\xi}_{i0}+\mathcal{F}_{i1}+\varrho_{i1}\mathcal{S}_{i1}, \tag{A.121}$$

$$\xi_{it}=\xi_{it-1}+\mathcal{F}_{it}+\varrho_{it}\mathcal{S}_{it}=\widehat{\xi}_{i0}+\sum_{\tau\in[t]}\mathcal{F}_{i\tau}+\sum_{\tau\in[t]}\varrho_{i\tau}\mathcal{S}_{i\tau},\ i\in[I],t\in[2:T], \tag{A.122}$$

其中, 变量 $\boldsymbol{\varrho}$ 属于如下集合:

$$\left\{\boldsymbol{\varrho}\in\mathbb{R}^{I\times T}\,\middle|\,\begin{array}{ll}\displaystyle\sum_{i\in[I]}|\varrho_{it}|\leqslant\Gamma_t^{\mathrm{Z}}, & t\in[T]\\ \displaystyle|\varrho_{it}|\leqslant\Gamma_i^{\mathrm{T}}-\mathbb{I}_{\{t\geqslant2\}}\left[\sum_{\tau\in[t-1]}|\varrho_{i\tau}|\right], & i\in[I],t\in[T]\\ \varrho_{it}\in[-1,1], & i\in[I],t\in[T]\end{array}\right\}. \tag{A.123}$$

对于任意 $i\in[I]$, 我们可以推出如下不等式:

$$|\varrho_{it}|\leqslant\Gamma_i^{\mathrm{T}}-\mathbb{I}_{\{t\geqslant2\}}\left[\sum_{\tau\in[t-1]}|\varrho_{i\tau}|\right],\ t\in[T]\Rightarrow\sum_{\tau=1}^{T}|\varrho_{i\tau}|\leqslant\Gamma_i^{\mathrm{T}}.$$

因此, $\boldsymbol{\xi}\in\mathcal{U}^{\Gamma}$. 对于每一个 $\boldsymbol{\xi}=(\boldsymbol{\xi}_1,\boldsymbol{\xi}_2,\cdots,\boldsymbol{\xi}_T)\in\mathcal{U}^{\Gamma}$, 满足

$$\sum_{t\in[T]}|\varrho_{it}|\leqslant\Gamma_i^{\mathrm{T}},\quad i\in[I].$$

因为 $|\varrho_{it}|\geqslant0,\forall i\in[I]$, 所以我们有如下不等式成立:

$$|\varrho_{it}|\leqslant\Gamma_i^{\mathrm{T}}-\mathbb{I}_{\{t\geqslant2\}}\left[\sum_{\tau\in[t-1]}|\varrho_{i\tau}|\right],\quad t\in[T],i\in[I].$$

因此, $\boldsymbol{\xi}$ 满足条件 (A.121)-(A.123), 即 $\boldsymbol{\xi}_t\in\mathcal{U}_t^{\Gamma}(\boldsymbol{\xi}_{t-1}),t\in[T]$. 综上命题 7.1 的等式 (7.14) 成立.

命题 7.2 的证明

根据命题 7.1, 问题 $\mathcal{Z}_{\text{NPV}}(\boldsymbol{x})$ 可以转化为如下最小最大化问题:

$$\min_{\boldsymbol{\xi}\in\mathcal{U}^{\Gamma}} \max_{\boldsymbol{y},\boldsymbol{z}} \quad \sum_{t\in[T]}\left[\beta^t\sum_{i\in[I]}\left(\sum_{j\in[J]} r_{ij}y_{ijt}-c_D z_{it}\right)-\beta^{t-1}\boldsymbol{c}^{\top}\boldsymbol{x}_t\right] \tag{A.124}$$

$$\text{s.t.} \quad \sum_{j\in[J]} y_{ijt}+z_{it}=\xi_{it},\ \forall i\in[I],t\in[T], \tag{A.125}$$

$$\sum_{i\in[I]} y_{ijt}\leqslant\sum_{\tau\in[t]} x_{j\tau}s_j,\ \forall j\in[J],t\in[T], \tag{A.126}$$

$$\boldsymbol{y}_t\in\mathbb{R}_+^{I\times J},\boldsymbol{z}_t\in\mathbb{R}_+^{I},\ \forall t\in[T]. \tag{A.127}$$

利用线性规划的强对偶定理, 对内层最大化问题做对偶, 可以得到问题 (A.124)-(A.127) 的对偶形式如下:

$$\min_{\boldsymbol{\xi}\in\mathcal{U}^{\Gamma}} \min_{\boldsymbol{f},\boldsymbol{g}} \quad \sum_{t\in[T]}\left[\sum_{i\in[I]}\xi_{it}f_{it}+\sum_{j\in[J]}\sum_{\tau\in[t]} x_{j\tau}s_j g_{jt}-\beta^{t-1}\boldsymbol{c}^{\top}\boldsymbol{x}_t\right] \tag{A.128}$$

$$\text{s.t.} \quad f_{it}+g_{jt}\geqslant\beta^t r_{ij},\ \forall i\in[I],j\in[J],t\in[T], \tag{A.129}$$

$$f_{it}\geqslant-\beta^t c_D,\ \forall i\in[I],t\in[T], \tag{A.130}$$

$$\boldsymbol{f}\in\mathbb{R}^{I\times T},\ \ \boldsymbol{g}\in\mathbb{R}_+^{J\times T}, \tag{A.131}$$

其中, $\boldsymbol{f},\boldsymbol{g}$ 为对偶变量. 接下来, 我们分三步推导问题 $\mathcal{Z}_{\text{NPV}}(\boldsymbol{x})$ 的混合整数规划形式.

第一步 首先假设对偶变量 $f_{it},i\in[I],t\in[T]$ 有一个有限常数上界 M_1, 即

$$f_{it}\leqslant M_1,\ \forall i\in[I],t\in[T]. \tag{A.132}$$

为了表述简洁, 记 OBJ 为问题 (A.128)-(A.131) 的最优目标函数值. 然后, 我们可以推出如下不等式:

$$\begin{aligned}&\text{OBJ}<M_0\\ &:=\sum_{t\in[T]}\left[-\sum_{i\in[I]}\xi_{it}^{+}\beta^t c_D+\sum_{j\in[J]}\sum_{\tau\in[t]}x_{j\tau}s_j\beta^t(r_{ij}+c_D)-\beta^{t-1}\boldsymbol{c}^{\top}\boldsymbol{x}_{t-1}+\Delta\right]>0,\end{aligned}$$

其中, Δ 是任意合适的正常数, ξ_{it}^{+} 是 ξ_{it} 的有限上界.

此外, 令

$$M_1 := \frac{M_0 + \sum_{t\in[T]} \beta^{t-1}\boldsymbol{c}^\top \boldsymbol{x}_t + \sum_{i\in[I]}\sum_{t\in[T]} \xi_{it}^{+}\beta^t c_D}{\min_{i\in[I],t\in[T]} \xi_{it}^{-}}, \tag{A.133}$$

其中, $\xi_{it}^{-} > 0$ 是 ξ_{it} 的一个有限下界. 给定 $\boldsymbol{f} \in \mathbb{R}^{I\times T}$, 如果存在任意的 $(i^*, t^*) \in [I] \times [T]$ 使得 $f_{i^*t^*} > M_1$, 对于任意的 $\boldsymbol{\xi} \in \mathcal{U}^\Gamma$, 则目标函数值必有

$$\begin{aligned}
&\sum_{t\in[T]}\left[\sum_{i\in[I]} \xi_{it} f_{it} + \sum_{j\in[J]}\sum_{\tau\in[0:t-1]} x_{j\tau} s_j g_{jt} - \beta^{t-1}\boldsymbol{c}^\top \boldsymbol{x}_t\right] \\
\geqslant& \sum_{t\in[T]}\sum_{i\in[I]} \xi_{it} f_{it} - \sum_{t\in[T]} \beta^{t-1}\boldsymbol{c}^\top \boldsymbol{x}_t \\
\geqslant& M_0 + \sum_{i\in[I]}\sum_{t\in[T]} \xi_{it}^{+}\beta^t c_D + \sum_{t\in[T]\setminus\{t^*\}}\sum_{i\in[I]\setminus\{i^*\}} \xi_{it} f_{it} \\
\geqslant& M_0 + \sum_{i\in[I]\setminus\{i^*\}}\sum_{t\in[T]\setminus\{t^*\}} \xi_{it}\left(\beta^t c_D + f_{it}\right) \\
\geqslant& M_0,
\end{aligned}$$

上述不等式成立, 由于 $\xi_{it} \geqslant 0, g_{jt} \geqslant 0, f_{it} \geqslant -\beta^t c_D$, $i \in [I], t \in [T]$. 因此, $\boldsymbol{f}$ 的最优解取值小于 M_1. 换句话说, 可以在不影响最优性的情况下, 将约束 (A.132) 加入到问题 (A.128)-(A.131) 中.

第二步 证明由式 (7.8) 定义的不确定集 $\mathcal{U}^\Gamma$ 等价于如下形式:

$$\mathcal{U}^\Gamma = \left\{ \boldsymbol{\xi} \in \mathbb{R}^{I\times T} \;\middle|\; \begin{array}{l}
\xi_{it} = \widehat{\xi}_{i0} + \sum_{\tau\in[t]} \mathcal{F}_{i\tau} + \sum_{\tau\in[t]} \left(\varrho_{i\tau}^{+} - \varrho_{i\tau}^{-}\right)\mathcal{S}_{i\tau}, \quad \forall i \in [I], t \in [T], \\
\sum_{i\in[I]} (\varrho_{it}^{+} + \varrho_{it}^{-}) \leqslant \Gamma_t^{\mathrm{Z}}, \ \forall t \in [T], \\
\sum_{t\in[T]} (\varrho_{it}^{+} + \varrho_{it}^{-}) \leqslant \Gamma_i^{\mathrm{T}}, \ \forall i \in [I], \\
\varrho_{it}^{+}, \varrho_{it}^{-} \in [0,1], \ \forall i \in [I], t \in [T]
\end{array} \right\}.$$

为了证明上述等价形式, 只需要证明以下两个集合等价:

$$\mathcal{P}_1 := \left\{ \boldsymbol{\varrho} \in \mathbb{R}^{I\times T} \;\middle|\; \begin{array}{l}
\sum_{i\in[I]} |\varrho_{it}| \leqslant \Gamma_t^{\mathrm{Z}}, \ \forall t \in [T], \\
\sum_{t\in[T]} |\varrho_{it}| \leqslant \Gamma_i^{\mathrm{T}}, \ \forall i \in [I], \\
\varrho_{it} \in [-1,1], \ \forall i \in [I], t \in [T]
\end{array} \right\}$$

和

$$\mathcal{P}_2 := \left\{ (\boldsymbol{\varrho}^+ - \boldsymbol{\varrho}^-) \in \mathbb{R}^{I\times T} \left| \begin{array}{l} \sum_{i\in[I]} (\varrho_{it}^+ + \varrho_{it}^-) \leqslant \Gamma_t^{\mathrm{Z}},\ \forall t\in[T], \\ \sum_{t\in[T]} (\varrho_{it}^+ + \varrho_{it}^-) \leqslant \Gamma_i^{\mathrm{T}},\ \forall i\in[I], \\ \varrho_{it}^+, \varrho_{it}^- \in [0,1],\ \forall i\in[I], t\in[T] \end{array} \right. \right\}.$$

一方面, 对于任意的 $\boldsymbol{\varrho}\in\mathcal{P}_1$, 总能找到 $\varrho_{it}^+, \varrho_{it}^- \in [0,1], i\in[I], t\in[T]$ 使得 $\varrho_{it} = \varrho_{it}^+ - \varrho_{it}^-, i\in[I], t\in[T]$. 此外, 如果 $|\varrho_{it}| = 0$, 则有 $\varrho_{it} = 0 = \varrho_{it}^+ = \varrho_{it}^-$, 这意味着 $\varrho_{it}^+ + \varrho_{it}^- = 0$. 否则, 对于任意的 $|\varrho_{it}| > 0$, 有 $\varrho_{it} = \varrho_{it}^+ - \varrho_{it}^- \neq 0$, 并且如果 $\varrho_{it}^+ > 0, \varrho_{it}^- > 0$, 那么总可以以相同的量缩小 ϱ_{it}^+ 和 ϱ_{it}^-, 直到其中一个为零, 同时保持 $\varrho_{it}^+ - \varrho_{it}^-$ 不变. 这实质上导致 ϱ_{it}^+ 和 ϱ_{it}^- 分别是 ϱ_{it} 的正负部分, 且 $\varrho_{it}^+ + \varrho_{it}^- = |\varrho_{it}|$. 因此,

$$\sum_{i\in[I]} (\varrho_{it}^+ + \varrho_{it}^-) = \sum_{i\in[I]} |\varrho_{it}| \leqslant \Gamma_t^{\mathrm{Z}}, \quad t\in[T],$$

并且

$$\sum_{t\in[T]} (\varrho_{it}^+ + \varrho_{it}^-) = \sum_{t\in[T]} |\varrho_{it}| \leqslant \Gamma_i^{\mathrm{T}}, \quad i\in[I].$$

综上所述, 总是可以找到 $\boldsymbol{\varrho} = \boldsymbol{\varrho}^+ - \boldsymbol{\varrho}^- \in \mathcal{P}_2$.

另一方面, 对于任意的 $\boldsymbol{\varrho}^+ - \boldsymbol{\varrho}^- \in \mathcal{P}_2$, 有 $\varrho_{it} := \varrho_{it}^+ - \varrho_{it}^- \in [-1,1]$ 且 $|\varrho_{it}| \leqslant \varrho_{it}^+ + \varrho_{it}^-, \forall i\in[I], t\in[T]$, 则有如下两个不等式:

$$\sum_{i\in[I]} |\varrho_{it}| \leqslant \sum_{i\in[I]} (\varrho_{it}^+ + \varrho_{it}^-) \leqslant \Gamma_t^{\mathrm{Z}}, \quad t\in[T]$$

和

$$\sum_{t\in[T]} |\varrho_{it}| \leqslant \sum_{t\in[T]} (\varrho_{it}^+ + \varrho_{it}^-) \leqslant \Gamma_i^{\mathrm{T}}, \quad i\in[I].$$

因此, $\boldsymbol{\varrho}^+ - \boldsymbol{\varrho}^- = \boldsymbol{\varrho} \in \mathcal{P}_1$.

第三步 推导问题 (A.128)-(A.131) 的混合整数规划等价形式. 给定任意可行的 $\boldsymbol{f}$, 考虑问题 (A.128)-(A.131) 的子问题:

$$\min_{\boldsymbol{\xi}\in\mathcal{U}^{\Gamma}} \sum_{t\in[T]} \sum_{i\in[I]} \xi_{it} f_{it}.$$

利用第二步的结果, 上述问题可以等价地转化为如下问题:

$$\min_{\boldsymbol{\varrho}^+, \boldsymbol{\varrho}^-} \quad \sum_{t\in[T]} \sum_{i\in[I]} f_{it} \left[\widehat{\xi}_{i0} + \sum_{\tau\in[t]} \mathcal{F}_{i\tau} \right] + \left[\sum_{\tau\in[t]} \left(\varrho_{i\tau}^+ f_{it} - \varrho_{i\tau}^- f_{it} \right) \mathcal{S}_{i\tau} \right] \quad \text{(A.134)}$$

$$\text{s.t.} \quad \sum_{i\in[I]}(\varrho_{it}^{+}+\varrho_{it}^{-}) \leqslant \Gamma_t^{\mathrm{Z}},\ \forall t\in[T], \tag{A.135}$$

$$\sum_{t\in[T]}(\varrho_{it}^{+}+\varrho_{it}^{-}) \leqslant \Gamma_i^{\mathrm{T}},\ \forall i\in[I], \tag{A.136}$$

$$\varrho_{it}^{+},\varrho_{it}^{-} \in [0,1],\ \forall i\in[I], t\in[T]. \tag{A.137}$$

接下来, 约束 (A.135)-(A.137) 可以写成如下矩阵形式:

$$\left\{ \begin{pmatrix} \boldsymbol{\varrho}^{+} \\ \boldsymbol{\varrho}^{-} \end{pmatrix} \in \mathbb{R}^{2\times I\times T} \;\middle|\; \begin{bmatrix} \boldsymbol{A} \\ \boldsymbol{I} \end{bmatrix} \begin{pmatrix} \boldsymbol{\varrho}^{+} \\ \boldsymbol{\varrho}^{-} \end{pmatrix} \leqslant \begin{pmatrix} \boldsymbol{\Omega}^{T} \\ \boldsymbol{\Omega}^{Z} \\ \mathbf{1} \end{pmatrix}, \begin{pmatrix} \boldsymbol{\varrho}^{+} \\ \boldsymbol{\varrho}^{-} \end{pmatrix} \geqslant \mathbf{0} \right\}.$$

注意到矩阵 $\boldsymbol{A}$ 满足以下性质: ① 矩阵 $\boldsymbol{A}$ 的每一列都包含两个符号相同的非零元素; ② 矩阵 $\boldsymbol{A}$ 的行可以被划分为两个子集, 每一列的两个非零元素都在不同的行子集里. 因此, $\boldsymbol{A}$ 是完全幺模的 (total unimodular) (Schrijver, 2003), 那么 $[\boldsymbol{A}^{\top},\boldsymbol{I}]^{\top}$ 也是完全单模的. 结合参数 $\Gamma_t^{\mathrm{T}},\Gamma_i^{\mathrm{Z}}, i\in[I], t\in[T]$ 都是整数, 意可得线性约束 (A.135)-(A.137) 组成可行域的极值点也是整数. 因此, 问题 (A.134)-(A.137) 可以转化为如下整数线性规划问题:

$$\min_{\boldsymbol{\pi}^{+},\boldsymbol{\pi}^{-}} \quad \sum_{t\in[T]}\sum_{i\in[I]} f_{it}\left[\widehat{\xi}_{i0}+\sum_{\tau\in[t]}\mathcal{F}_{i\tau}\right]+\left[\sum_{\tau\in[t]}\left(\pi_{i\tau}^{+}f_{it}-\pi_{i\tau}^{-}f_{it}\right)\mathcal{S}_{i\tau}\right] \tag{A.138}$$

$$\text{s.t.} \quad \sum_{i\in[I]}(\pi_{it}^{+}+\pi_{it}^{-}) \leqslant \Gamma_t^{\mathrm{Z}},\ \forall t\in[T], \tag{A.139}$$

$$\sum_{t\in[T]}(\pi_{it}^{+}+\pi_{it}^{-}) \leqslant \Gamma_i^{\mathrm{T}},\ \forall i\in[I], \tag{A.140}$$

$$\pi_{it}^{+},\pi_{it}^{-} \in \{0,1\},\ \forall i\in[I], t\in[T]. \tag{A.141}$$

在第一步中, 我们证明了 $f_{it}\leqslant M_1,\ \forall\, i\in[I], t\in[T]$, 利用替换变量, 令 $q_{it}=f_{it}+\beta^t c_D$, 可以得到 $\pi_{i\tau}^{+}f_{it}=\pi_{i\tau}^{+}q_{it}-\pi_{i\tau}^{+}\beta^t c_D$, $\pi_{i\tau}^{-}f_{it}=\pi_{i\tau}^{-}q_{it}-\pi_{i\tau}^{-}\beta^t c_D$, 以及

$$0\leqslant q_{it}\leqslant M:=M_1+\beta^t c_D, \quad \forall\, i\in[I], t\in[T].$$

利用标准线性化方法, 可以得到乘积项:

$$h_{it\tau}^{+}=\pi_{i\tau}^{+}q_{it}, \quad \forall\, i\in[I], t\in[T], \tau\in[1:t],$$

等价于如下线性约束系统:

$$\left\{\begin{array}{l} h_{it\tau}^{+}\geqslant 0,\ \forall i\in[I],t\in[T],\tau\in[1:t] \\ h_{it\tau}^{+}\leqslant q_{it},\ \forall i\in[I],t\in[T],\tau\in[1:t] \\ h_{it\tau}^{+}\geqslant q_{it}+(\pi_{i\tau}^{+}-1)M,\ \forall i\in[I],t\in[T],\tau\in[1:t] \\ h_{it\tau}^{+}\leqslant \pi_{i\tau}^{+}M,\ \forall i\in[I],t\in[T],\tau\in[1:t] \end{array}\right\}.$$

利用类似的方法, 可以将 $\pi_{i\tau}^{-}q_{it}$ 为 $h_{it\tau}^{-}$ 对于任意的 $i\in[I],t\in[T],\tau\in[1:t]$ 进行线性化. 通过上述三步证明, 可以将问题 $\mathcal{Z}_{\mathrm{NPV}}(\boldsymbol{x})$ 转化为一个混合整数线性规划问题 (7.15).

命题 7.3 的证明

该命题可以直接从命题 7.2 的证明中直接得到. 因此, 省略了证明细节.

命题 7.4 的证明

首先, 处理问题 (7.19) 内层最小最大化问题:

$$\min_{\boldsymbol{\xi}\in\mathcal{U}^{\Gamma}}\max_{(\boldsymbol{y},\boldsymbol{z})\in\mathcal{Y}(\boldsymbol{x},\boldsymbol{\xi})}\phi(\boldsymbol{x},\boldsymbol{y},\boldsymbol{z}). \tag{A.142}$$

根据命题 7.2 的证明过程, 可以得到问题 (A.142) 内层最大化问题的对偶问题:

$$\min_{\boldsymbol{\xi}\in\mathcal{U}^{\Gamma}}\min_{\boldsymbol{f},\boldsymbol{g}}\quad\sum_{t\in[T]}\left[\sum_{i\in[I]}\xi_{it}f_{it}+\sum_{j\in[J]}\sum_{\tau\in[t]}x_{j\tau}s_{j}g_{jt}-\beta^{t-1}\boldsymbol{c}^{\top}\boldsymbol{x}_{t}\right] \tag{A.143}$$

$$\text{s.t.}\quad f_{it}+g_{jt}\geqslant\beta^{t}r_{ij},\ \forall i\in[I],j\in[J],t\in[T], \tag{A.144}$$

$$f_{it}\geqslant-\beta^{t}c_{D},\ \forall i\in[I],t\in[T], \tag{A.145}$$

$$\boldsymbol{f}\in\mathbb{R}^{I\times T},\ \ \boldsymbol{g}\in\mathbb{R}_{+}^{J\times T}, \tag{A.146}$$

其中, $\boldsymbol{f}\in\mathbb{R}^{T\times I}$ 和 $\boldsymbol{g}\in\mathbb{R}_{+}^{T\times J}$ 是对偶变量. 其次, 从命题 7.2 的证明过程可知, 当不确定集合 $\mathcal{U}^{\Gamma}$ 替换为集合 $\mathcal{V}(\Gamma)$ 时, 上述问题的最优目标值不会改变, 即有

$$\min_{\boldsymbol{\xi}\in\mathcal{V}(\Gamma)}\min_{\boldsymbol{f},\boldsymbol{g}}\left\{\sum_{t\in[T]}\left[\sum_{i\in[I]}\xi_{it}f_{it}+\sum_{j\in[J]}\sum_{\tau=1}^{t}x_{j\tau}s_{j}g_{jt}-\beta^{t-1}\boldsymbol{c}^{\top}\boldsymbol{x}_{t}\right]\ \middle|\ \text{(A.144)-(A.146)}\right\}.$$

此外, 对于每个 $\boldsymbol{\xi}\in\mathcal{V}(\Gamma)$, 对上述问题内层最小化问题做对偶, 可得到问题 (A.143)-(A.143)(或原始的最小最大化问题 (A.142)) 的等价形式:

$$\min_{\boldsymbol{\xi}\in\mathcal{V}(\Gamma)}\max_{(\boldsymbol{y},\boldsymbol{z})\in\mathcal{Y}(\boldsymbol{x},\boldsymbol{\xi})}\phi(\boldsymbol{x},\boldsymbol{y},\boldsymbol{z}). \tag{A.147}$$

我们用上境图形式 (epigraph form) 将问题 (A.147) 转化为如下问题:

$$\begin{aligned}\max\quad & \gamma\\ \text{s.t.}\quad & \gamma \leqslant \max_{(\boldsymbol{y},\boldsymbol{z})\in\mathcal{Y}(\boldsymbol{x},\boldsymbol{\xi})} \phi(\boldsymbol{x},\boldsymbol{y},\boldsymbol{z}),\ \boldsymbol{\xi}\in\mathcal{V}(\Gamma)\,.\end{aligned}$$

因为 $\mathcal{V}(\Gamma)$ 是一个离散集合, 所以上述可以转化为如下线性规划问题:

$$\begin{aligned}\max\quad & \gamma\\ \text{s.t.}\quad & \left.\begin{array}{l}\gamma \leqslant \phi(\boldsymbol{x},\boldsymbol{y},\boldsymbol{z}),\\ (\boldsymbol{y},\boldsymbol{z})\in\mathcal{Y}(\boldsymbol{x},\boldsymbol{\xi}),\end{array}\right.\quad \boldsymbol{\xi}\in\mathcal{V}(\Gamma)\,,\end{aligned}$$

进一步, 利用由式 (7.21) 定义的 $\mathcal{K}(\boldsymbol{x},\boldsymbol{\xi})$, 上述问题可以等价地转化为

$$\max\quad \gamma \tag{A.148}$$

$$\text{s.t.}\quad \gamma\in\mathcal{K}(\boldsymbol{x},\boldsymbol{\xi}),\ \boldsymbol{\xi}\in\mathcal{V}(\Gamma)\,. \tag{A.149}$$

现在, 把固定成本项代入问题 (A.148)-(A.149)中, 则可以将问题 (7.19) 等价地转化为如下混合整数线性规划问题:

$$\begin{aligned}\max_{\boldsymbol{x}}\quad & \sum_{t\in[T]}\left[-\beta^{t-1}\boldsymbol{c}^{\top}\boldsymbol{x}_t\right]+\gamma\\ \text{s.t.}\quad & \gamma\in\mathcal{K}(\boldsymbol{x},\boldsymbol{\xi}),\ \boldsymbol{\xi}\in\mathcal{V}(\Gamma)\,,\\ & \boldsymbol{x}\in\mathcal{X},\end{aligned}$$

因此, 命题 7.4 的结论成立.

命题 7.5 的证明

给定任意 $(\boldsymbol{x},\boldsymbol{p},\boldsymbol{q})\in\{(\boldsymbol{x},\boldsymbol{p},\boldsymbol{q})\mid$ 问题 (7.27) 可行域$\}$ 以及 $\boldsymbol{\xi}\in\mathcal{U}^{\Gamma}$, 问题 (7.27) 内层最大化问题具有如下形式:

$$\begin{aligned}&\max_{(\boldsymbol{y},\boldsymbol{z})\in\mathcal{Y}(\boldsymbol{x},\boldsymbol{\xi})\cap\boldsymbol{\Lambda}(\boldsymbol{p},\boldsymbol{q},\boldsymbol{\xi})}\phi(\boldsymbol{x},\boldsymbol{y},\boldsymbol{z})\\ &=\begin{cases}\phi_{\mathrm{NPV}}(\boldsymbol{x},[\boldsymbol{p}_j^{\top}\boldsymbol{\xi}]_J,\boldsymbol{q}^{\top}\boldsymbol{\xi}), & \text{如果 } \mathcal{Y}(\boldsymbol{x},\boldsymbol{\xi})\cap\boldsymbol{\Lambda}(\boldsymbol{p},\boldsymbol{q},\boldsymbol{\xi})\neq\varnothing,\\ -\infty, & \text{其他.}\end{cases}\end{aligned}$$

因此, 对于每个 $(\boldsymbol{x},\boldsymbol{p},\boldsymbol{q})\in\{(\boldsymbol{x},\boldsymbol{p},\boldsymbol{q})\mid$ 问题 (7.27) 可行域$\}$, 可以将问题 (7.27) 的内层最小最大化问题转化为如下形式:

$$\min_{\boldsymbol{\xi}\in\mathcal{U}^{\Gamma}}\left[\max_{(\boldsymbol{y},\boldsymbol{z})\in\mathcal{Y}(\boldsymbol{x},\boldsymbol{\xi})\cap\boldsymbol{\Lambda}(\boldsymbol{p},\boldsymbol{q},\boldsymbol{\xi})}\phi(\boldsymbol{x},\boldsymbol{y},\boldsymbol{z})\right]$$

$$= \begin{cases} -\infty, & \text{如果} \ \exists\, \boldsymbol{\xi} \in \mathcal{U}^{\Gamma}, \text{ s.t. } \mathcal{Y}(\boldsymbol{x},\boldsymbol{\xi}) \cap \boldsymbol{\Lambda}(\boldsymbol{p},\boldsymbol{q},\boldsymbol{\xi}) = \varnothing, \\ \min\limits_{\boldsymbol{\xi}\in\mathcal{U}^{\Gamma}} \phi_{\mathrm{NPV}}(\boldsymbol{x}, [\boldsymbol{p}_j^{\top}\boldsymbol{\xi}]_J, \boldsymbol{q}^{\top}\boldsymbol{\xi}), & \text{其他}. \end{cases}$$

此外, 因为约束 $\bigcap_{\boldsymbol{\xi}\in\mathcal{U}^{\Gamma}} \mathcal{Y}(\boldsymbol{x},\boldsymbol{\xi}) \cap \boldsymbol{\Lambda}(\boldsymbol{p},\boldsymbol{q},\boldsymbol{\xi}) \neq \varnothing$ 与约束 $\boldsymbol{p}_{tj}^{\top}\boldsymbol{\xi}_t \leqslant \sum_{\tau\in[t]} x_{j\tau}s_j$, $\forall \boldsymbol{\xi} \in \mathcal{U}^{\Gamma}, t\in[T], j\in[J]$, 所以对于每个 $(\boldsymbol{x},\boldsymbol{p},\boldsymbol{q}) \in \{(\boldsymbol{x},\boldsymbol{p},\boldsymbol{q}) \mid \text{问题 (7.27) 可行域}\}$, 问题 (7.27) 可以等价地转化为如下形式:

$$\max_{\boldsymbol{x},\boldsymbol{p},\boldsymbol{q}} \quad \min_{\boldsymbol{\xi}\in\mathcal{U}^{\Gamma}} \sum_{t\in[T]} \left[\beta^t \sum_{i\in[I]} \left(\sum_{j\in[J]} r_{ij}p_{ijt} - c_D q_{it} \right) \xi_{it} - \beta^{t-1}\boldsymbol{c}^{\top}\boldsymbol{x}_t \right] \tag{A.150}$$

$$\text{s.t.} \quad \sum_{i\in[I]} p_{ijt}\xi_{it} \leqslant \sum_{\tau\in[t]} x_{j\tau}s_j, \ \forall \boldsymbol{\xi} \in \mathcal{U}^{\Gamma}, j\in[J], t\in[T], \tag{A.151}$$

$$(\boldsymbol{x},\boldsymbol{p},\boldsymbol{q}) \in \{(\boldsymbol{x},\boldsymbol{p},\boldsymbol{q}) \mid \text{问题 (7.27) 可行域}\}, \tag{A.152}$$

其中, 目标函数 (A.150) 中最小化问题:

$$\min_{\boldsymbol{\xi}\in\mathcal{U}^{\Gamma}} \sum_{t\in[T]} \sum_{i\in[I]} \beta^t \left[\sum_{j\in[J]} r_{ij}p_{ijt} - c_D q_{it} \right] \xi_{it},$$

可以转化为如下线性规划问题:

$$\begin{aligned} \min_{\boldsymbol{\xi},\boldsymbol{\lambda},\boldsymbol{\varrho}} \quad & \sum_{t\in[T]} \sum_{i\in[I]} \beta^t \left[\sum_{j\in[J]} r_{ij}p_{ijt} - c_D q_{it} \right] \xi_{it} \\ \text{s.t.} \quad & \xi_{it} - \sum_{\tau\in[t]} \varrho_{i\tau}\mathcal{S}_{i\tau} = \widehat{\xi}_{i0} + \sum_{\tau\in[t]} \mathcal{F}_{i\tau}, \ \forall i\in[I], t\in[T], \\ & \sum_{i\in[I]} \lambda_{it} \leqslant \Gamma_t^{\mathrm{Z}}, \ \forall t\in[T], \\ & \sum_{t\in[T]} \lambda_{it} \leqslant \Gamma_i^{\mathrm{T}}, \ \forall i\in[I], \\ & \lambda_{it} - \varrho_{it} \geqslant 0, \ \forall i\in[I], t\in[T], \\ & \lambda_{it} + \varrho_{it} \geqslant 0, \ \forall i\in[I], t\in[T], \\ & \varrho_{it} \geqslant -1, \ \forall i\in[I], t\in[T], \\ & \varrho_{it} \leqslant 1, \ \forall i\in[I], t\in[T], \\ & \boldsymbol{\xi}, \boldsymbol{\lambda}, \boldsymbol{\varrho} \in \mathbb{R}^{I\times T}. \end{aligned} \tag{A.153}$$

利用线性规划强对偶定理, 可以将上述问题转化为如下等价对偶问题:

$$\max \quad \sum_{t\in[T]} \sum_{i\in[I]} \left[\widehat{\xi}_{i0} + \sum_{\tau\in[t]} \mathcal{F}_{i\tau} \right] d_{it}^{o} + \sum_{t\in[T]} \Gamma_t^{\mathrm{Z}} b_t + \sum_{i\in[I]} \Gamma_i^{\mathrm{T}} \varphi_i + \sum_{t\in[T]} \sum_{i\in[I]} \left[\psi_{it} - \nu_{it} \right]$$

$$\begin{aligned}
\text{s.t.}\quad & \beta^t\left[\sum_{j\in[J]} r_{ij}p_{ijt} - c_D q_{it}\right] - d_{it}^o = 0,\ \forall i\in[I], t\in[T],\\
& b_t + \varphi_i^o + \gamma_{it}^o + \xi_{it}^o = 0,\ \forall i\in[I], t\in[T],\\
& \psi_{it}^o + \nu_{it}^o + \xi_{it}^o - \gamma_{it}^o - \sum_{\tau\in[t:T]} d_{i\tau}^o \mathcal{S}_{i\tau} = 0,\ \forall i\in[I], t\in[T],\\
& \boldsymbol{d}^o\in\mathbb{R}^{I\times T}, \boldsymbol{b}^o\in\mathbb{R}_-^T, \boldsymbol{\varphi}^o\in\mathbb{R}_-^I, \boldsymbol{\gamma}^o, \boldsymbol{\xi}^o, \boldsymbol{\nu}^o\in\mathbb{R}_+^{I\times T}, \boldsymbol{\psi}^o\in\mathbb{R}_-^{I\times T}.
\end{aligned}$$

最后, 对于每个 $j\in[J], t\in[T]$, 将鲁棒约束 (A.151) 做对偶, 并结合上述所得结果, 可以将模型 (7.27) 转化为一个混合整数线性规划问题 (7.28).

A.5.2 基于情景树的多阶段随机规划模型

基于不确定参数的情景 (或场景) 来构建多阶段随机规划模型是一种常用的多阶段建模方法, 其中不确定参数 (本章的原料产量) 的情景作为数据输入到模型中. 然而, 多阶段随机规划建模的难点之一是对非预期性 (non-anticipativity) 约束进行建模, 这要求在决策过程的任何阶段所做出的决策不依赖于不确定参数的未来实现值或未来的决策 (Shapiro et al., 2021). 在基于不确定参数的情景的多阶段随机规划模型中, 这意味着如果第 t 阶段之前的两个不确定情景路径是相同的, 那么依赖于这两个情景路径的决策必须是相同的. 非预期性在很大程度上会增加多阶段模型中约束的数量, 从而增加计算负担. 为此, 本节在多阶段资源回收设施选址规划问题中采用情景树 (scenario tree) 方法来模拟原料产量的不确定性, 该方法利用树状结构自动将相同的情景沿路径分组, 因此可以保证非预期性, 而无需在模型中增加任何约束. 情景树方法是建模多阶段或具有随机过程结构问题中参数不确定性的一种常用方法.

为了简洁地介绍基于原料产量情景树的多阶段资源回收设施选址模型, 本节假设对于每一阶段 t 不同居民区 i 的原料产量 ξ_{it} 的情景数量是相同的. 对于 $t\in\{0\}\cup[T]$, 记 $[N_t]$ 为第 t 阶段的原料产量节点 (情景) 集, 节点数记为 N_t, 其中, $[N_0] := \{1\}$ 用于 $\boldsymbol{x}_0$ 来保证表述的一致性. 此外, 记 $\mathcal{P}(n)$ 为非根节点 n 的所有父节点的集合. 根据定义的节点 (情景) 下标集, 可以把下标 t 从原料产量 ξ_{it} 中移到集合 $[N_t]$ 中, 来更清楚地表示居民区 i 的原料产量节点 (情景) 为 $\xi_i^n, n\in[N_t], t\in[T]$, 它的一种等价表示为 $\xi_{ti}^{k_t}$, 代表居民区 i 在第 t 阶段第 k_t 个原料产量节点 (情景). 同样地, 可以将选址和原料分配决策表示为 y_{ij}^n, z_i^n 和 x_j^n, 对于任意 $n\in[N_t], t\in[T]$.

利用上述定义, 可以构建一个基于原料产量情景树的多阶段风险规避随机资源回收设施选址模型, 其目标是以 $1-\alpha\in(0,1)$ 的置信度最小化条件在险价值

(CV@R) Rockafellar et al., 2000):

$$
\begin{aligned}
\min \quad & \sum_{t\in[T]} \mathrm{CV@R}_{t,\alpha} \\
\text{s.t.} \quad & \sum_{j\in[J]} y_{ij}^n + z_i^n = \xi_i^n,\ \forall n\in[N_t], t\in[T], i\in[I], \\
& \sum_{i\in[I]} y_{ij}^n \leqslant \sum_{m\in\mathcal{P}(n)} x_j^m s_j,\ \forall n\in[N_t], t\in[T], j\in[J], \\
& \sum_{m\in\mathcal{P}(n)} x_j^m \leqslant 1,\ \forall n\in[N_t], j\in[J], \\
& x_j^n \in\{0,1\},\ \forall n\in[N_{t-1}], t\in[T], j\in[J], \\
& y_{ij}^n, z_i^n\in\mathbb{R}_+,\ n\in[N_t],\ \forall t\in[T], i\in[I], j\in[J],
\end{aligned}
$$

其中,

$$
\begin{aligned}
\mathrm{CV@R}_{t,\alpha} = \sum_{t\in[T]} \Bigg\{ \theta_t + \frac{1}{\alpha}\Bigg[& \frac{\beta^{t-1}}{N_{t-1}} \sum_{n\in[N_{t-1}]}\sum_{j\in[J]} c_j x_j^n \\
& - \frac{\beta^t}{N_t} \sum_{n\in[N_t]}\sum_{j\in[J]}\sum_{i\in[I]} \left(r_{ij} y_{ij}^n - c_D z_{ij}^n\right) - \theta_t \Bigg]^+ \Bigg\}.
\end{aligned}
$$

可以看出, 上述模型本质上是一个混合整数线性规划问题, 当问题规模 (使用的场景数量、规划时间段数量以及备选地址数量) 很大时, 求解该问题将非常耗时.

A.6 第 8 章的证明

命题 8.1 的证明

对于任意 $\mathbb{P}\in\mathcal{F}(\theta)$, 有

$$
\mathbb{E}_{\mathbb{P}}\left[\boldsymbol{c}_t^\top \boldsymbol{x}_t + \mathrm{Q}_t(\boldsymbol{X}_t,\tilde{\boldsymbol{u}}_t)\right] = \mathbb{E}_{\mathbb{P}_1}\left[\mathbb{E}_{\mathbb{P}_2}\left[\cdots\mathbb{E}_{\mathbb{P}_t}\left[\boldsymbol{c}_t^\top \boldsymbol{x}_t + \mathrm{Q}_t(\boldsymbol{X}_t,\tilde{\boldsymbol{u}}_t) | \tilde{\boldsymbol{U}}_{t-1}\right]\cdots|\tilde{\boldsymbol{U}}_1\right]\right],
$$

其中, $\mathbb{P}_1\in\mathcal{F}_1(\theta)$ 以及 $\mathbb{P}_t\in\mathcal{F}_{t|\tilde{\boldsymbol{U}}_{t-1}}(\theta)$, $\forall t\in[2:T]$ 构成联合分布 $\mathbb{P}$. 接下来, 证明上述等式在概率分布取极大值后依然成立. 一方面, 对任意 $\mathbb{P}\in\mathcal{F}(\theta)$, 均存在 $\mathbb{P}_1\in\mathcal{F}_1(\theta)$ 和 $\mathbb{P}_t\in\mathcal{F}_{t|\tilde{\boldsymbol{U}}_{t-1}}(\theta)$, $\forall t\in[2:T]$ 构成联合分布 $\mathbb{P}$. 因此,

$$
\mathrm{Z}_t(\boldsymbol{X},\theta) = \sup_{\mathbb{P}\in\mathcal{F}(\theta)} \mathbb{E}_{\mathbb{P}}\left[\boldsymbol{c}_t^\top \boldsymbol{x}_t + \mathrm{Q}_t(\boldsymbol{X}_t,\tilde{\boldsymbol{u}}_t)\right]
$$

$$\leqslant \sup_{\mathbb{P}_1\in\mathcal{F}_1(\theta)} \mathbb{E}_{\mathbb{P}_1}\left[\sup_{\mathbb{P}_2\in\mathcal{F}_{2|\tilde{\boldsymbol{U}}_1}(\theta)} \mathbb{E}_{\mathbb{P}_2}\left[\cdots \sup_{\mathbb{P}_t\in\mathcal{F}_{t|\tilde{\boldsymbol{U}}_{t-1}}(\theta)} \mathbb{E}_{\mathbb{P}_t}\left[\boldsymbol{c}_t^\top \boldsymbol{x}_t \right.\right.\right.$$

$$\left.\left.\left. + \mathrm{Q}_t(\boldsymbol{X}_t, \tilde{\boldsymbol{u}}_t)\middle|\tilde{\boldsymbol{U}}_{t-1}\right]\cdots\middle|\tilde{\boldsymbol{U}}_1\right]\right].$$

另一方面, 任意 $\mathbb{P}_1 \in \mathcal{F}_1(\theta)$ 和 $\mathbb{P}_t \in \mathcal{F}_{t|\tilde{\boldsymbol{U}}_{t-1}}(\theta)$, $\forall t \in [2:T]$ 构成的分布 $\mathbb{P} \in \mathcal{F}(\theta)$, 进而推出相反的不等式. 因此, 命题 8.1 成立.

命题 8.2 的证明

首先, 建立随机需求与随机噪声 (ERN) 之间的等式关系. 从第 1 阶段到第 t 阶段, 依次将前期的需求代入估计模型 (8.3), 即

$$\tilde{\boldsymbol{u}}_1 = \underbrace{\boldsymbol{u}_1^0}_{=\boldsymbol{\eta}_1} + \underbrace{\boldsymbol{I}_K}_{=\boldsymbol{B}_1^1}\tilde{\boldsymbol{\xi}}_1,$$

$$\tilde{\boldsymbol{u}}_2 = \underbrace{\boldsymbol{u}_2^0 + \boldsymbol{A}_2^1\boldsymbol{\eta}_1}_{=\boldsymbol{\eta}_2} + \underbrace{\boldsymbol{A}_2^1}_{=\boldsymbol{B}_2^1}\tilde{\boldsymbol{\xi}}_1 + \underbrace{\boldsymbol{I}_K}_{=\boldsymbol{B}_2^2}\tilde{\boldsymbol{\xi}}_2,$$

$$\vdots$$

$$\tilde{\boldsymbol{u}}_t = \underbrace{\boldsymbol{u}_t^0 + \sum_{s\in[t-1]} \hat{\boldsymbol{A}}_t^s\boldsymbol{\eta}_s}_{=\boldsymbol{\eta}_t} + \sum_{s\in[t-1]} \underbrace{\left(\sum_{s<i_1<i_2<\cdots<i_s<t-1} \boldsymbol{A}_{i_1}^s\boldsymbol{A}_{i_2}^{i_1}\cdots\boldsymbol{A}_t^{i_s}\right)}_{=\boldsymbol{B}_t^s}\tilde{\boldsymbol{\xi}}_s + \underbrace{\boldsymbol{I}_K}_{=\boldsymbol{B}_t^t}\tilde{\boldsymbol{\xi}}_t,$$

其中, $\boldsymbol{I}_K$ 是纬度 $K\times K$ 的单位矩阵, 对于每个 $t\in[T]$, $\boldsymbol{\eta}_t$ 和 $\boldsymbol{B}_t^s$, $\forall s\in[t]$ 可利用时间序列模型和历史样本计算得到. 因此, 对于每个 $t\in[T]$, 我们可以将需求表示为 $\tilde{\boldsymbol{u}}_t = \boldsymbol{\eta}_t + \sum_{s\in[t]} \boldsymbol{B}_t^s\tilde{\boldsymbol{\xi}}_s$.

其次, 将需求 $\tilde{\boldsymbol{u}}_t$ 的表达式代入式 (8.6) 中, 可得极值概率分布下的期望成本等价于如下形式:

$$\mathrm{Z}_t(\boldsymbol{X}, \theta)$$

$$= \sup_{\mathbb{P}_1\in\mathcal{F}_1(\theta)} \mathbb{E}_{\mathbb{P}_1}\left[\sup_{\mathbb{P}_2\in\mathcal{F}_{2|\tilde{\boldsymbol{U}}_1}(\theta)} \mathbb{E}_{\mathbb{P}_2}\right.$$

$$\left.\cdot\left[\cdots \sup_{\mathbb{P}_t\in\mathcal{F}_{t|\tilde{\boldsymbol{U}}_{t-1}}(\theta)} \mathbb{E}_{\mathbb{P}_t}\left[\boldsymbol{c}_t^\top\boldsymbol{x}_t + \mathrm{Q}_t(\boldsymbol{X}_t, \tilde{\boldsymbol{u}}_t)\middle|\tilde{\boldsymbol{U}}_{t-1}\right]\cdots\middle|\tilde{\boldsymbol{U}}_1\right]\right]$$

$$= \sup_{\mathbb{Q}_1\in\mathcal{B}_1(\theta)} \mathbb{E}_{\mathbb{Q}_1}\left[\sup_{\mathbb{Q}_2\in\mathcal{B}_2(\theta)} \mathbb{E}_{\mathbb{Q}_2}\left[\cdots \sup_{\mathbb{Q}_t\in\mathcal{B}_t(\theta)} \mathbb{E}_{\mathbb{Q}_t}\right.\right.$$

$$\cdot\left[\boldsymbol{c}_t^\top \boldsymbol{x}_t + \left(\boldsymbol{\eta}_t + \sum_{s\in[t]} \boldsymbol{B}_t^s \tilde{\boldsymbol{\xi}}_s\right)^\top \boldsymbol{f}_t(\boldsymbol{X}_t) \middle| \tilde{\boldsymbol{\xi}}_{[t-1]}\right] \cdots \middle| \tilde{\boldsymbol{\xi}}_1\right]\Bigg]$$

$$= \sup_{\mathbb{Q}_1\in\mathcal{B}_1(\theta)} \mathbb{E}_{\mathbb{Q}_1}\left[\sup_{\mathbb{Q}_2\in\mathcal{B}_2(\theta)} \mathbb{E}_{\mathbb{Q}_2}\left[\cdots \sup_{\mathbb{Q}_t\in\mathcal{B}_t(\theta)} \mathbb{E}_{\mathbb{Q}_t}\left[\boldsymbol{c}_t^\top \boldsymbol{x}_t + \left(\boldsymbol{\eta}_t + \sum_{s\in[t]} \boldsymbol{B}_t^s \tilde{\boldsymbol{\xi}}_s\right)^\top \boldsymbol{f}_t(\boldsymbol{X}_t)\right]\right]\right]$$

$$= \sup_{\mathbb{Q}_{[t]}\in\mathcal{B}_{[t]}(\theta)} \mathbb{E}_{\mathbb{Q}_{[t]}}\left[\boldsymbol{c}_t^\top \boldsymbol{x}_t + \boldsymbol{\eta}_t^\top \boldsymbol{f}_t(\boldsymbol{X}_t) + \sum_{s\in[t]} [\boldsymbol{B}_t^s \tilde{\boldsymbol{\xi}}_s]^\top \boldsymbol{f}_t(\boldsymbol{X}_t)\right].$$

其中, 第三个等式成立是因为条件期望是确定性期望. 换句话说, 对任意随机噪声 $\tilde{\boldsymbol{\xi}}_{[s-1]}$ 实现值, Wasserstein 球 $\mathcal{B}_s(\theta)$ 中的极值概率分布仅取决于参数 $[\boldsymbol{B}_t^s]^\top \boldsymbol{f}_t(\boldsymbol{X}_t)$, 即极值概率分布总是同一个分布. 因此, 命题 8.2 的结果成立.

定理 8.1 的证明

根据命题 8.2, 极值概率分布下的期望成本 $\mathrm{Z}_t(\boldsymbol{X},\theta)$ 具有如下形式:

$$\mathrm{Z}_t(\boldsymbol{X},\theta) = \sup_{\mathbb{Q}_{[t]}\in\mathcal{B}_{[t]}(\theta)} \mathbb{E}_{\mathbb{Q}_{[t]}}\left[\boldsymbol{c}_t^\top \boldsymbol{x}_t + \boldsymbol{\eta}_t^\top \boldsymbol{f}_t(\boldsymbol{X}_t) + \sum_{s\in[t]} [\boldsymbol{B}_t^s \tilde{\boldsymbol{\xi}}_s]^\top \boldsymbol{f}_t(\boldsymbol{X}_t)\right]$$

$$= \boldsymbol{c}_t^\top \boldsymbol{x}_t + \boldsymbol{\eta}_t^\top \boldsymbol{f}_t(\boldsymbol{X}_t) + \sum_{s\in[t]} \sup_{\mathbb{Q}_s\in\mathcal{B}_s(\theta)} \mathbb{E}_{\mathbb{Q}_s}\left[\tilde{\boldsymbol{\xi}}_s^\top [\boldsymbol{B}_t^s]^\top \boldsymbol{f}_t(\boldsymbol{X}_t)\right]$$

$$= \boldsymbol{c}_t^\top \boldsymbol{x}_t + \boldsymbol{\eta}_t^\top \boldsymbol{f}_t(\boldsymbol{X}_t) + \sum_{s\in[t]} \mathbb{E}_{\hat{\mathbb{Q}}_s}\left[\tilde{\boldsymbol{\xi}}_s^\top [\boldsymbol{B}_t^s]^\top \boldsymbol{f}_t(\boldsymbol{X}_t)\right] + \theta \left\|[\boldsymbol{B}_t^s]^\top \boldsymbol{f}_t(\boldsymbol{X}_t)\right\|_*$$

$$= \mathbb{E}_{\hat{\mathbb{Q}}_{[t]}}\left[\boldsymbol{c}_t^\top \boldsymbol{x}_t + \tilde{\boldsymbol{u}}_t^\top \boldsymbol{f}_t(\boldsymbol{X}_t)\right] + \sum_{s\in[t]} \theta \left\|[\boldsymbol{B}_t^s]^\top \boldsymbol{f}_t(\boldsymbol{X}_t)\right\|_*,$$

其中, 利用 (Mohajerin Esfahani and Kuhn, 2018) 中定理 6.3 的正则化结果可得第三个等式成立, 最后一个等式成立是因为 $\tilde{\boldsymbol{u}}_t = \boldsymbol{\eta}_t + \sum_{s\in[t]} \boldsymbol{B}_t^s \tilde{\boldsymbol{\xi}}_s$. 因此, $\mathrm{Z}_t(\boldsymbol{X},\theta)$ 的梯度为 $\nabla_\theta \mathrm{Z}_t(\boldsymbol{X},\theta) = \sum_{s\in[t]} \left\|[\boldsymbol{B}_t^s]^\top \boldsymbol{f}_t(\boldsymbol{X}_t)\right\|_*$. 此外, 利用上述关于 $\mathrm{Z}_t(\boldsymbol{X},\theta)$ 的等价形式及其梯度, 可以将容量充足条件下预算驱动型多阶段枢纽选址模型 (8.2) 等价地转化为如下形式:

$$\theta^\star = \max_{\boldsymbol{X}\in\mathcal{X}} \min_{t\in[T]} \left\{ \frac{\left(\tau_t - \mathbb{E}_{\hat{\mathbb{Q}}_{[t]}}\left[\boldsymbol{c}_t^\top \boldsymbol{x}_t + \tilde{\boldsymbol{u}}_t^\top \boldsymbol{f}_t(\boldsymbol{X}_t)\right]\right)^+}{\nabla_\theta \mathrm{Z}_t(\boldsymbol{X},\theta)} \right\}.$$

推论 8.1 的证明

利用命题 8.2 中向量 $\boldsymbol{\eta}_t$ 和矩阵 $\boldsymbol{B}_t^1, \cdots, \boldsymbol{B}_t^t, t \in [T]$ 的定义, 可以得到第 t 期成本的线性表达式 $\boldsymbol{c}_t^\top \boldsymbol{x}_t + \boldsymbol{\eta}_t^\top \boldsymbol{f}_t(\boldsymbol{X}_t) + \sum_{s\in[t]} [\boldsymbol{B}_t^s \boldsymbol{\xi}_s]^\top \boldsymbol{f}_t(\boldsymbol{X}_t)$. 因此, 对每个 $s \in [t]$, 在经验分布 $\hat{\mathbb{Q}}_s = \mathcal{N}(\mathbf{0}, \boldsymbol{\Sigma})$ 下总成本的方差为

$$\mathbb{V}_{\hat{\mathbb{Q}}_s}[[\boldsymbol{B}_t^s \tilde{\boldsymbol{\xi}}_s]^\top \boldsymbol{f}_t(\boldsymbol{X}_t)] = [\boldsymbol{f}_t(\boldsymbol{X}_t)]^\top \boldsymbol{B}_t^s \boldsymbol{\Sigma} [\boldsymbol{B}_t^s]^\top \boldsymbol{f}_t(\boldsymbol{X}_t).$$

此外, 利用定义在协方差矩阵 $\boldsymbol{\Sigma}$ 上的 Mahalanobis 范数, 可以将 $Z_t(\boldsymbol{X}, \theta)$ 在 θ 处的梯度转化为

$$\begin{aligned}\nabla_\theta Z_t(\boldsymbol{X}, \theta) &= \sum_{s\in[t]} \left\| [\boldsymbol{B}_t^s]^\top \boldsymbol{f}_t(\boldsymbol{X}_t) \right\|_{\mathrm{M}}^* = \sum_{s\in[t]} \sqrt{[\boldsymbol{f}_t(\boldsymbol{X}_t)]^\top \boldsymbol{B}_t^s \boldsymbol{\Sigma} [\boldsymbol{B}_t^s]^\top \boldsymbol{f}_t(\boldsymbol{X}_t)} \\ &= \sum_{s\in[t]} \sqrt{\mathbb{V}_{\hat{\mathbb{Q}}_s}[[\boldsymbol{B}_t^s \tilde{\boldsymbol{\xi}}_s]^\top \boldsymbol{f}_t(\boldsymbol{X}_t)]}.\end{aligned}$$

特别地, 在第一阶段, 即 $t = 1$ 时, 可以推出

$$\nabla_\theta Z_1(\boldsymbol{X}, \theta) = \sqrt{\mathbb{V}_{\hat{\mathbb{Q}}_1}[[\boldsymbol{B}_1^1 \tilde{\boldsymbol{\xi}}_1]^\top \boldsymbol{f}_1(\boldsymbol{x}_1)]} = \sqrt{\mathbb{V}_{\hat{\mathbb{Q}}_1}[\boldsymbol{c}_1^\top \boldsymbol{x}_1 + \tilde{\boldsymbol{u}}_1^\top \boldsymbol{f}_1(\boldsymbol{x}_1)]}.$$

定理 8.2 的证明

根据定理 8.1, 可得模型 (8.2) 的可行集为

$$\left\{ (\boldsymbol{X}, \theta) \,\middle|\, \begin{aligned} &\mathbb{E}_{\hat{\mathbb{Q}}_{[t]}} \left[\boldsymbol{c}_t^\top \boldsymbol{x}_t + \tilde{\boldsymbol{u}}_t^\top \boldsymbol{f}_t(\boldsymbol{X}_t) \right] \\ &+ \sum_{s\in[t]} \theta \| [\boldsymbol{B}_t^s]^\top \boldsymbol{f}_t(\boldsymbol{X}_t) \|_* \leqslant \tau_t,\ \forall t \in [T],\ \boldsymbol{X} \in \mathcal{X},\ \theta \geqslant 0 \end{aligned} \right\}.$$

通过提升变量 $\boldsymbol{h} = (h_{kt})_{k\in[K], t\in[T]} \in \mathbb{R}_+^{KT}$ 和 $\boldsymbol{\lambda} = (\lambda_t)_{t\in[T]} \in \mathbb{R}_+^T$, 将 $\boldsymbol{f}_t(\boldsymbol{X}_t) = (f_{kt}(\boldsymbol{X}_t))_{k\in[K]}$ 引入上述约束中, 则该可行集等价于一个混合整数锥约束系统 (8.9).

定理 8.3 的证明

当模型 (8.2) 只有唯一的最优解 $\boldsymbol{X}^\star$ 时, 则 $\theta^\star(\boldsymbol{\tau}) = \max_{\boldsymbol{X}\in\mathcal{X}} \min_{t\in[T]} \gamma_t(\boldsymbol{X})$ 严格大于 $\min_{t\in[T]} \gamma_t(\boldsymbol{X})$ $\boldsymbol{X} \in \mathcal{X} \setminus \{\boldsymbol{X}^\star\}$. 因此, 对于任意 $t \in [T]$, 如果 $\Delta\tau_t$ 足够小, 预算驱动型模型 (8.2) 的最优解集仍然是 $\{\boldsymbol{X}^\star\}$. 接下来, 在给定最优解 $\boldsymbol{X}^\star$ 下, 推导成本预算 $\boldsymbol{\tau}$ 对最优鲁棒性水平的影响. 如果 $\mathcal{T}^\star(\boldsymbol{\tau}, \boldsymbol{X}^\star) = \{t^\star\}$, 当成本

预算在 $\boldsymbol{\tau}$ 的充分小邻域内变化时, 对应的最优鲁棒性水平为

$$\theta^\star(\boldsymbol{\tau}) = \frac{\left(\tau_{t^\star} - \mathbb{E}_{\hat{\mathbb{Q}}_{[t^\star]}}\left[\boldsymbol{c}_{t^\star}^\top \boldsymbol{x}_{t^\star}^\star + \tilde{\boldsymbol{u}}_{t^\star}^\top \boldsymbol{f}_{t^\star}(\boldsymbol{X}_{t^\star}^\star)\right]\right)^+}{\sum_{s\in[t^\star]} \|[\boldsymbol{B}_{t^\star}^s]^\top \boldsymbol{f}_{t^\star}(\boldsymbol{X}_{t^\star}^\star)\|_*}.$$

则定理 8.3 的结论显然成立. 否则, 减少第 $t^\star \in \mathcal{T}^\star(\boldsymbol{\tau}, \boldsymbol{X}^\star)$ 阶段的预算, 相应的分量 $\gamma_{t^\star}(\boldsymbol{X}^\star)$ 仍然在所有阶段中最小. 因此,

$$\nabla_{\tau_{t^\star}}^- \theta^\star(\boldsymbol{\tau}) = \frac{1}{\sum_{s\in[t^\star]} \|[\boldsymbol{B}_{t^\star}^s]^\top \boldsymbol{f}_{t^\star}(\boldsymbol{X}_{t^\star}^\star)\|_*}.$$

略微减少第 $t \notin \mathcal{T}^\star(\boldsymbol{\tau}, \boldsymbol{X}^\star)$ 阶段的预算, 对应的 $\gamma_t(\boldsymbol{X}^\star)$ 依旧不是最小的. 因此, $\nabla_{\tau_t}^- \theta^\star(\boldsymbol{\tau}) = 0$. 提高第 $t \in [T]$ 阶段内预算, 因为 $\mathcal{T}^\star(\boldsymbol{\tau}, \boldsymbol{X}^\star) \setminus \{t\} \neq \varnothing$, 所以最优的鲁棒性水平保持不变, 进而推出 $\nabla_{\tau_t}^+ \theta^\star(\boldsymbol{\tau}) = 0,\ \forall t \in [T]$.

命题 8.3 的证明

利用命题 8.2 中向量 $\boldsymbol{\eta}_t$ 和矩阵 $\boldsymbol{B}_t^1, \cdots, \boldsymbol{B}_t^t, t \in [T]$ 的定义, 随机需求满足 $\tilde{\boldsymbol{u}}_t = \boldsymbol{\eta}_t + \sum_{s\in[t]} \boldsymbol{B}_t^s \tilde{\boldsymbol{\xi}}_s$. 值得注意的是, 第 t 阶段需求的随机性由 $\tilde{\boldsymbol{\xi}}_1, \cdots, \tilde{\boldsymbol{\xi}}_t$ 决定. 因此, 我们可以用随机噪声 $\tilde{\boldsymbol{\xi}}_{[t]}$ 替换决策规则 $\boldsymbol{z}(\cdot)$ 中的需求. 利用命题 8.2 相同的证明逻辑, 我们可以得到命题 8.3 的结论成立.

定理 8.4 的证明

将拓展的线性决策规则 (8.12) 代入近似问题 $\bar{\mathrm{Y}}_t(\boldsymbol{X}, \theta)$ 中, 并分别根据随机变量 $\boldsymbol{\xi}_{[t]}$, $\boldsymbol{\zeta}_{[t]}$, 和 $\pi_{[t]}$, 将目标函数和约束中的项重新排列, 可以得到如下等价问题:

$$\begin{array}{lll}
\min & \boldsymbol{c}_t^\top \boldsymbol{x}_t + \boldsymbol{w}_t^\top \boldsymbol{\eta}_t + \bar{\boldsymbol{v}}_t^\top \boldsymbol{z}_0 & \\
& \quad + \displaystyle\sum_{s\in[t]} \sup_{\mathbb{Q}_s \in \mathcal{G}_s(\theta)} \mathbb{E}_{\mathbb{Q}_s}\left[(\bar{\boldsymbol{v}}_t^\top \boldsymbol{Z}_s + \boldsymbol{w}_t^\top \boldsymbol{B}_s^t)\boldsymbol{\xi}_s + \bar{\boldsymbol{v}}_t^\top \boldsymbol{W}_s \boldsymbol{\zeta}_s + \bar{\boldsymbol{v}}_t^\top \boldsymbol{Y}_s \pi_s\right] & \\
\text{s.t.} & \left.\begin{array}{l}
\displaystyle\sum_{s\in[t]} (\boldsymbol{1}^\top \boldsymbol{Z}_s - \boldsymbol{B}_t^s)\boldsymbol{\xi}_s + \boldsymbol{1}^\top \boldsymbol{W}_s \boldsymbol{\zeta}_s + \boldsymbol{1}^\top \boldsymbol{Y}_s \pi_s \leqslant \boldsymbol{\eta}_t - \boldsymbol{1}^\top \boldsymbol{z}_0, \\
\displaystyle\sum_{s\in[t]} \boldsymbol{D}\left(\boldsymbol{Z}_s \boldsymbol{\xi}_s + \boldsymbol{W}_s \boldsymbol{\zeta}_s + \boldsymbol{Y}_s \pi_s\right) \leqslant \boldsymbol{\Gamma}_t \boldsymbol{X}_t - \boldsymbol{D}\boldsymbol{z}_0, \\
\displaystyle\sum_{s\in[t]} -\boldsymbol{Z}_s \boldsymbol{\xi}_s - \boldsymbol{W}_s \boldsymbol{\zeta}_s - \boldsymbol{Y}_s \pi_s \leqslant \boldsymbol{z}_0,
\end{array}\right\} & \begin{array}{l}\forall (\boldsymbol{\xi}_s, \boldsymbol{\zeta}_s, \pi_s) \in \mathcal{W}_{sn}, \\ \forall s \in [t],\ n \in [N],\end{array} \\
& \boldsymbol{z}_0,\ \boldsymbol{Y}_s \in \mathbb{R}^{I^2 K},\ \boldsymbol{Z}_s,\ \boldsymbol{W}_s \in \mathbb{R}^{K \times I^2 K}, & \forall s \in [t],
\end{array}$$

其中, 利用 (Bertsimas et al., 2022) 中的命题 4, 分布鲁棒约束可转化为鲁棒约束.

接下来, 我们处理目标函数中极值概率分布下的期望成本 (由 s 索引). 为了表述简洁, 令 $g(\boldsymbol{\xi}_s, \boldsymbol{\zeta}_s, \pi_s) = (\bar{\boldsymbol{v}}_t^\top \boldsymbol{Z}_s + \boldsymbol{w}_t^\top \boldsymbol{B}_s^t)\boldsymbol{\xi}_s + \bar{\boldsymbol{v}}_t^\top \boldsymbol{W}_s \boldsymbol{\zeta}_s + \bar{\boldsymbol{v}}_t^\top \boldsymbol{Y}_s \pi_s$, 并且对每个

$n \in [N]$, 假设 $\mathbb{Q}_s^n$ 是给定 $\hat{\boldsymbol{\xi}}_s^n$ 的条件分布, 我们可以推出如下等式:

$$
\begin{aligned}
&\sup_{\mathbb{Q}_s \in \mathcal{G}_s(\theta)} \mathbb{E}_{\mathbb{Q}_s}[g(\tilde{\boldsymbol{\xi}}_s, \tilde{\boldsymbol{\zeta}}_s, \tilde{\pi}_s]\\
&= \left\{ \begin{array}{ll} \sup\limits_{\mathbb{Q}_s^n, n\in[N]} & \dfrac{1}{N} \sum\limits_{n\in[N]} \displaystyle\int_{\mathcal{W}_{sn}} g(\boldsymbol{\xi}_s, \boldsymbol{\zeta}_s, \pi_s) \mathbb{Q}_s^n(d\boldsymbol{\xi}_s, d\boldsymbol{\zeta}_s, d\pi_s) \\ \text{s.t.} & \dfrac{1}{N} \sum\limits_{n\in[N]} \displaystyle\int_{\mathcal{W}_{sn}} \pi_s \mathbb{Q}_s^n(d\boldsymbol{\xi}_s, d\boldsymbol{\zeta}_s, d\pi_s) \leqslant \theta \end{array} \right. \\
&= \left\{ \begin{array}{l} \sup\limits_{\mathbb{Q}_s^n, n\in[N]} \inf\limits_{\lambda_s \geqslant 0} \quad \dfrac{1}{N} \sum\limits_{n\in[N]} \displaystyle\int_{\mathcal{W}_{sn}} g(\boldsymbol{\xi}_s, \boldsymbol{\zeta}_s, \pi_s) \mathbb{Q}_s^n(d\boldsymbol{\xi}_s, d\boldsymbol{\zeta}_s, d\pi_s) \\ \qquad\qquad + \lambda_s \left(\theta - \dfrac{1}{N} \sum\limits_{n\in[N]} \displaystyle\int_{\mathcal{W}_{sn}} \pi_s \mathbb{Q}_s^n(d\boldsymbol{\xi}_s, d\boldsymbol{\zeta}_s, d\pi_s) \right) \end{array} \right. \\
&= \inf_{\lambda_s \geqslant 0} \sup_{\mathbb{Q}_s^n, n\in[N]} \left\{ \lambda_s \theta + \frac{1}{N} \sum_{n\in[N]} \int_{\mathcal{W}_{sn}} (g(\boldsymbol{\xi}_s, \boldsymbol{\zeta}_s, \pi_s) - \lambda_s \pi_s) \, \mathbb{Q}_s^n(d\boldsymbol{\xi}_s, d\boldsymbol{\zeta}_s, d\pi_s) \right\} \\
&= \inf_{\lambda_s \geqslant 0} \left\{ \lambda_s \theta + \frac{1}{N} \sum_{n\in[N]} \sup_{(\boldsymbol{\xi}_s, \boldsymbol{\zeta}_s, \pi_s)\in\mathcal{W}_{sn}} \{ g(\boldsymbol{\xi}_s, \boldsymbol{\zeta}_s, \pi_s) - \lambda_s \tilde{\pi}_s \} \right\} \\
&= \left\{ \begin{array}{ll} \inf\limits_{\lambda_s \geqslant 0} & \lambda_s \theta + \dfrac{1}{N} \sum\limits_{n\in[N]} \sigma_s^n \\ \text{s.t.} & g(\boldsymbol{\xi}_s, \boldsymbol{\zeta}_s, \pi_s) - \lambda_s \tilde{\pi}_s \leqslant \sigma_s^n \quad \forall (\boldsymbol{\xi}_s, \boldsymbol{\zeta}_s, \pi_s) \in \mathcal{W}_{sn}, \ n \in [N], \end{array} \right.
\end{aligned}
$$

其中, 第一个等式由 $\mathcal{G}_s(\theta)$ 的定义可得, 第三个等式利用最小最大化定理推出 (Sion, 1958). 因此, 定理 8.4 的结果成立.

附录 B　凸优化基础

B.1　凸 集 与 锥

定义 B.1 (凸集)　集合 $\boldsymbol{C}$ 被称为凸集, 如果集合 $\boldsymbol{C}$ 中任意两点间的线段都位于该集合中, 即对于任意的 $\boldsymbol{x}_1, \boldsymbol{x}_2 \in \boldsymbol{C}$ 以及 $\lambda \in [0,1]$, 我们有 $\lambda \boldsymbol{x}_1 + (1-\lambda)\boldsymbol{x}_2 \in \boldsymbol{C}$.

凸集有以下重要性质:

(1) (任意个) 凸集的交集是凸集, 即若对于任意的 $i \in [I]$, $\boldsymbol{C}_i$ 都是凸集, 那么 $\bigcap_{i\in[I]} \boldsymbol{C}_i$ 也是凸集, 但凸集的并集不一定是凸集.

(2) 凸集在线性函数映射下的象也是凸集, 即若 $\boldsymbol{C}$ 是凸集, 那么集合 $\{\boldsymbol{A}\boldsymbol{x}+b \mid \boldsymbol{x} \in \boldsymbol{C}\}$ 也是凸集; 凸集在线性函数映射下的原象也是凸集, 即若 $\boldsymbol{C}$ 是凸集, 那么集合 $\{\boldsymbol{x} \mid \boldsymbol{A}\boldsymbol{x} + \boldsymbol{b} \in \boldsymbol{C}\}$ 也是凸集.

(3) 凸集在它的某些维度的投影也是凸集, 即 $\boldsymbol{C}$ 是凸集当且仅当集合 $\{\boldsymbol{x} \mid \exists \boldsymbol{y} : (\boldsymbol{x}, \boldsymbol{y}) \in \boldsymbol{C}\}$ 是凸集.

下面, 我们介绍一些重要的凸集, 本书后续部分将多次涉及这些凸集.

例 24 (子空间)　若 n 维欧氏空间中的非空集合 $\boldsymbol{S}$ 满足: 对于集合 $\boldsymbol{S}$ 中任意两元素 $\boldsymbol{x}, \boldsymbol{y}$ 以及任意两实数 a, b, $a\boldsymbol{x} + b\boldsymbol{y} \in \boldsymbol{S}$, 则称集合 $\boldsymbol{S}$ 是 n 维欧氏空间中的子空间.

可以看出集合 $\boldsymbol{S}_1 = \left\{(x,0)^\top \mid x \in \mathbb{R}\right\} \subseteq \mathbb{R}^2$ 是凸集, 且是子空间. 而 $\boldsymbol{S}_2 = \boldsymbol{S}_1 + (0,1)^\top = \{(x,1) \mid x \in \mathbb{R}\} \subseteq \mathbb{R}^2$ 是凸集, 但不是子空间. 给定矩阵 $\boldsymbol{A} \in \mathbb{R}^{m\times n}$, 那么零空间与值域空间

$$\mathcal{N}(\boldsymbol{A}) := \{\boldsymbol{x} \in \mathbb{R}^n \mid \boldsymbol{A}\boldsymbol{x} = \boldsymbol{0}\}, \quad \mathcal{R}\left(\boldsymbol{A}^\top\right) := \left\{\boldsymbol{A}^\top \boldsymbol{z} \in \mathbb{R}^n \mid \boldsymbol{z} \in \mathbb{R}^m\right\}$$

都是 $\mathbb{R}^n$ 中的子空间.

例 25 (超平面与半空间)　若集合 $\boldsymbol{C}$ 可表示为 $\{\boldsymbol{x} \mid \boldsymbol{a}^\top \boldsymbol{x} = b\}$ 的形式, 其中 $\boldsymbol{a} \in \mathbb{R}^n$, $\boldsymbol{a} \neq 0$ 且 $b \in \mathbb{R}$, 我们称集合 $\boldsymbol{C}$ 为超平面; 若集合 $\boldsymbol{C}$ 可表示为 $\{\boldsymbol{x} \mid \boldsymbol{a}^\top \boldsymbol{x} \leqslant b\}$ 的形式, 我们称集合 $\boldsymbol{C}$ 为半空间.

例 26 (支持向量机)　给定 n 维欧氏空间中的点 $\boldsymbol{x}$, 该点到超平面 $\{\boldsymbol{x} \mid \boldsymbol{a}^\top \boldsymbol{x} = b\}$ 的距离为

$$(\boldsymbol{x} - \boldsymbol{x}_0)^\top \frac{\boldsymbol{a}}{\|\boldsymbol{a}\|} = \frac{\boldsymbol{a}^\top \boldsymbol{x} - b}{\|\boldsymbol{a}\|},$$

其中 $\boldsymbol{x}_0$ 满足 $\boldsymbol{a}^\top\boldsymbol{x}_0=b$. 假设训练集 $\{(\hat{\boldsymbol{x}}_i,\hat{y}_i)\}$ 是可分的, 其中 $\hat{\boldsymbol{x}}_i$ 为样本点, $\hat{y}_i$ 为与之对应的标签, 取值为 $\{-1,+1\}$, 那么对于任意的 $i\in[N]$, 有

$$\hat{y}_i\left(\frac{\boldsymbol{a}^\top\hat{\boldsymbol{x}}_i-b}{\|\boldsymbol{a}\|}\right)>0.$$

因此, 根据最大化可分类间隔的优化目标, 可写出支持向量机模型如下:

$$\max_{\boldsymbol{a},b}\left\{M:\hat{y}_i\left(\frac{\boldsymbol{a}^\top\hat{\boldsymbol{x}}_i-b}{\|\boldsymbol{a}\|}\right)\geqslant M,\forall i\in[N]\right\}.$$

例 27 若集合 $\boldsymbol{C}$ 可表达为 $\{\boldsymbol{x}\in\mathbb{R}^n\mid\boldsymbol{A}\boldsymbol{x}\geqslant\boldsymbol{b}\}$, 其中 $\boldsymbol{A}\in\mathbb{R}^{m\times n}$ 且 $\boldsymbol{b}\in\mathbb{R}^m$, 则我们称集合 $\boldsymbol{C}$ 为多面体 (polyhedron). 可以看出, 多面体是有限个半空间和超平面的交集, 且多面体是凸集.

定义 B.2 (凸包) 有限集合 $\{\boldsymbol{x}_1,\boldsymbol{x}_2,\cdots,\boldsymbol{x}_n\}$ 的凸包 (convex hull) 由这 n 个元素的凸组合所构成, 记为 $\operatorname{conv}\{\boldsymbol{x}_1,\boldsymbol{x}_2,\cdots,\boldsymbol{x}_n\}$, 即

$$\operatorname{conv}\{\boldsymbol{x}_1,\boldsymbol{x}_2,\cdots,\boldsymbol{x}_n\}:=\left\{\sum_{i=1}^n\lambda_i\boldsymbol{x}_i\mid\lambda_i\geqslant 0,i=1,2,\cdots,n,\ \sum_{i=1}^n\lambda_i=1\right\}.$$

顾名思义, 一个集合的凸包总是凸的. 事实上, 可以证明集合 $\{\boldsymbol{x}_1,\boldsymbol{x}_2,\cdots,\boldsymbol{x}_n\}$ 的凸包是包含 $\{\boldsymbol{x}_1,\boldsymbol{x}_2,\cdots,\boldsymbol{x}_n\}$ 最小的凸集. 也就是说, 如果 $\boldsymbol{B}$ 是任一包含 $\{\boldsymbol{x}_1,\boldsymbol{x}_2,\cdots,\boldsymbol{x}_n\}$ 的凸集, 那么有 $\operatorname{conv}\{\boldsymbol{x}_1,\boldsymbol{x}_2,\cdots,\boldsymbol{x}_n\}\subseteq\boldsymbol{B}$. 下面给出有界多面体的凸包描述.

定理 B.1 任意有界多面体 $\{\boldsymbol{x}\in\mathbb{R}^n\mid\boldsymbol{A}\boldsymbol{x}\geqslant\boldsymbol{b}\}$ 都可以写成由有限个向量所构成的集合的凸包.

定义 B.3 (仿射包) 集合 $\boldsymbol{C}$ 的仿射包 (affine hull) 由 $\boldsymbol{C}$ 中所有的点的仿射组合所构成, 记为 $\operatorname{aff}(\boldsymbol{C})$, 即

$$\operatorname{aff}(\boldsymbol{C}):=\left\{\sum_{i=1}^n\lambda_i\boldsymbol{x}_i\mid\boldsymbol{x}_i\in\boldsymbol{C},i=1,2,\cdots,n,\ \sum_{i=1}^n\lambda_i=1\right\}.$$

我们定义集合 $\boldsymbol{C}$ 的仿射维数为其仿射包的维数. 如果集合 $\boldsymbol{C}\subseteq\mathbb{R}^n$ 的仿射维数小于 n, 那么这个集合在其仿射包中.

定义 B.4 (相对内部) 定义 $\boldsymbol{C}$ 的相对内部 (relative interior) 为 $\operatorname{aff}(\boldsymbol{C})$ 的内部, 记为 $\operatorname{ri}(\boldsymbol{C})$, 即

$$\operatorname{ri}(\boldsymbol{C}):=\left\{\boldsymbol{x}\in\boldsymbol{C}\mid\text{对于某些}r>0,\mathbb{B}(\boldsymbol{x},r)\cap\operatorname{aff}(\boldsymbol{C})\subseteq\boldsymbol{C}\right\},$$

其中 $\mathbb{B}(\boldsymbol{x},r)=\{\boldsymbol{y}\mid\|\boldsymbol{y}-\boldsymbol{x}\|\leqslant r\}$.

定义 B.5 (锥与凸锥)　集合 $\boldsymbol{C}$ 被称为锥 (cone), 如果对于任意的 $\boldsymbol{x}\in\boldsymbol{C}$ 以及 $\lambda\geqslant 0$, $\lambda\boldsymbol{x}\in\boldsymbol{C}$. 此外, 如果锥 $\boldsymbol{C}$ 还是凸的, 我们称 $\boldsymbol{C}$ 是一个凸锥, 即对于任意的 $\boldsymbol{x}_1,\boldsymbol{x}_2\in\boldsymbol{C}$ 和 $\lambda\in[0,1]$, $\lambda\boldsymbol{x}_1+(1-\lambda)\boldsymbol{x}_2\in\boldsymbol{C}$.

对于任意 $\lambda_i\geqslant 0$, $i=1,\cdots,n$, 具有 $\lambda_1\boldsymbol{x}_1+\cdots+\lambda_n\boldsymbol{x}_n$ 形式的点称为 $\{\boldsymbol{x}_1,\boldsymbol{x}_2,\cdots,\boldsymbol{x}_n\}$ 的锥组合 (或非负线性组合). 如果 $\boldsymbol{x}_i$ 均属于锥 $\boldsymbol{C}$, 那么 $\boldsymbol{x}_i$ 的每一个锥组合也在 $\boldsymbol{C}$ 中.

定义 B.6 (锥包)　有限集合 $\{\boldsymbol{x}_1,\boldsymbol{x}_2,\cdots,\boldsymbol{x}_n\}$ 的锥包 (conic hull) 由这 n 个元素的锥组合所组成, 记为 $\operatorname{cone}\{\boldsymbol{x}_1,\boldsymbol{x}_2,\cdots,\boldsymbol{x}_n\}$, 即

$$\operatorname{cone}\{\boldsymbol{x}_1,\boldsymbol{x}_2,\cdots,\boldsymbol{x}_n\}:=\left\{\sum_{i=1}^n\lambda_i\boldsymbol{x}_i\mid\lambda_i\geqslant 0,i=1,2,\cdots,n\right\}.$$

同样地, 我们可以证明集合 $\{\boldsymbol{x}_1,\boldsymbol{x}_2,\cdots,\boldsymbol{x}_n\}$ 的锥包是包含 $\{\boldsymbol{x}_1,\boldsymbol{x}_2,\cdots,\boldsymbol{x}_n\}$ 的最小凸锥. 下面, 我们给出一些常用的锥.

例 28　非负象限 $\mathbb{R}^n_+$ 以及洛伦兹锥 (二阶锥) 是两个常用的凸锥, 其中后者形式如下:

$$\mathcal{L}^m:=\{\boldsymbol{x}=(x_1,x_2,\cdots,x_m)^\top\in\mathbb{R}^m\mid\|(x_1,x_2,\cdots,x_{m-1})\|_2\leqslant x_m\}.$$

另一个著名的凸锥是半正定锥 $\mathcal{S}^m_+$. 半正定锥是由所有对称半正定矩阵所构成的集合.

例 29 (正常锥)　一个锥 $\mathcal{K}\in\mathbb{R}^n$ 被称为正常锥 (proper cone), 如果 (1)$\mathcal{K}$ 是凸集; (2)$\mathcal{K}$ 是闭集; (3)$\mathcal{K}$ 有非空的内部; (4)$\mathcal{K}$ 不含有直线 (即若 $\boldsymbol{x}\in\mathcal{K}$ 且 $-\boldsymbol{x}\in\mathcal{K}$, 则 $\boldsymbol{x}=\boldsymbol{0}$).

正常锥 $\mathcal{K}$ 可以用来定义广义不等式, 进而构造了 $\mathbb{R}^n$ 空间上的一个偏序关系, 并且与常用的 “$\geqslant$” 有许多类似性质. 具体而言, 给定正常锥 $\mathcal{K}$, 可以定义偏序关系 “$\succeq_{\mathcal{K}}$” 如下:

$$\boldsymbol{x}\succeq_{\mathcal{K}}\boldsymbol{y}\iff\boldsymbol{x}-\boldsymbol{y}\in\mathcal{K}.$$

广义不等式 “$\succeq_{\mathcal{K}}$” 满足许多常用的性质, 如传递性、反身性等.

定义 B.7 (对偶锥)　假设 $\mathcal{K}\in\mathbb{R}^n$ 是一个锥, 那么集合 $\mathcal{K}^*=\{\boldsymbol{y}\in\mathbb{R}^n\mid\boldsymbol{x}^\top\boldsymbol{y}\geqslant 0,\forall\boldsymbol{x}\in\mathcal{K}\}$ 称为 $\mathcal{K}$ 的对偶锥.

根据对偶锥的定义不难看出, $\mathcal{K}^*$ 是闭凸集. 同时如果 $\mathcal{K}_1\subseteq\mathcal{K}_2$, 那么 $\mathcal{K}_2^*\subseteq\mathcal{K}_1^*$. 此外, 若 $\mathcal{K}$ 是一个正常锥, 那么它的对偶锥 $\mathcal{K}^*$ 也是正常锥, 且 $(\mathcal{K}^*)^*=\mathcal{K}$.

例 30　如下为一些常见对偶锥的表达形式.

(1) 非负象限、二阶锥以及半正定锥是自对偶 (self-dual) 的, 即 $\mathcal{K}^*=\mathcal{K}$.

(2) 指数锥 $\mathcal{K}_{\exp} = \mathrm{cl}\{(x_1, x_2, x_3) \in \mathbb{R}^3 \mid x_1 \geqslant x_2 \exp(x_3/x_2),\ x_1, x_2 > 0\}$ 的对偶锥为

$$\mathcal{K}_{\exp}^* = \mathrm{cl}\left\{(y_1, y_2, y_3) \in \mathbb{R}^3 \mid -y_1 \log(-y_1/y_3) + y_1 \leqslant y_2, y_1 < 0, y_3 > 0\right\}.$$

(3) 假设 $\boldsymbol{\alpha} > 0$ 且 $\sum_{i\in[n]} \alpha_i = 1$, 那么幂锥 $\mathcal{K}_{\boldsymbol{\alpha}} = \mathrm{cl}\{(\boldsymbol{x}, y) \in \mathbb{R}^{n+1} \mid |y| \leqslant x_1^{\alpha_1} \cdots x_n^{\alpha_n}, \boldsymbol{x} \geqslant \boldsymbol{0}\}$ 的对偶锥为

$$\mathcal{K}_{\boldsymbol{\alpha}}^* = \left\{(\boldsymbol{u}, v) \in \mathbb{R}^{n+1} \mid |v| \leqslant (u_1/\alpha_1)^{\alpha_1} \cdots (u_1/\alpha_n)^{\alpha_n}, \boldsymbol{u} \geqslant \boldsymbol{0}\right\}.$$

(4) 给定 $\mathbb{R}^n$ 中的范数 $\|\cdot\|$, 那么范数锥 $\mathcal{K}_{\|\cdot\|} = \{(\boldsymbol{x}, t) \in \mathbb{R}^{n+1} \mid \|\boldsymbol{x}\| \leqslant t\}$ 的对偶锥为

$$\mathcal{K}_{\|\cdot\|}^* = \left\{(\boldsymbol{u}, v) \in \mathbb{R}^{n+1} \mid \|\boldsymbol{u}\|_* \leqslant v\right\}.$$

例 31 多面体 $\{\boldsymbol{x} \mid \boldsymbol{A}\boldsymbol{x} \geqslant 0\}$ 也是一个凸锥, 叫做多面体锥. 给定多面体 $\boldsymbol{P} = \{\boldsymbol{y} \mid \boldsymbol{A}\boldsymbol{y} \geqslant \boldsymbol{b}\}$, 那么其回收锥 (recession cone) 为

$$\{\boldsymbol{x} \in \mathbb{R}^n \mid \boldsymbol{A}(\boldsymbol{y} + \lambda \boldsymbol{x}) \geqslant \boldsymbol{b}, \forall \lambda \geqslant 0\} = \{\boldsymbol{x} \in \mathbb{R}^n \mid \boldsymbol{A}\boldsymbol{x} \geqslant \boldsymbol{0}\}.$$

下面, 我们给出任意多面体的锥包描述.

定理 B.2 任何一个多面体锥 $\boldsymbol{C} = \{\boldsymbol{x} \in \mathbb{R}^n \mid \boldsymbol{A}\boldsymbol{x} \geqslant 0\}$ 都可以由有限个向量生成. 换言之, 存在有限集合 $\{\boldsymbol{b}_1, \boldsymbol{b}_2, \cdots, \boldsymbol{b}_m\} \subseteq \mathbb{R}^n$ 使得 $\boldsymbol{C} = \mathrm{cone}\{\boldsymbol{b}_1, \boldsymbol{b}_2, \cdots, \boldsymbol{b}_m\}$, 即

$$\{\boldsymbol{x} \mid \boldsymbol{A}\boldsymbol{x} \geqslant \boldsymbol{0}\} = \left\{\sum_{i=1}^m \lambda_i \boldsymbol{b}_i,\ \lambda_i \geqslant 0, i = 1, 2, \cdots, m\right\}.$$

凸分析中最基本的结果之一是分离超平面定理. 我们首先给出单点集和闭凸集的分离形式.

定理 B.3 (分离超平面定理) 假定 $\boldsymbol{S}$ 是 n 维欧氏空间上的非空闭凸集, 点 $\boldsymbol{x}^\star$ 为集合 $\boldsymbol{S}$ 外一点. 那么存在向量 $\boldsymbol{c} \in \mathbb{R}^n$, 使得对于任意的 $\boldsymbol{x} \in \boldsymbol{S}$, 我们有 $\boldsymbol{c}^\top \boldsymbol{x}^\star < \boldsymbol{c}^\top \boldsymbol{x}$.

分离超平面定理的一个直接应用是 Farkas 引理.

定理 B.4 (Farkas 引理) 假定 $\boldsymbol{A} \in \mathbb{R}^{m\times n}$, $\boldsymbol{b} \in \mathbb{R}^m$. 那么以下两个断言有且仅有一个成立:

(1) 存在 $\boldsymbol{x} \geqslant 0$ 使得 $\boldsymbol{A}\boldsymbol{x} = \boldsymbol{b}$.

(2) 存在向量 $\boldsymbol{p}$ 使得 $\boldsymbol{p}^\top \boldsymbol{A} \geqslant \boldsymbol{0}^\top$ 且 $\boldsymbol{p}^\top \boldsymbol{b} < 0$.

定理 B.5 假设 $\boldsymbol{C}$ 和 $\boldsymbol{D}$ 是两个非空且不相交的凸集, 那么存在 $\boldsymbol{a} \neq 0$ 以及 $b \in \mathbb{R}$ 使得

$$\boldsymbol{a}^\top \boldsymbol{x} \leqslant b,\ \forall \boldsymbol{x} \in \boldsymbol{C};\quad \boldsymbol{a}^\top \boldsymbol{x} \geqslant b,\ \forall \boldsymbol{x} \in \boldsymbol{D}.$$

设 $\boldsymbol{C} \subseteq \mathbb{R}^n$, 且 $\boldsymbol{x}_0$ 是其边界上的一点. 如果 $\boldsymbol{a} \neq 0$, 并且对任意 $\boldsymbol{x} \in \boldsymbol{C}$ 满足 $\boldsymbol{a}^\top \boldsymbol{x} \leqslant \boldsymbol{a}^\top \boldsymbol{x}_0$, 那么称超平面 $\left\{\boldsymbol{x} \mid \boldsymbol{a}^\top \boldsymbol{x} = \boldsymbol{a}^\top \boldsymbol{x}_0\right\}$ 为集合 $\boldsymbol{C}$ 在点 $\boldsymbol{x}_0$ 处的支撑超平面, 这等于说点 $\boldsymbol{x}_0$ 与集合 $\boldsymbol{C}$ 被超平面所分离 $\left\{\boldsymbol{x} \mid \boldsymbol{a}^\top \boldsymbol{x} = \boldsymbol{a}^\top \boldsymbol{x}_0\right\}$. 接下来, 我们给出另一种形式的分离超平面定理.

定理 B.6 假设 $\boldsymbol{C}$ 是非空凸集, $\boldsymbol{x}_0$ 是集合 $\boldsymbol{C}$ 边界上的点, 那么存在 $\boldsymbol{x}_0$ 处的支撑超平面, 即集合 $\boldsymbol{S} = \{\boldsymbol{x} \mid \boldsymbol{a}^\top(\boldsymbol{x} - \boldsymbol{x}_0) = 0\}$ 在点 $\boldsymbol{x}_0$ 处与集合 $\boldsymbol{C}$ 相切, 且对于集合 $\boldsymbol{C}$ 中的任意一点, 有 $\boldsymbol{a}^\top(\boldsymbol{x} - \boldsymbol{x}_0) \leqslant 0$.

利用上述分离超平面定理, 我们可以得出另外一个重要的结果——支撑超平面定理.

定理 B.7 (支撑超平面定理) 假设 $\boldsymbol{C}$ 是非空凸集, 那么对于任意给定的集合 $\boldsymbol{C}$ 的边界点 $\boldsymbol{x}_0$, 存在一个点 $\boldsymbol{x}_0$ 处的支撑超平面 $\boldsymbol{S} = \{\boldsymbol{x} \mid \boldsymbol{a}^\top(\boldsymbol{x} - \boldsymbol{x}_0) = 0\}$ 与集合 $\boldsymbol{C}$ 相切于 $\boldsymbol{x}_0$, 并且

$$\boldsymbol{a}^\top(\boldsymbol{x} - \boldsymbol{x}_0) \leqslant 0, \quad \forall \boldsymbol{x} \in \boldsymbol{C}.$$

B.2 凸 函 数

定义 B.8 假设 $\boldsymbol{C} \subseteq \mathbb{R}^n$ 是凸集, 那么实值函数 $f(\boldsymbol{x}) : \boldsymbol{C} \mapsto \mathbb{R}$ 是凸的, 如果对于任意的 $\boldsymbol{x}_1, \boldsymbol{x}_2 \in \boldsymbol{C}$ 以及 $\lambda \in [0, 1]$, 如下不等式成立:

$$f(\lambda \boldsymbol{x}_1 + (1 - \lambda)\boldsymbol{x}_2) \leqslant \lambda f(\boldsymbol{x}_1) + (1 - \lambda) f(\boldsymbol{x}_2).$$

若当 $\boldsymbol{x}_1 \neq \boldsymbol{x}_2$ 且 $0 < \lambda < 1$ 时, 上述不等式严格成立, 我们称函数 f 是严格凸函数. 此外, 如果对于任意的 $\boldsymbol{x}_1, \boldsymbol{x}_2 \in \boldsymbol{C}$ 以及 $\lambda \in [0, 1]$, 有 $f(\lambda \boldsymbol{x}_1 + (1 - \lambda)\boldsymbol{x}_2) \geqslant \lambda f(\boldsymbol{x}_1) + (1 - \lambda) f(\boldsymbol{x}_2)$, 那么我们称函数 f 是凹函数. 不难看出, $f(\boldsymbol{x})$ 是凸函数当且仅当 $-f(\boldsymbol{x})$ 是凹函数. 常见的凸函数有多项式函数、绝对值函数、范数函数等.

定义 B.9 (对偶范数) 假设 $\|\cdot\|$ 是 $\mathbb{R}^n$ 上的范数, 这包括 ℓ_1, ℓ_2 或 ℓ_∞ 范数. 我们记 $\|\cdot\|^*$ 为 $\|\cdot\|$ 的对偶范数, 其中 $\|\cdot\|^*$ 定义如下:

$$\|\boldsymbol{y}\|^* = \sup_{\|\boldsymbol{x}\| \leqslant 1} \boldsymbol{y}^\top \boldsymbol{x}.$$

关于对偶范数, 常用的结论如下:

$$\|\boldsymbol{y}\|_1^* = \sup_{\|\boldsymbol{x}\|_1 \leqslant 1} \boldsymbol{y}^\top \boldsymbol{x} = \|\boldsymbol{y}\|_\infty, \quad \|\boldsymbol{y}\|_2^* = \sup_{\|\boldsymbol{x}\|_2 \leqslant 1} \boldsymbol{y}^\top \boldsymbol{x} = \|\boldsymbol{y}\|_2,$$

$$\|\boldsymbol{y}\|_\infty^* = \sup_{\|\boldsymbol{x}\|_\infty \leqslant 1} \boldsymbol{y}^\top \boldsymbol{x} = \|\boldsymbol{y}\|_1.$$

此外, 利用 Hölder 不等式可知: 对于任意给定的 $p \in (1, \infty)$, $\|\boldsymbol{y}\|_p^* = \|\boldsymbol{y}\|_q$, 其中 $1/p + 1/q = 1$.

定义 B.10 (梯度) 定义在 $\mathbb{R}^n$ 上的函数 $f(\boldsymbol{x})$ 在点 $\hat{\boldsymbol{x}}$ 处的梯度向量, 记作 $\nabla f(\hat{\boldsymbol{x}})$, 是由一阶偏导数所组成的向量:

$$\nabla f(\hat{\boldsymbol{x}}) := \left(\frac{\partial f}{\partial x_1}(\hat{\boldsymbol{x}}), \frac{\partial f}{\partial x_2}(\hat{\boldsymbol{x}}), \cdots, \frac{\partial f}{\partial x_n}(\hat{\boldsymbol{x}})\right)^\top.$$

函数 $f(\boldsymbol{x})$ 在点 $\hat{\boldsymbol{x}}$ 处的黑塞矩阵, 记作 $\mathbf{H}_f(\hat{\boldsymbol{x}})$, 是由二阶偏导数所构成的矩阵:

$$\mathbf{H}_f(\hat{\boldsymbol{x}})_{ij} = \frac{\partial^2 f}{\partial x_i \partial x_j}(\hat{\boldsymbol{x}}).$$

例 32 假设 $f(\boldsymbol{x}) = \dfrac{1}{2}\boldsymbol{x}^\top \mathbf{M}\boldsymbol{x}$, 其中 $\mathbf{M}$ 是对称矩阵, 那么 $\nabla f(\boldsymbol{x}) = \mathbf{M}\boldsymbol{x}$ 且 $\mathbf{H}_f(\boldsymbol{x}) = \mathbf{M}$.

例 33 (最小二乘估计) 给定样本 $\{\boldsymbol{x}_i, y_i\}_{i=1}^n$, 且各个样本间相互独立. 现我们想找到一条直线来拟合数据 $\{\boldsymbol{x}_i, y_i\}_{i=1}^n$, 那么我们可以写出相应的回归方程为 $\hat{y}_i(\boldsymbol{\beta}) = \boldsymbol{x}_i^\top \boldsymbol{\beta}$, $i \in [n]$, 其中 $\boldsymbol{\beta}$ 为待估计参数. 而参数 $\boldsymbol{\beta}$ 的最小二乘估计量 $\hat{\boldsymbol{\beta}}$ 可以通过求解以下二阶锥规划问题得出

$$\min_{\boldsymbol{\beta}} \sum_{i=1}^n (y_i - \hat{y}_i(\boldsymbol{\beta}))^2 = \sum_{i=1}^n (y_i - \boldsymbol{x}_i^\top \boldsymbol{\beta})^2 = \|\boldsymbol{y} - \boldsymbol{X}\boldsymbol{\beta}\|_2^2,$$

其中 $\boldsymbol{X} = (\boldsymbol{x}_1, \cdots, \boldsymbol{x}_n)^\top$, $\boldsymbol{y} = (y_1, \cdots, y_n)$. 不难得出上述优化问题的法方程 (normal equation) 为 $\boldsymbol{X}^\top \boldsymbol{X}\boldsymbol{\beta} = \boldsymbol{X}^\top \boldsymbol{y}$, 求解得到

$$\boldsymbol{\beta}^* = (\boldsymbol{X}^\top \boldsymbol{X})^{-1} \boldsymbol{X}^\top \boldsymbol{y}.$$

这里我们假设对称矩阵 $\boldsymbol{X}^\top \boldsymbol{X}$ 是非奇异的.

定理 B.8 (一阶条件) 假定函数 $f(\boldsymbol{x}) : \mathbb{R}^n \longmapsto \mathbb{R}$ 的定义域 $\mathrm{dom}(f)$ 是凸集, 且在其定义域上可微. 那么 $f(\boldsymbol{x})$ 是凸函数当且仅当对任意的 $\boldsymbol{x}, \boldsymbol{y} \in \mathrm{dom}(f)$,

$$f(\boldsymbol{y}) \geqslant f(\boldsymbol{x}) + \nabla f(\boldsymbol{x})^\top (\boldsymbol{y} - \boldsymbol{x}).$$

定理 B.9 (二阶条件) 假设函数 $f(\boldsymbol{x}) : \mathbb{R}^n \longmapsto \mathbb{R}$ 的定义域 $\mathrm{dom}(f)$ 是凸集, 且在其定义域上二次可微. 那么 $f(\boldsymbol{x})$ 是凸函数当且仅当其黑塞矩阵 $\mathbf{H}_f(\boldsymbol{x})$ 为半正定矩阵, 即对任意给定 $\boldsymbol{x}$,

$$\boldsymbol{z}^\top \mathbf{H}_f(\boldsymbol{x})\boldsymbol{z} \geqslant 0, \quad \forall \boldsymbol{z} \in \mathbb{R}^n.$$

例 34 (二次函数)　假定 $\boldsymbol{M}$ 是对称矩阵, 且函数 $f(\boldsymbol{x})=\frac{1}{2}\boldsymbol{x}^\top\boldsymbol{M}\boldsymbol{x}+\boldsymbol{b}^\top\boldsymbol{x}+c$, 那么 f 是凸函数当且仅当 $\boldsymbol{M}$ 是半正定矩阵.

例 35 (方差函数)　假定 $\boldsymbol{\xi}=(\xi_1,\xi_2,\cdots,\xi_n)^\top$ 为 n 个资产的回报率, 而向量 $\boldsymbol{x}$ 为投资者在 n 个资产上的投资比例, 那么投资总回报的方差函数 $f(\boldsymbol{x})=\mathrm{Var}(\boldsymbol{\xi}^\top\boldsymbol{x})$ 是凸函数. 事实上,

$$\mathrm{Var}(\boldsymbol{\xi}^\top\boldsymbol{x})=\mathbb{E}\Big[(\boldsymbol{\xi}-\boldsymbol{\mu})^\top\boldsymbol{x}\Big]^2=\boldsymbol{x}^\top\underbrace{\mathbb{E}\Big[(\boldsymbol{\xi}-\boldsymbol{\mu})(\boldsymbol{\xi}-\boldsymbol{\mu})^\top\Big]}_{\boldsymbol{\Sigma}}\boldsymbol{x},$$

其中 $\boldsymbol{\mu}$, $\boldsymbol{\Sigma}$ 分别为资产回报率的均值以及协方差矩阵. 由于协方差矩阵是半正定的, 利用凸函数的二阶条件判定定理, 即可证明方差函数 $f(\boldsymbol{x})=\mathrm{Var}(\boldsymbol{\xi}^\top\boldsymbol{x})$ 为凸函数.

凸函数有以下重要性质:

(1) 如果 $f_1(\boldsymbol{x})$ 以及 $f_2(\boldsymbol{x})$ 都是凸函数, 那么对于任意给定 $a,b\geqslant 0$, $af_1(\boldsymbol{x})+bf_2(\boldsymbol{x})$ 也是凸函数.

(2) 如果 $f(\boldsymbol{x})$ 是凸函数, 令 $\boldsymbol{x}=\boldsymbol{A}\boldsymbol{y}+\boldsymbol{b}$, 那么函数 $g(\boldsymbol{y}):=f(\boldsymbol{A}\boldsymbol{y}+\boldsymbol{b})$ 关于 $\boldsymbol{y}$ 也是凸函数.

(3) 如果对于任意的 $i\in\mathcal{I}$, $f_i(\boldsymbol{x})$ 都是凸函数, 那么 $g(\boldsymbol{x}):=\max_{i\in\mathcal{I}}f_i(\boldsymbol{x})$ 也是凸函数.

(4) 假定函数 $f(\boldsymbol{x},\boldsymbol{y})$ 关于 $(\boldsymbol{x},\boldsymbol{y})$ 是联合凸函数且集合 $\boldsymbol{C}$ 是凸集, 则函数 $g(\boldsymbol{x})=\inf_{\boldsymbol{y}\in\boldsymbol{C}}f(\boldsymbol{x},\boldsymbol{y})$ 是凸函数.

(5) 如果函数 $f(\boldsymbol{x})$ 在集合 $\boldsymbol{C}$ 上是凸函数, 那么它的透视函数 (perspective function)$g(\boldsymbol{x},t)=tf(\boldsymbol{x}/t)$ 在定义域 $\{(\boldsymbol{x},t)\mid \boldsymbol{x}/t\in\boldsymbol{C},\ t>0\}$ 上是凸函数.

定义 B.11 (上境图)　假定 $f(\boldsymbol{x})$ 是定义在凸集 $\boldsymbol{C}\subseteq\mathbb{R}^n$ 上的实值函数, 那么函数 f 的上境图 (epigraph) 定义为

$$\mathrm{epi}(f):=\{(\boldsymbol{x},t)\in\boldsymbol{C}\times\mathbb{R}\mid f(\boldsymbol{x})\leqslant t\}.$$

事实上, 凸集与凸函数之间的联系可以通过上境图来建立.

定理 B.10　给定任意凸集 $\boldsymbol{C}$, 函数 $f(\boldsymbol{x}):\boldsymbol{C}\mapsto\mathbb{R}$ 是凸函数当且仅当其上境图 $\mathrm{epi}(f)$ 是凸集.

例 36 (从上境图的观点解释一阶条件)　由于上境图 $\mathrm{epi}(f)$ 是凸集, 根据支撑超平面定理, 给定上境图 $\mathrm{epi}(f)$ 中边界上任意一点 $(\boldsymbol{x},f(\boldsymbol{x}))$, 我们有

$$\begin{bmatrix}\nabla f(\boldsymbol{x})\\-1\end{bmatrix}^\top\left(\begin{bmatrix}\boldsymbol{y}\\t\end{bmatrix}-\begin{bmatrix}\boldsymbol{x}\\f(\boldsymbol{x})\end{bmatrix}\right)\leqslant 0,\quad\forall(\boldsymbol{y},t)\in\mathrm{epi}(f),$$

其中向量 $(\nabla f(\boldsymbol{x}), -1)^\top$ 定义了上境图 epi(f) 在点 $(\boldsymbol{x}, f(\boldsymbol{x}))$ 处的支撑超平面. 这也等价于

$$t \geqslant f(\boldsymbol{x}) + [\nabla f(\boldsymbol{x})]^\top (\boldsymbol{y} - \boldsymbol{x}), \forall \boldsymbol{x} \in \mathrm{dom}(f),\ \forall (\boldsymbol{y}, t) \in \mathrm{epi}(f).$$

下面, 我们考虑如下凸优化问题并给出其最优性条件:

$$\min_{\boldsymbol{x}} \{f(\boldsymbol{x}) : \boldsymbol{x} \in \mathcal{X}\},$$

其中 $f(\boldsymbol{x})$ 是凸函数且可行域 $\mathcal{X}$ 是凸集.

定理 B.11 (局部最优等价于全局最优) 假设 $f : \mathbb{R}^n \mapsto (-\infty, \infty]$ 是凸函数且可行域 $\mathcal{X}$ 是非空凸集, 那么函数 $f(\boldsymbol{x})$ 的任一局部极小值点一定是全局极小值点.

我们称点 $\boldsymbol{z}$ 是函数 $f(\boldsymbol{x})$ 的局部极小值点, 如果 $\exists r > 0$ 使得

$$f(\boldsymbol{z}) = \inf\{f(\boldsymbol{x}) : \boldsymbol{x} \in \mathcal{X}, \|\boldsymbol{x} - \boldsymbol{z}\|_2 \leqslant r\}.$$

定理 B.12 (可微凸函数的最优性条件) 假设 $f : \mathbb{R}^n \mapsto (-\infty, \infty]$ 是可微凸函数且可行域 $\mathcal{X}$ 是非空凸集, 那么

$$\boldsymbol{x}^* \in \operatorname*{argmin}_{\boldsymbol{x} \in \mathcal{X}} f(\boldsymbol{x}) \iff \nabla f(\boldsymbol{x}^*)^\top (\boldsymbol{x} - \boldsymbol{x}^*) \geqslant 0, \forall \boldsymbol{x} \in \mathcal{X}.$$

定理 B.2 的几何意义如图 B.1 所示.

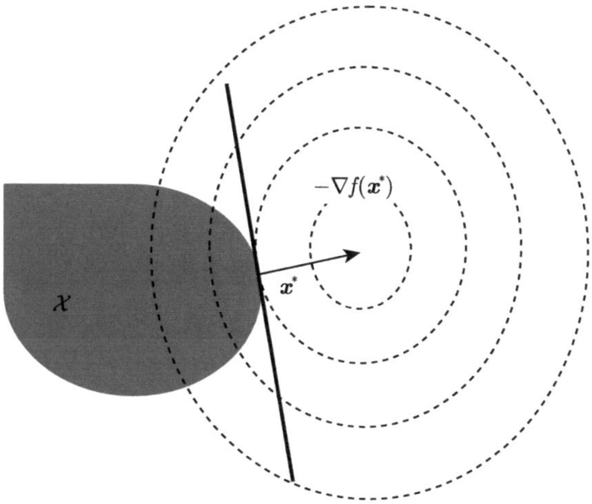

图 B.1 如果 $\nabla f(\boldsymbol{x}^*) \neq 0$, 那么 $-\nabla f(\boldsymbol{x}^*)$ 定义了可行域 $\mathcal{X}$ 在点 $\boldsymbol{x}^*$ 处的一个支撑超平面

例 37 (线性约束下的最优性条件) 考虑仅有线性等式约束下的优化问题, 即

$$\begin{aligned}\min_{\boldsymbol{x}} \quad & f_0(\boldsymbol{x}) \\ \text{s.t.} \quad & \boldsymbol{A}\boldsymbol{x} = \boldsymbol{b},\end{aligned}$$

其中 $\boldsymbol{A} \in \mathbb{R}^{m\times n}$. 那么上述优化问题的最优性条件可写为

$$\left\{\nabla f_0(\boldsymbol{x}) + \boldsymbol{A}^\top \boldsymbol{\nu} = 0, \boldsymbol{\nu} \in \mathbb{R}^m\right\} \neq \varnothing.$$

事实上, 对于任一可行的 $\boldsymbol{x}$, 其最优性条件变为

$$\nabla f_0(\boldsymbol{x})^\top(\boldsymbol{y} - \boldsymbol{x}) \geqslant \boldsymbol{0}, \quad \forall \boldsymbol{y} : A\boldsymbol{y} = \boldsymbol{b}.$$

由于 $\boldsymbol{x}$ 是可行解, 那么对于每一个可行的 $\boldsymbol{y}$, 都存在 $\boldsymbol{v} \in \mathcal{N}(A)$, 使得 $\boldsymbol{y} = \boldsymbol{x} + \boldsymbol{v}$. 因此, 最优性条件进一步可重写为

$$\nabla f_0(\boldsymbol{x})^\top \boldsymbol{v} \geqslant \boldsymbol{0}, \quad \forall \boldsymbol{v} \in \mathcal{N}(A),$$

并且我们注意到, 如果线性函数在子空间上是非负的, 那么该函数在子空间上的取值恒为 0. 因此对于任意的 $\boldsymbol{v} \in \mathcal{N}$, 我们有 $\nabla f_0(\boldsymbol{x})^\top \boldsymbol{v} = 0$, 或者

$$\nabla f_0(\boldsymbol{x}) \perp \mathcal{N}(A).$$

此外, 由于 $\mathcal{N}^\perp = \mathcal{R}(A^\top)$, 上述最优性条件可重新表述为 $\nabla f_0(\boldsymbol{x}) \in \mathcal{R}\left(A^\top\right)$, 也就是, 存在 $\boldsymbol{\nu} \in \mathbb{R}^m$, 使得

$$\nabla f_0(\boldsymbol{x}) + A^\top \boldsymbol{\nu} = 0.$$

定理 B.12 给出了可微凸函数的最优性条件, 现在我们考察一般凸函数的最优性条件. 在此之前, 给出一般函数凸性的判定定理.

定理 B.13 (一般函数凸性的判断) 函数 $f : \mathbb{R}^n \mapsto \mathbb{R}$ 是凸的当且仅当对于任意 $\boldsymbol{z} \in \mathbb{R}^n$, 存在向量 $\boldsymbol{s}(\boldsymbol{z}) \in \mathbb{R}^n$ 使得对于任意的 $\boldsymbol{x} \in \mathbb{R}^n$, 有

$$f(\boldsymbol{x}) \geqslant f(\boldsymbol{z}) + \boldsymbol{s}(\boldsymbol{z})^\top(\boldsymbol{x} - \boldsymbol{z}),$$

其中, $\boldsymbol{s}(\boldsymbol{z})$ 称为函数 f 在点 $\boldsymbol{z}$ 处的次梯度 (subgradient).

函数 f 在点 $\boldsymbol{z}$ 处次梯度所构成的集合称为函数 f 在点 $\boldsymbol{z}$ 处的次微分 (subdifferential), 记作 $\partial f(\boldsymbol{z})$. 特别地, 如果函数 f 是可微的, 那么此时任意一点处函数 f 的次微分退化到函数 f 在该点的梯度, 即 $\partial f(\boldsymbol{z}) = \{\nabla f(\boldsymbol{z})\}$.

命题 B.1 (一般凸函数的最优性条件) 假设 $f : \mathbb{R}^n \mapsto (-\infty, \infty]$ 是一个正常 (proper) 凸函数且其可行域 $\mathcal{X}$ 为非空凸集, 那么

$$\boldsymbol{x}^* \in \underset{\boldsymbol{x}\in\mathcal{X}}{\operatorname{argmin}} f(\boldsymbol{x}) \iff \boldsymbol{0} \in \partial f(\boldsymbol{x}^*).$$

B.3 凸规划与对偶理论

考虑如下优化问题:

$$\begin{aligned}\min\ & f_0(\boldsymbol{x})\\ \text{s.t. } & f_i(\boldsymbol{x})\leqslant 0,\quad i=1,\cdots,m,\\ & \boldsymbol{a}_i^\top\boldsymbol{x}=b_i,\quad i=1,\cdots,p.\end{aligned}$$

若函数 $f_i(\boldsymbol{x}), i=0,1,\cdots,m$ 均为凸函数, 则上述问题是一个凸优化问题. 相较于一般优化问题, 凸优化问题有三个要求: ① 目标函数必须是凸的; ② 不等式约束函数必须是凸的; ③ 等式约束函数必须是仿射 (affine) 的. 由于不等式约束对应的集合是凸集, 等式约束对应的超平面也是凸集, 而凸集的交集为凸集, 因此我们立即注意到凸优化问题的可行域为凸集.

1. 拉格朗日对偶

考虑如下标准形式的优化问题:

$$\begin{aligned}\min\ & f_0(\boldsymbol{x})\\ \text{s.t. } & f_i(\boldsymbol{x})\leqslant 0,\quad i=1,\cdots,m,\\ & h_i(\boldsymbol{x})=0,\quad i=1,\cdots,p,\\ & \boldsymbol{x}\in\mathcal{X}.\end{aligned}\tag{B.1}$$

假定上述问题的可行域非空且目标函数最优值为 $p^\star$. 定义上述问题的拉格朗日函数为

$$L(\boldsymbol{x},\boldsymbol{\lambda},\boldsymbol{\nu})=f_0(\boldsymbol{x})+\sum_{i=1}^m\lambda_i f_i(\boldsymbol{x})+\sum_{i=1}^p\nu_i h_i(\boldsymbol{x}),$$

这里 λ_i 称为第 i 个不等式约束对应的拉格朗日乘子, ν_i 称为第 i 个等式约束对应的拉格朗日乘子. 那么可以定义拉格朗日对偶函数如下:

$$g(\boldsymbol{\lambda},\boldsymbol{\nu})=\inf_{\boldsymbol{x}\in\mathcal{X}}L(\boldsymbol{x},\boldsymbol{\lambda},\boldsymbol{\nu})=\inf_{x\in\mathcal{D}}\left(f_0(\boldsymbol{x})+\sum_{i=1}^m\lambda_i f_i(\boldsymbol{x})+\sum_{i=1}^p\nu_i h_i(\boldsymbol{x})\right).$$

注意到, 对偶函数 $g(\boldsymbol{\lambda},\boldsymbol{\nu})$ 是一族关于 $(\boldsymbol{\lambda},\boldsymbol{\nu})$ 的仿射函数的逐点下确界, 这意味着无论原问题(B.1)是不是凸的, 对偶函数 $g(\boldsymbol{\lambda},\boldsymbol{\nu})$ 总是凹函数.

此外, 对于任意一组 $(\boldsymbol{\lambda},\boldsymbol{\nu})$, 其中 $\boldsymbol{\lambda}\geqslant 0$, 拉格朗日对偶函数给出了优化问题 (B.1) 的最优值的一个下界, 即对于任意的 $\boldsymbol{\lambda}\geqslant\mathbf{0}$ 以及 $\boldsymbol{\nu}$, 我们有 $g(\boldsymbol{\lambda},\boldsymbol{\nu})\leqslant p^\star$.

由于该下界与参数 $\boldsymbol{\lambda}, \boldsymbol{\nu}$ 相关, 那么一个自然的问题是, 从拉格朗日函数能够得到的最好下界是什么？事实上, 这个问题可以表述为下述优化问题

$$\begin{aligned} &\max\ g(\boldsymbol{\lambda}, \boldsymbol{\nu}) \\ &\text{s.t. } \boldsymbol{\lambda} \geqslant 0. \end{aligned} \tag{B.2}$$

该问题称为问题 (B.1) 的拉格朗日对偶问题. 由于极大化的目标函数是凹函数且约束集合是凸集, 拉格朗日对偶问题 (B.2) 是一个凸优化问题. 因此, 对偶问题的凸性和原问题 (B.1) 是否为凸优化问题无关.

拉格朗日对偶问题的最优值, 用 $d^\star$ 表示, 根据定义, 这是通过拉格朗日函数得到的原问题最优值 $p^\star$ 的最好下界. 特别地, 有如下重要不等式:

$$d^\star \leqslant p^\star. \tag{B.3}$$

即使原问题不是凸问题, 上述不等式亦成立. 这个性质称为弱对偶性. 当 $d^\star$ 和 $p^\star$ 无限时, 弱对偶性不等式 (B.3) 同样成立. 例如, 如果原问题(B.1)无下界, 即 $p^\star = -\infty$, 弱对偶性说明 $d^\star = -\infty$, 这意味着拉格朗日对偶问题(B.2)不存在可行解. 反过来, 若对偶问题无上界, 即 $d^\star = \infty$, 弱对偶性说明 $p^\star = \infty$, 即原问题(B.1)不存在可行解.

将差值 $p^\star - d^\star$ 定义为原问题的最优对偶间隙. 它给出了原问题最优值以及通过拉格朗日对偶函数所能得到的最好下界之间的差值. 如果等式

$$d^\star = p^\star$$

成立, 那么强对偶性成立. 这说明从拉格朗日对偶函数得到的最好的下界是紧的. 对于一般情况, 强对偶性不成立. 然而, 如果原问题 (B.1) 是凸问题且可以表述为如下形式:

$$\begin{aligned} &\min\ f_0(\boldsymbol{x}) \\ &\text{s.t. } f_i(\boldsymbol{x}) \leqslant 0, \quad i = 1, \cdots, m, \\ &\qquad Ax = b, \end{aligned}$$

其中函数 $f_0, \cdots, f_m$ 是凸函数, 那么强对偶性通常成立. 有很多研究成果给出了强对偶性成立的条件, 这些条件称为约束准则.

一个常用的约束准则是 Slater 条件: 存在一点 $\boldsymbol{x} \in \operatorname{relint} \mathcal{D}$ 使得下列严格不等式成立

$$f_i(\boldsymbol{x}) < 0, \quad i = 1, \cdots, m, \quad \boldsymbol{A}\boldsymbol{x} = \boldsymbol{b}.$$

满足上述条件的点也称为严格可行点. 当 Slater 条件成立且原问题是凸问题时, 强对偶性成立.

此外, 如果不等式约束函数 f_i 中有一些是仿射函数, 那么 Slater 条件可以进一步改进. 假设钱 k 个约束函数 $f_1, \cdots, f_k$ 是仿射的, 则若存在 $\boldsymbol{x} \in \operatorname{relint} \mathcal{D}$ 使得

$$f_i(\boldsymbol{x}) \leqslant 0, \quad i=1,\cdots,k, \quad f_i(\boldsymbol{x}) < 0, \quad i=k+1,\cdots,m, \quad \boldsymbol{A}\boldsymbol{x}=\boldsymbol{b},$$

强对偶性成立. 换言之, 仿射不等式不需要严格成立. 注意到当所有约束条件都是线性等式或不等式且 $\operatorname{dom}(f_0)$ 是开集时, 上述改进的 Slater 条件就是可行性条件.

2. 锥对偶

给定任意正常锥 $\mathcal{K}$, 考虑如下标准形式的锥规划问题:

$$\begin{aligned} &\min\ \boldsymbol{c}^\top \boldsymbol{x} \\ &\text{s.t.}\ \boldsymbol{A}\boldsymbol{x}-\boldsymbol{b} \in \mathcal{K}. \end{aligned} \tag{B.4}$$

记 $\mathcal{K}^*$ 为 $\mathcal{K}$ 的对偶锥, 引入对偶变量 $\boldsymbol{y} \in \mathcal{K}^*$, 上述锥规划的对偶问题为

$$\begin{aligned} &\max\ \boldsymbol{b}^\top \boldsymbol{y} \\ &\text{s.t.}\ \boldsymbol{A}^\top \boldsymbol{y} = \boldsymbol{c}, \\ &\qquad \boldsymbol{y} \in \mathcal{K}^*. \end{aligned} \tag{B.5}$$

下述定理描述了锥规划问题中的弱对偶性质、强对偶性质及其所需条件.

定理 B.14 (Shapiro, 2001) 考虑锥规划问题 (B.4) 以及其对偶问题 (B.5).

(1) 问题 (B.5) 的对偶问题等价于原问题 (B.4).

(2) 对于原问题 (B.4) 中任意可行的 $\boldsymbol{x}$ 以及对偶问题 (B.5) 中任意可行的 $\boldsymbol{y}$, 我们有 $\boldsymbol{c}^\top \boldsymbol{x} \geqslant \boldsymbol{y}^\top \boldsymbol{b}$.

(3) 若问题 (B.4) 是有下界的, 即最优值大于 $-\infty$, 且存在 $\boldsymbol{x}$ 使得 $\boldsymbol{A}\boldsymbol{x}-\boldsymbol{b} \in \operatorname{int} \mathcal{K}$, 那么问题 (B.5) 是可解的且其最优值与问题 (B.4) 的最优值相等.

(4) 如果问题 (B.4) 或 (B.5) 中至少有一个是有界的且含有严格可行的解, 那么任意的原始-对偶解 $(\boldsymbol{x}, \boldsymbol{y})$ 是最优的当且仅当 $\boldsymbol{c}^\top \boldsymbol{x} = \boldsymbol{b}^\top \boldsymbol{y}$ 或者 $\boldsymbol{y}^\top(\boldsymbol{A}\boldsymbol{x}-\boldsymbol{b}) = 0$.

下面, 我们给出一个简单的例子来更好地阐述上述定理.

例 38 我们考虑如下二阶锥规划问题:

$$\begin{aligned} &\min\ x_2 \\ &\text{s.t.}\ \sqrt{x_1^2 + x_2^2} \leqslant x_1. \end{aligned}$$

不难看出, 其最优值为 0 且最优解为 $(x_1,0)$, 其中 x_1 为任意实数. 同时, 我们可以求出上述问题的对偶问题为

$$\begin{aligned}&\max\ 0\\&\text{s.t.}\begin{bmatrix}y_0+y_1\\y_2\end{bmatrix}=\begin{bmatrix}0\\1\end{bmatrix},\\&\sqrt{y_1^2+y_2^2}\leqslant y_0.\end{aligned}$$

但是我们发现, 对偶问题无解. 此时尽管原问题最优值有限, 但强对偶不一定成立. 下面, 我们将原问题等价地写成二阶锥规划的形式

$$\begin{aligned}&\min\ x_2\\&\text{s.t.}\begin{bmatrix}1&0\\1&0\\0&1\end{bmatrix}\begin{bmatrix}x_1\\x_2\end{bmatrix}\in\mathcal{L}^3,\end{aligned}$$

其中 $\mathcal{L}^3$ 是洛伦兹锥. 利用上述锥对偶定理, 我们可写出其对偶问题为

$$\begin{aligned}&\max\ 0\\&\text{s.t.}\begin{bmatrix}1&1&0\\0&0&1\end{bmatrix}\begin{bmatrix}y_0\\y_1\\y_2\end{bmatrix}=\begin{bmatrix}0\\1\end{bmatrix},\\&\boldsymbol{y}\in\mathcal{L}^3.\end{aligned}$$

3. Fenchel 对偶

定义 B.12　假设 f 和 g 是 $\mathbb{R}^n$ 上的实值函数, 且 f 和 g 的定义域分别为 $\operatorname{dom}(f)=\{\boldsymbol{x}\mid f(\boldsymbol{x})<\infty\}$, $\operatorname{dom}(g)=\{\boldsymbol{x}\mid g(\boldsymbol{x})>-\infty\}$. 那么, 函数 f 的凸共轭 (convex conjugate) 函数 f^* 为

$$f^*(y)=\sup_{\boldsymbol{x}\in\operatorname{dom}(f)}\{\boldsymbol{x}^\top\boldsymbol{y}-f(\boldsymbol{x})\},$$

函数 g 的凹共轭函数 g_* 为

$$g_*(y)=\inf_{\boldsymbol{x}\in\operatorname{dom}(g)}\{\boldsymbol{x}^\top\boldsymbol{y}-g(\boldsymbol{x})\}.$$

注意到, $f^*(y)$ 总是一个凸函数, 而 $g_*(y)$ 总是一个凹函数. 此外, 当 f 和 g 是下半连续 (lower semi-continuous) 的闭函数时, 可以证明 (Bertsekas, 2009)

$$f^{**}=f,\quad g_{**}=g.$$

例 39 定义集合 $\boldsymbol{S}$ 的示性函数 (indicator function)$\delta(\cdot\mid\boldsymbol{S})$ 如下:

$$\delta(\boldsymbol{x}\mid\boldsymbol{S})=\begin{cases}0, & \boldsymbol{x}\in\boldsymbol{S}\\ \infty, & \boldsymbol{x}\notin\boldsymbol{S},\end{cases}$$

那么利用定义可知, $\delta(\cdot\mid\boldsymbol{S})$ 的共轭函数为集合 $\boldsymbol{S}$ 的支撑函数, 即 $\delta^*(\boldsymbol{y}\mid S)=\sup_{\boldsymbol{x}\in S}\boldsymbol{x}^\top\boldsymbol{y}$.

例 40 下面, 我们给出一些常用函数的共轭函数.

(1) 指数函数 $f(x)=\mathrm{e}^x$ 的共轭函数为 $f^*(y)=y\log y-y$.

(2) 给定正定矩阵 $\boldsymbol{Q}\in\boldsymbol{S}_{++}^n$, 那么二次函数 $f(\boldsymbol{x})=\dfrac{1}{2}\boldsymbol{x}^\top\boldsymbol{Q}\boldsymbol{x}$ 的共轭函数为 $f^*(\boldsymbol{y})=\dfrac{1}{2}\boldsymbol{y}^\top\boldsymbol{Q}^{-1}\boldsymbol{y}$.

(3) 函数 $f(\boldsymbol{x})=\log\left(\sum_{i\in[n]}\mathrm{e}^{x_i}\right)$ 的共轭函数为

$$f^*(\boldsymbol{y})=\begin{cases}\displaystyle\sum_{i\in n}y_i\log y_i, & \boldsymbol{y}\geqslant\boldsymbol{0}\text{ 且 }\boldsymbol{1}^\top\boldsymbol{y}=1\\ \infty, & \text{其他情况}.\end{cases}$$

(4) 范数函数 $f(\boldsymbol{x})=||\boldsymbol{x}||$ 的共轭函数为

$$f^*(\boldsymbol{y})=\begin{cases}0, & ||\boldsymbol{y}||_*\leqslant 1\\ \infty, & ||\boldsymbol{y}||_*>1.\end{cases}$$

接下来, 我们给出 Fenchel 对偶定理的主要形式.

定理 B.15 (Ben-Tal et al., 2015) 假设 f 和 g 分别为 $\mathbb{R}^n$ 上正常的闭凸函数和闭凹函数, 且其定义域分别为 $\operatorname{dom}(f)=\{\boldsymbol{x}\mid f(\boldsymbol{x})<\infty\}$, $\operatorname{dom}(g)=\{\boldsymbol{x}\mid g(\boldsymbol{x})>-\infty\}$. 考虑如下优化问题

$$\inf\{f(\boldsymbol{x})-g(\boldsymbol{x})\mid\boldsymbol{x}\in\operatorname{dom}(f)\cap\operatorname{dom}(g)\},\tag{P}$$

以及它的 Fenchel 对偶问题

$$\sup\left\{g_*(\boldsymbol{y})-f^*(\boldsymbol{y})\mid\boldsymbol{y}\in\operatorname{dom}\left(g_*\right)\cap\operatorname{dom}\left(f^*\right)\right\}.\tag{D}$$

那么,

(1) 如果 $\operatorname{ri}(\operatorname{dom}(f))\cap\operatorname{ri}(\operatorname{dom}(g))\neq\varnothing$, 则问题 (P) 的最优值与问题 (D) 的最优值相等, 且问题 (D) 的最优值是可达的.

(2) 如果 $\operatorname{ri}(\operatorname{dom}(f^*))\cap\operatorname{ri}(\operatorname{dom}(g_*))\neq\varnothing$, 则问题 (P) 的最优值与问题 (D) 的最优值相等, 且问题 (P) 的最优值是可达的.

参考文献

ACIMOVIC J, GRAVES S C, 2014. Making better fulfillment decisions on the fly in an online retail environment. Manufacturing & Service Operations Management, 17(1): 34-51.

AFIFY B, RAY S, SOEANU A, et al., 2019. Evolutionary learning algorithm for reliable facility location under disruption. Expert Systems with Applications, 115: 223-244.

ALUMUR S, KARA B Y. 2008. Network hub location problems: The state of the art. European Journal of Operational Research, 190(1): 1-21.

ALUMUR S A, NICKEL S, SALDANHA-DA GAMA F, 2012. Hub location under uncertainty. Transportation Research Part B: Methodological, 46(4): 529-543.

ARDESTANI-JAAFARI A, DELAGE E, 2018. The value of flexibility in robust location-transportation problems. Transportation Science, 52(1): 189-209.

ATAMTÜRK A, ZHANG M, 2007. Two-stage robust network flow and design under demand uncertainty. Operations Research, 55(4): 662-673.

BALCIK B, BEAMON B M, 2008. Facility location in humanitarian relief. International Journal of Logistics, 11(2): 101-121.

BANKS C, 2009. Optimising anaerobic digestion. University of Southampton, England.

BARON O, MILNER J, NASERALDIN H, 2011. Facility location: a robust optimization approach. Production and Operations Management, 20(5): 772-785.

BASCIFTCI B, AHMED S, SHEN S, 2021. Distributionally robust facility location problem under decision-dependent stochastic demand. European Journal of Operational Research, 292(2): 548-561.

BAYRAKSAN G, LOVE D K, 2015. Data-driven stochastic programming using phi-divergences. The Operations Research Revolution: 1-19.

BEIGL P, LEBERSORGER S, SALHOFER S, 2008. Modelling municipal solid waste generation: a review. Waste Management, 28(1): 200-214.

BEN-TAL A, DEN HERTOG D, DE WAEGENAERE A, et al., 2013. Robust solutions of optimization problems affected by uncertain probabilities. Management Science, 59(2): 341-357.

BEN-TAL A, DEN HERTOG D, VIAL J-P, 2015. Deriving robust counterparts of nonlinear uncertain inequalities. Mathematical Programming, 149(1): 265-299.

BEN-TAL A, EL GHAOUI L, NEMIROVSKI A, 2009. Robust optimization. Princeton and Oxford: Princeton University Press.

BEN-TAL A, GORYASHKO A, GUSLITZER E, et al., 2004. Adjustable robust solutions of uncertain linear programs. Mathematical Programming, 99(2):351-376.

BEN-TAL A, NEMIROVSKI A, 1998. Robust convex optimization. Mathematics of Operations Research, 23(4): 769-805.

BEN-TAL A, NEMIROVSKI A, 1999. Robust solutions of uncertain linear programs. Operations Research Letters, 25(1): 1-13.

BEN-TAL A, NEMIROVSKI A, 2000. Robust solutions of linear programming problems contaminated with uncertain data. Mathematical Programming, 88: 411-424.

BEN-TAL A, NEMIROVSKI A, 2002. Robust optimization-methodology and applications. Mathematical Programming, 92: 453-480.

BENKO D, COROIAN D, 2018, A new angle on the fermat-torricelli point. The College Mathematics Journal, 49(3): 195-199.

BERMAN O, WANG J, 2006. The 1-median and 1-antimedian problems with continuous probabilistic demand weights. INFOR Information Systems and Operational Research, 44(4): 267-283.

BERTSEKAS D, 2009. Convex optimization theory: vol 1. Nashua, NH: Athena Scientific.

BERTSIMAS D, BROWN D B, CARAMANIS C, 2011. Theory and applications of robust optimization. SIAM Review, 53(3): 464-501.

BERTSIMAS D, DE RUITER F, 2016. Duality in two-stage adaptive linear optimization: Faster computation and stronger bounds. INFORMS Journal on Computing, 28(3): 500-511.

BERTSIMAS D, DOAN X V, NATARAJAN K, et al., 2010. Models for minimax stochastic linear optimization problems with risk aversion. Mathematics of Operations Research, 35(3): 580-602.

BERTSIMAS D, GOYAL V, 2012. On the power and limitations of affine policies in two-stage adaptive optimization. Mathematical Programming, 134(2): 491-531.

BERTSIMAS D, GUPTA V, KALLUS N, 2018. Data-driven robust optimization. Mathematical Programming, 167(2):235-292.

BERTSIMAS D, SHTERN S, 2018. A scalable algorithm for two-stage adaptive linear optimization. arXiv preprint arXiv:1807.02812.

BERTSIMAS D, SHTERN S, STURT B, 2023. A data-driven approach to multi-stage stochastic linear optimization. Management Science, 69(1): 51-74.

BERTSIMAS D, SIM M, 2004. The price of robustness. Operations Research, 52(1): 35-53.

BERTSIMAS D, SIM M, ZHANG M, 2019. Adaptive distributionally robust optimization. Management Science, 65(2): 604-618.

BERTSIMAS D, TSITSIKLIS J N, 1997. Introduction to linear optimization. Nashua, NH: Athena Scientific.

BIRGE J R, LOUVEAUX F, 2011. Introduction to stochastic programming. New York: Springer Science & Business Media.

BLANCHET J, KANG Y, MURTHY K, 2019. Robust Wasserstein profile inference and applications to machine learning. Journal of Applied Probability, 56(3): 830-857.

BLANCHET J, MURTHY K, 2019. Quantifying distributional model risk via optimal transport. Mathematics of Operations Research, 44(2): 565-600.

BOUKANI F H, MOGHADDAM B F, PISHVAEE M S, 2016. Robust optimization approach to capacitated single and multiple allocation hub location problems. Computational and Applied Mathematics, 35(1): 45-60.

BOWERMAN B L, O'CONNELL R T, KOEHLER A B, 2005. Forecasting, time series, and regression: an applied approach. South-Western Pub.

BRIMBERG J, DREZNER Z, 2013. A new heuristic for solving the p-median problem in the plane. Computers & Operations Research, 40(1): 427-437.

BROWN D B, SIM M, 2009. Satisficing measures for analysis of risky positions. Management Science, 55(1): 71-84.

BRYAN D L, O'KELLY M E, 1999. Hub-and-spoke networks in air transportation: an analytical review. Journal of Regional Science, 39(2): 275-295.

BUTLER J, HOOPER P, 2000. Factors determining the post-consumer waste recycling burden. Journal of Environmental Planning and Management, 43(3): 407-432.

CALAFIORE G C, EL GHAOUI L, 2006. On distributionally robust chance-constrained linear programs. Journal of Optimization Theory and Applications, 130(1): 1-22.

CAMPBELL J F, ERNST A T, KRISHNAMOORTHY M, 2005. Hub arc location problems: part i-introduction and results. Management Science, 51(10): 1540-1555.

CAMPBELL J F, O'KELLY M E, 2012. Twenty-five years of hub location research. Transportation Science, 46(2): 153-169.

CARBONE R, 1974. Public facilities location under stochastic demand. Information Systems and Operational Research, 12(3): 261-270.

CARDIN M A, XIE Q, NG T S, et al., 2017. An approach for analyzing and managing flexibility in engineering systems design based on decision rules and multistage stochastic programming. IISE Transactions, 49(1): 1-12.

CARELLO G, DELLA CROCE F, GHIRARDI M, et al., 2004. Solving the hub location problem in telecommunication network design: a local search approach. Networks: An International Journal, 44(2): 94-105.

ÇETINER S, SEPIL C, SÜRAL H, 2010. Hubbing and routing in postal delivery systems. Annals of Operations Research, 181: 109-124.

CHARNES A, COOPER W W, 1959. Chance-constrained programming. Management Science, 6(1): 73-79.

CHEN Z, KUHN D, WIESEMANN W, 2024. Data-driven chance constrained programs over Wasserstein balls. Operations Research, 72(1): 410-424.

CHEN Z, KUHN D, WIESEMANN W, 2023. On approximations of data-driven chance constrained programs over Wasserstein balls. Operations Research Letters, 51(3): 226-233.

CHEN Z, SIM M, XIONG P, 2020. Robust stochastic optimization made easy with RSOME. Management Science, 66(8): 3329-3339.

CHEN Z, SIM M, XU H, 2019. Distributionally robust optimization with infinitely constrained ambiguity sets. Operations Research, 67(5): 1328-1344.

CHENG C, ADULYASAK Y, ROUSSEAU L-M, 2021. Robust facility location under demand uncertainty and facility disruptions. Omega, 103: 102429.

COINTREAU-LEVINE S, 1994. Private sector participation in municipal solid waste services in developing countries: vol. 1. The Formal Sector. The World Bank.

CONTRERAS I, CORDEAU J-F, LAPORTE G, 2011. Stochastic uncapacitated hub location. European Journal of Operational Research, 212(3): 518-528.

CONTRERAS I, O'Kelly M, 2019. Hub location problems//LAPORTE G S, NICKEL F S, ed, Location science, chapter 12, pages 327-363. Springer International Publishing, 2nd ed.

COOPER L, 1963. Location-allocation problems. Operations Research, 11(3): 331-343.

COOPER L, 1974. A random locational equilibrium problem. Journal of Regional Science, 14(1): 47-54.

CORREIA I, NICKEL S, SALDANHA-DA-GAMA F, 2018. A stochastic multi-period capacitated multiple allocation hub location problem. Formulation and inequalities. Omega, 74: 122-134.

CROCI D, JABALI O, MALUCELLI F, 2023. The balanced p-median problem with unitary demand. Computers & Operations Research, 155: 106242.

CUI T, OUYANG Y, SHEN Z-J M, 2010. Reliable facility location design under the risk of disruptions. Operations Research, 58(4): 998-1011.

DAHLÉN L, VUKICEVIC S, MEIJER J-E, et al., 2007. Comparison of different collection systems for sorted household waste in sweden. Waste Management, 27(10): 1298-1305.

DASKIN M S, COULLARD C R, SHEN Z-J M, 2002. An inventory-location model: formulation, solution algorithm and computational results. Annals of Operations Research, 110(1-4): 83-106.

DE CAMARGO R S, MIRANDA G, Jr., LUNA H P, 2008. Benders decomposition for the uncapacitated multiple allocation hub location problem. Computers & Operations Research, 35(4): 1047-1064.

DE MAESSCHALCK R, JOUAN-RIMBAUD D, MASSART D L, 2000. The Mahalanobis distance. Chemometrics and Intelligent Laboratory Systems, 50(1): 1-18.

DE RUITER F, 2018. Primal and dual approaches to adjustable robust optimization. PhD thesis, Tilburg University, School of Economics and Management.

DELAGE E, YE Y, 2010. Distributionally robust optimization under moment uncertainty with application to data-driven problems. Operations Research, 58(3): 595-612.

DOU X, ANITESCU M, 2019. Distributionally robust optimization with correlated data from vector autoregressive processes. Operations Research Letters, 47(4): 294-299.

DREZNER Z, HAMACHER H W, 2004. Facility location: applications and theory. Berlin, Heidelberg: Springer Science & Business Media.

DREZNER Z, SALHI S, 2017. Incorporating neighborhood reduction for the solution of the planar p-median problem. Annals of Operations Research, 258: 639-654.

DÖNMEZ Z, KARA B Y, KARSU O, et al., 2021. Humanitarian facility location under uncertainty: Critical review and future prospects. Omega, 102: 102393.

EBERY J, KRISHNAMOORTHY M, ERNST A, et al., 2000. The capacitated multiple allocation hub location problem: Formulations and algorithms. European Journal of Operational Research, 120: 614-631.

EL GHAOUI L, OUSTRY F, LEBRET H, 1998. Robust solutions to uncertain semidefinite programs. SIAM Journal on Optimization, 9(1): 33-52.

ERKUT E, KARAGIANNIDIS A, PERKOULIDIS G, et al., 2008. A multicriteria facility location model for municipal solid waste management in North Greece. European Journal of Operational Research, 187(3): 1402-1421.

ESCUDERO L F, LANDETE M, MARÍN A, 2009. A branch-and-cut algorithm for the winner determination problem. Decision Support Systems, 46(3): 649-659.

FARAHANI R Z, FALLAH S, RUIZ R, et al., 2019. OR models in urban service facility location: a critical review of applications and future developments. European Journal of Operational Research, 276(1): 1-27.

FARAHANI R Z, HEKMATFAR M, ARABANI A B, et al., 2013. Hub location problems: a review of models, classification, solution techniques, and applications. Computers & Industrial Engineering, 64(4): 1096-1109.

FEARON D, 2006. Alfred weber: theory of the location of industries, 1909. Ŕ Center for Spatially Integrated Social Science.

FERNÁNDEZ E, LANDETE M, 2019. Fixed-charge facility location problems//LAPORTE G, NICKEL S, SALDANHA-DA-GAMA F, ed. Location science (chapter 4, pages 67-98). 2nd ed. Switzerland: Springer International Publishing.

FRANK H, 1966. Optimum locations on a graph with probabilistic demands. Operations Research, 14(3): 409-421.

FUJII M, FUJITA T, OHNISHI S, et al., 2014. Regional and temporal simulation of a smart recycling system for municipal organic solid wastes. Journal of Cleaner Production, 78: 208-215.

GABREL V, LACROIX M, MURAT C, et al., 2014. Robust location transportation problems under uncertain demands. Discrete Applied Mathematics, 164: 100-111.

GAMBELLA C, GHADDAR B, NAOUM-SAWAYA J. 2021. Optimization problems for machine learning: a survey. European Journal of Operational Research, 290: 807-828.

GAMBELLA C, MAGGIONI F, VIGO D, 2019. A stochastic programming model for a tactical solid waste management problem. European Journal of Operational Research, 273(2): 684-694.

GAO R, 2023. Finite-sample guarantees for Wasserstein distributionally robust optimization: Breaking the curse of dimensionality. Operations Research, 71(6): 2291-2360.

GAO R, CHEN X, KLEYWEGT A J, 2022. Wasserstein distributionally robust optimization and variation regularization. Operations Research, 72(3): 1177-1191.

GAO R, KLEYWEGT A, 2023. Distributionally robust stochastic optimization with Wasserstein distance. Mathematics of Operations Research, 48(2): 603-655.

GAO Y, QIN Z, 2016. A chance constrained programming approach for uncertain p-hub center location problem. Computers & Industrial Engineering, 102: 10-20.

GEISSER S, 2017. Predictive inference. New York: Chapman and Hall/CRC.

GELAREH S, NICKEL S, 2008. A benders decomposition for hub location problems arising in public transport. Operations Research Proceedings 2007: Selected Papers of the Annual International Conference of the German Operations Research Society (GOR) Saarbrücken, September 5-7, 2007, Pages 129-134. Springer.

GELAREH S, PISINGER D, 2011. Fleet deployment, network design and hub location of liner shipping companies. Transportation Research Part E: Logistics and Transportation Review, 47(6): 947-964.

GHAFFARI-NASAB N, GHAZANFARI M, TEIMOURY E, 2015. Robust optimization approach to the design of hub-and-spoke networks. The International Journal of Advanced Manufacturing Technology, 76(5-8): 1091-1110.

GHIANI G, LAGANÀ D, MANNI E, et al., 2014. Operations research in solid waste management: a survey of strategic and tactical issues. Computers & Operations Research, 44: 22-32.

GIVENS C R, SHORTT R M, 1984. A class of Wasserstein metrics for probability distributions. Michigan Mathematical Journal, 31(2): 231-240.

GOH J, SIM M, 2010. Distributionally robust optimization and its tractable approximations. Operations Research, 58(4): 902-917.

GOLLOWITZER S, LJUBIĆ I, 2011. Mip models for connected facility location: a theoretical and computational study. Computers & Operations Research, 38(2): 435-449.

GORISSEN B L, YANIKOĞLU İ, DEN HERTOG D, 2015. A practical guide to robust optimization. Omega, 53: 124-137.

GÖRNER S, KANZOW C, 2016. On Newton's method for the Fermat-Weber location problem. Journal of Optimization Theory and Applications, 170: 107-118.

GOTOH J-Y, KIM M J, LIM A E, 2018. Robust empirical optimization is almost the same as mean–variance optimization. Operations Research Letters, 46(4): 448-452.

GUERRERO L A, MAAS G, HOGLAND W, 2013. Solid waste management challenges for cities in developing countries. Waste Management, 33(1): 220-232.

GÜLPINAR N, PACHAMANOVA D, ÇANAKOĞLU E, 2013. Robust strategies for facility location under uncertainty. European Journal of Operational Research, 225(1): 21-35.

GUNDUPALLI S P, HAIT S, THAKUR A, 2017. A review on automated sorting of source-separated municipal solid waste for recycling. Waste Management, 60: 56-74.

HAIXIANG G, YIJING L, SHANG J, et al., 2017. Learning from class-imbalanced data: Review of methods and applications. Expert Systems with Applications, 73: 220-239.

HAMACHER H W, LABBÉ M, NICKEL S, et al., 2004. Adapting polyhedral properties from facility to hub location problems. Discrete Applied Mathematics, 145(1): 104-116.

HANASUSANTO G A, KUHN D, 2018. Conic programming reformulations of two-stage distributionally robust linear programs over Wasserstein balls. Operations Research, 66(3): 849-869.

HANASUSANTO G A, ROITCH V, KUHN D, et al., 2017. Ambiguous joint chance constraints under mean and dispersion information. Operations Research, 65(3): 751-767.

HANNAN M, AL MAMUN M A, HUSSAIN A, et al., 2015. A review on technologies and their usage in solid waste monitoring and management systems: Issues and challenges. Waste Management, 43: 509-523.

HAO Z, HE L, HU Z, et al., 2020. Robust vehicle pre-allocation with uncertain covariates. Production and Operations Management, 29(4): 955-972.

HODDER J E, JUCKER J V, 1985. A simple plant-location model for quantity-setting firms subject to price uncertainty. European Journal of Operational Research, 21(1): 39-46.

HU J, CHEN Z, WANG S, 2022. Budget-driven multi-period hub location: a robust time series approach. Available at SSRN 4221971.

HU Z, HONG J, 2013. Kullback-Leibler divergence constrained distributionally robust optimization. Optimization Online [2023-11-10].

JIA H, ORDÓÑEZ F, DESSOUKY M, 2007. A modeling framework for facility location of medical services for large-scale emergencies. IIE Transactions, 39(1): 41-55.

JIANG N, XIE W, 2022. Also-x and also-x+: Better convex approximations for chance constrained programs. Operations Research, 70(6): 3581-3600.

JIANG N, XIE W, 2023. Also-x#: Better convex approximations for distributionally robust chance constrained programs. arXiv preprint arXiv:2302.01737.

JIANG R, GUAN Y, 2016. Data-driven chance constrained stochastic program. Mathematical Programming, 158(1-2): 291-327.

JIANG R, GUAN Y, 2018. Risk-averse two-stage stochastic program with distributional ambiguity. Operations Research, 66(5): 1390-1405.

JUCKER J V, CARLSON R C, 1976. The simple plant-location problem under uncertainty. Operations Research, 24(6): 1045-1055.

KANNAN R, BAYRAKSAN G, LUEDTKE J R, 2020. Residuals-based distributionally robust optimization with covariate information. arXiv preprint arXiv:2012.01088.

KIM H, O'KELLY M E, 2009. Reliable p-hub location problems in telecommunication networks. Geographical Analysis, 41(3): 283-306.

KIM Y, OH I, LEE J-D, 2015. Economic impact assessment of public-private matching fund programs using firm-level data. The Singapore Economic Review, 60(4): 1550060.

KLINCEWICZ J G, 1998. Hub location in backbone/tributary network design: a review. Location Science, 6(1-4): 307-335.

KLOSE A, DREXL A, 2005. Facility location models for distribution system design. European Journal of Operational Research, 162(1): 4-29.

KOSTKA G, FIEDLER J, 2016. Large infrastructure projects in Germany: between ambition and realities. London, UK: Palgrave Macmillan.

KUBY M J, GRAY R G, 1993. The hub network design problem with stopovers and feeders: The case of federal express. Transportation Research Part A: Policy and Practice, 27(1): 1-12.

KÜÇÜKYAVUZ S, JIANG R, 2022. Chance-constrained optimization under limited distributional information: a review of reformulations based on sampling and distributional robustness. EURO Journal on Computational Optimization, 10: 100030.

KUHN D, ESFAHANI P M, NGUYEN V A, et al., 2019. Wasserstein distributionally robust optimization: Theory and applications in machine learning. Operations research & management science in the age of analytics, pages 130-166. INFORMS.

LAM S-W, NG T S, SIM M, et al., 2013. Multiple objectives satisficing under uncertainty. Operations Research, 61(1): 214-227.

LAPORTE G, NICKEL S, SALDANHA-DA-GAMA F, 2019. Location science. 2nd ed. New York: Springer International Publishing.

LAUNHARDT C, 1900. The principles of location: The theory of the trace. part i-the commercial trace//Bewley A, trans., 1900.

LEE C, WILHELM W, 2010. On integrating theories of international economics in the strategic planning of global supply chains and facility location. International Journal of Production Economics, 124(1): 225-240.

LI Y, HUANG G, 2012. Robust interval quadratic programming and its application to waste management under uncertainty. Environmental Systems Research, 1(1): 7.

LI Y, HUANG G H, NIE X, et al., 2008. A two-stage fuzzy robust integer programming approach for capacity planning of environmental management systems. European Journal of Operational Research, 189(2): 399-420.

LIMBOURG S, JOURQUIN B, 2009. Optimal rail-road container terminal locations on the european network. Transportation Research Part E: Logistics and Transportation Review, 45(4): 551-563.

LISTEŞ O, DEKKER R, 2005. A stochastic approach to a case study for product recovery network design. European Journal of Operational Research, 160(1): 268-287.

LIU F, CHEN Z, WANG S, 2023. Globalized distributionally robust counterpart. Journal on Computing, 35(5): 1120-1142.

LIU K, LI Q, ZHANG Z, 2019. Distributionally robust optimization of an emergency medical service station location and sizing problem with joint chance constraints. Transportation Research Part B: Methodological, 119: 79-101.

LIU T, LI G, 2021. Robust recycling facility location with clustering. Computers & Operations Research, 136: 105466.

LIU T, SALDANHA-DA-GAMA F, WANG S, et al., 2022. Robust stochastic facility location: sensitivity analysis and exact solution. INFORMS Journal on Computing, 34(5):2776-2803.

LIU Y, XU H, YANG S-J S, et al., 2018. Distributionally robust equilibrium for continuous games: Nash and Stackelberg models. European Journal of Operational Research, 265(2):

631-643.
LOUVEAUX F, 1993. Stochastic location analysis. Location Science, 1: 127-154.
LOUVEAUX F V, 1986. Discrete stochastic location models. Annals of Operations Research, 6: 21-34.
LU M, RAN L, SHEN Z-J M, 2015. Reliable facility location design under uncertain correlated disruptions. Manufacturing & Service Operations Management, 17(4): 445-455.
LÜTKEPOHL H, 2005. New introduction to multiple time series analysis. Berlin: Springer Science & Business Media.
MALIK N K A, ABDULLAH S H, MANAF L A, 2015. Community participation on solid waste segregation through recycling programmes in Putrajaya. Procedia Environmental Sciences, 30: 10-14.
MCCONNELL V P, 2014. Economics of WtE. Renewable Energy Focus, 15(6): 38-41.
MELO M T, NICKEL S, SALDANHA-DA-GAMA F, 2009. Facility location and supply chain management: a review. European Journal of Operational Research, 196(2): 401-412.
MERAKLI M, YAMAN H, 2016. Robust intermodal hub location under polyhedral demand uncertainty. Transportation Research Part B: Methodological, 86: 66-85.
MERAKLI M, YAMAN H, 2017. A capacitated hub location problem under hose demand uncertainty. Computers & Operations Research, 88: 58-70.
MIEHLE W, 1958. Link-length minimization in networks. Operations Research, 6(2): 232-243.
MIRCHANDANI P B, OUDJIT A, WONG R T, 1985. "Multidimensional" extensions and a nested dual approach for the m-median problem. European Journal of Operational Research, 21(1): 121-137.
MIŠKOVIĆ S, STANIMIROVIĆ Z, GRUJIČIĆ I, 2017. Solving the robust two-stage capacitated facility location problem with uncertain transportation costs. Optimization Letters, 11: 1169-1184.
MOHAJERIN ESFAHANI P, KUHN D, 2018. Data-driven distributionally robust optimization using the Wasserstein metric: Performance guarantees and tractable reformulations. Mathematical Programming, 171(1): 115-166.
MULVEY J M, VANDERBEI R J, ZENIOS S A, 1995. Robust optimization of large-scale systems. Operations Research, 43(2): 264-281.
NAGY G, SALHI S, 2007. Location-routing: Issues, models and methods. European Journal of Operational Research, 177(2): 649-672.
NAVARRO-ESBRI J, DIAMADOPOULOS E, GINESTAR D, 2002. Time series analysis and forecasting techniques for municipal solid waste management. Resources, Conservation and Recycling, 35(3): 201-214.
NEMIROVSKI A, SHAPIRO A, 2006. Scenario approximations of chance constraints. Probabilistic and randomized methods for design under uncertainty, pages 3-47. Springer.
NESTEROV Y, NEMIROVSKII A, 1994. Interior-point polynomial algorithms in convex

programming. SIAM.
NIKOOFAL M E, SADJADI S J, 2010. A robust optimization model for p-median problem with uncertain edge lengths. The International Journal of Advanced Manufacturing Technology, 50: 391-397.
O'KELLY M E, BRYAN D, 1998. Hub location with flow economies of scale. Transportation Research Part B: Methodological, 32(8): 605-616.
OMRAN A, MAHMOOD A, ABDUL AZIZ H, et al., 2009. Investigating households attitude toward recycling of solid waste in Malaysia: a case study. International Journal of Environmental Research, 3(2): 275-288.
PELS E, 2021. Optimality of the hub-spoke system: a review of the literature, and directions for future research. Transport Policy, 104: A1-A10.
PERAKIS G, SIM M, TANG Q, et al., 2022. Robust pricing and production with information partitioning and adaptation. Management Science, 69(3): 1398-1419.
PFLUG G, WOZABAL D, 2007. Ambiguity in portfolio selection. Quantitative Finance, 7(4): 435-442.
QI M, CAO Y, SHEN Z-J M, 2022. Distributionally robust conditional quantile prediction with fixed design. Management Science, 68(3): 1639-1658.
RAHIMIAN H, MEHROTRA S, 2019. Distributionally robust optimization: a review. arXiv preprint arXiv:1908.05659.
ROCKAFELLAR R T, URYASEV S, et al., 2000. Optimization of conditional value-at-risk. Journal of Risk, 2: 21-42.
SADKA E, 2007. Public-private partnerships: a public economics perspective. CESifo Economic Studies, 53(3): 466-490.
SAIF A, DELAGE E, 2021. Data-driven distributionally robust capacitated facility location problem. European Journal of Operational Research, 291(3): 995-1007.
SALDANHA-DA-GAMA F, 2022. Facility location in logistics and transportation: an enduring relationship. Transportation Research Part E: Logistics and Transportation Review, 166:102903.
SAMANTA M, WELSH A H, 2013. Bootstrapping for highly unbalanced clustered data. Computational Statistics & Data Analysis, 59: 70-81.
SCHRIJVER A, 2003. Combinatorial optimization: polyhedra and efficiency: vol 24. New York: Springer.
SHAFIEEZADEH-ABADEH S, KUHN D, ESFAHANI P M, 2019. Regularization via mass transportation. Journal of Machine Learning Research, 20(103): 1-68.
SHAHABI M, UNNIKRISHNAN A, 2014. Robust hub network design problem. Transportation Research Part E: Logistics and Transportation Review, 70: 356-373.
SHAPIRO A, 2001. On duality theory of conic linear problems. Semi-infinite Programming. Kluwer Academic Publishers, Dordrecht, the Netherlands. 135-165.
SHAPIRO A, DENTCHEVA D, RUSZCZYNSKI A, 2021. Lectures on stochastic programming: modeling and theory. SIAM.

SHARAFI A, TALEIZADEH A A, AMALNICK M S, 2018. Fair allocation in financial disputes between public-private partnership stakeholders using game theory. Service Science, 10(1): 1-11.

SHEHADEH K S, SANCI E, 2021. Distributionally robust facility location with bimodal random demand. Computers & Operations Research, 134: 105257.

SHEN H, JIANG R, 2021. Convex chance-constrained programs with Wasserstein ambiguity. arXiv preprint arXiv:2111.02486.

SHEN Z-J M, ZHAN R, ZHANG J, 2011. The reliable facility location problem: Formulations, heuristics, and approximation algorithms. INFORMS Journal on Computing, 23: 470-482.

SHIODE S, DREZNER Z, 2003. A competitive facility location problem on a tree network with stochastic weights. European Journal of Operational Research, 149(1): 47-52.

SIM T, LOWE T J, THOMAS B W, 2009. The stochastic p-hub center problem with service-level constraints. Computers & Operations Research, 36(12): 3166-3177.

SINGH K N, 2008. The uncapacitated facility location problem: some applications in scheduling and routing. International Journal of Operations Research, 5(1): 36-43.

SION M, 1958. On general minimax theorems. Pacific Journal of Mathematics, 8(1): 171-176.

SNYDER L V, 2006. Facility location under uncertainty: A review. IIE Transactions, 38(7): 547-564.

ŠOMPLÁK R, FERDAN T, PAVLAS M, et al., 2013. Waste-to-energy facility planning under uncertain circumstances. Applied Thermal Engineering, 61(1): 106-114.

SONG J, SONG D, ZHANG X, et al., 2013. Risk identification for PPP waste-to-energy incineration projects in China. Energy Policy, 61: 935-962.

SOYSTER A L, 1973. Technical note: convex programming with set-inclusive constraints and applications to inexact linear programming. Operations Research, 21(5): 1154-1157.

SRIVASTAVA A K, NEMA A K, 2012. Fuzzy parametric programming model for multi-objective integrated solid waste management under uncertainty. Expert Systems with Applications, 39(5): 4657-4678.

SUN W, HUANG G, LV Y, et al., 2012. Waste management under multiple complexities: Inexact piecewise-linearization-based fuzzy flexible programming. Waste Management, 32(6): 1244-1257.

TAI J, ZHANG W, CHE Y, et al., 2011. Municipal solid waste source-separated collection in China: a comparative analysis. Waste Management, 31(8): 1673-1682.

TAN Q, HUANG G, CAI Y, 2010. Waste management with recourse: an inexact dynamic programming model containing fuzzy boundary intervals in objectives and constraints. Journal of Environmental Management, 91(9): 1898-1913.

UNNIKRISHNAN A, VALSARAJ V, WALLER S T, 2009. Stochastic and dynamic shipper carrier network design problem. Networks and Spatial Economics, 9(4): 525.

VERTER V, DINCER M C, 1992. An integrated evaluation of facility location, capacity acquisition, and technology selection for designing global manufacturing strategies. European Journal of Operational Research, 60(1): 1-18.

VILLANI C, et al., 2009. Optimal transport: old and new: vol 338. New York: Springer.

WAGNER M R, BHADURY J, PENG S, 2009. Risk management in uncapacitated facility location models with random demands. Computers & Operations Research, 36(4): 1002-1011.

WANG S, CHEN Z, LIU T, 2020. Distributionally robust hub location. Transportation Science, 54(5): 1189-1210.

WANG S, HUANG G H, YANG B, 2012. An interval-valued fuzzy-stochastic programming approach and its application to municipal solid waste management. Environmental Modelling & Software, 29(1): 24-36.

WANG S, NG T-S, 2019. Robustness of resource recovery systems under feedstock uncertainty. Production and Operations Management, 28(3): 628-649.

WANG S, NG T S, WONG M, 2016. Expansion planning for waste-to-energy systems using waste forecast prediction sets. Naval Research Logistics (NRL), 63(1): 47-70.

WANG X, KUO Y-H, SHEN H, et al., 2021a. Target-oriented robust location-transportation problem with service-level measure. Transportation Research Part B: Methodological, 153: 1-20.

WANG Z, YOU K, WANG Z, et al., 2021b. Multi-period facility location and capacity planning under ∞-Wasserstein joint chance constraints in humanitarian logistics. arXiv preprint arXiv:2111.15057.

WIESEMANN W, KUHN D, SIM M, 2014. Distributionally robust convex optimization. Operations Research, 62(6): 1358-1376.

World Bank Report, 1999. Decision makers' guide to municipal solid waste incineration. The World Bank. Washington, D.C.

XIE W, 2020. Tractable reformulations of two-stage distributionally robust linear programs over the type-∞ wasserstein ball. Operations Research Letters, 48(4): 513-523.

XIE W, 2021. On distributionally robust chance constrained programs with Wasserstein distance. Mathematical Programming, 186(1-2): 115-155.

XIE W, AHMED S, 2018. On deterministic reformulations of distributionally robust joint chance constrained optimization problems. SIAM Journal on Optimization, 28(2): 1151-1182.

XIONG J, NG T S A, WANG S, 2016. An optimization model for economic feasibility analysis and design of decentralized waste-to-energy systems. Energy, 101: 239-251.

XU Y, HUANG G, QIN X, et al., 2009. SRCCP: a stochastic robust chance-constrained programming model for municipal solid waste management under uncertainty. Resources, Conservation and Recycling, 53(6): 352-363.

XU Y, HUANG G, QIN X, et al., 2010. An interval-parameter stochastic robust optimization model for supporting municipal solid waste management under uncertainty. Waste management, 30(2): 316-327.

YAMAN H, CARELLO G, 2005. Solving the hub location problem with modular link capacities. Computers & Operations Research, 32(12): 3227-3245.

YANG T-H, 2009. Stochastic air freight hub location and flight routes planning. Applied Mathematical Modelling, 33(12): 4424-4430.

YANIKOGLU I, GORISSEN B, DEN HERTOG D, 2019. A survey of adjustable robust optimization. European Journal of Operational Research, 277(3): 799-813.

YEOMANS J S, 2007. Solid waste planning under uncertainty using evolutionary simulation-optimization. Socio-Economic Planning Sciences, 41(1): 38-60.

YILMAZ S, SELIM H, 2013. A review on the methods for biomass to energy conversion systems design. Renewable and Sustainable Energy Reviews, 25: 420-430.

YU G, HASKELL W B, LIU Y, 2017. Resilient facility location against the risk of disruptions. Transportation Research Part B: Methodological, 104: 82-105.

YU X, SHEN S, AHMED S, 2021. On the value of multistage stochastic facility location with risk aversion. arXiv preprint arXiv:2105.11005.

ZENG B, ZHAO L, 2013. Solving two-stage robust optimization problems using a column-and-constraint generation method. Operations Research Letters, 41(5): 457-461.

ZETINA C A, CONTRERAS I, CORDEAU J-F, et al., 2017. Robust uncapacitated hub location. Transportation Research Part B: Methodological, 106: 393-410.

ZHANG L, YANG J, GAO R, 2023. Optimal robust policy for feature-based newsvendor. Management Science.

ZHAO C, GUAN Y, 2015. Data-driven risk-averse two-stage stochastic program with ζ-structure probability metrics. Available on Optimization Online, 2(5): 1-40.

ZHAO C, GUAN Y, 2018. Data-driven risk-averse stochastic optimization with Wasserstein metric. Operations Research Letters, 46(2): 262-267.

ZHEN J, DEN HERTOG D, SIM M, 2018. Adjustable robust optimization via Fourier-Motzkin elimination. Operations Research, 66(4): 1086-1100.

ZHU H, HUANG G, 2011. Slfp: a stochastic linear fractional programming approach for sustainable waste management. Waste Management, 31(12): 2612-2619.

ZHU N, FU C, MA S, 2018. Data-driven distributionally robust optimization approach for reliable travel-time-information-gain-oriented traffic sensor location model. Transportation Research Part B: Methodological, 113: 91-120.

ZYMLER S, KUHN D, RUSTEM B, 2013. Distributionally robust joint chance constraints with second-order moment information. Mathematical Programming, 137(1-2): 167-198.